THE MASTER ARCHITECT SERIES

EISENMAN ARCHITECTS

Selected and Current Works

世界建筑大师优秀作品集锦

埃森曼建筑师事务所

王丽娜　译

中国建筑工业出版社

著作权合同登记图字：01－2003－0646号

图书在版编目（CIP）数据

埃森曼建筑师事务所／澳大利亚 Images 出版集团编；王丽娜译.
—北京：中国建筑工业出版社，2004
（世界建筑大师优秀作品集锦）
ISBN 7－112－06695－6

Ⅰ.埃... Ⅱ.①澳...②王... Ⅲ.建筑设计－作品集－英国－现代 Ⅳ.TU206

中国版本图书馆 CIP 数据核字（2004）第086699号

责任编辑：程素荣
责任设计：郑秋菊
责任校对：赵明霞

世界建筑大师优秀作品集锦
埃森曼建筑师事务所
王丽娜　译

中国建筑工业出版社出版、发行（北京西郊百万庄）
新　华　书　店　经　销
北京嘉泰利德公司制版
恒美印务有限公司印刷
*
开本：787×1092毫米　1/10　印张：25⅗　字数：600千字
2005年1月第一版　2005年1月第一次印刷
定价：**218.00**元
ISBN 7－112－06695－6
TU·5849（12649）

本社网址：http://www.china－abp.com.cn
网上书店：http://www.china－building.com.cn

Contents

目　录

Introduction

导　　言

The Eisenman Wave

By Sanford Kwinter

埃森曼冲击波

斯坦福·昆特

我们很难说清以下哪一项是埃森曼更为重要的职业成就：是在30年间推出一个不断更新完善的令人震撼的学说，亦或是舍弃了几乎所有的既定体系的便利条件和传统赋予的不可避免的自满，以及那条达到“成功”的老路——设计“伟大”的作品或者发展个人化的建筑风格。在这两个方面，彼得·埃森曼不仅与他同辈的建筑师有所不同（毕竟，可以毫不客气地说他的同道“纽约五”现在已经堕落到手法主义的地步），而且卓然独立于几乎所有当代的建筑师。

当埃森曼在20世纪60年代早期开始他的职业生涯的时候，建筑师还是一个相当老成深厚的职业，就今天而言，情况依然没变：建筑师从他们的敌人——古典主义中获得力量来源，马上他们要反过来激烈地反对古典主义。埃森曼和他的追随者们总是能够运用自由的客体比如历史事件（广岛原子弹爆炸），特殊境况（上帝之死，家庭生活及社会习俗的转变）以及思想学说（转换语言学，结构主义，概念论，反人类学）来对抗已经发展成熟的关于秩序和场所的学说，并且试图扭转古老的中产阶级式的观念：重视空间价值胜过时间价值。埃森曼的任务已经变成了一种与现代主义相抗衡的新的（实践），它从福科和尼采那里借鉴表述方式，从而带来建筑学业外学科的信息。这种实践产生于充满变化的无物之阵，在虚空流体的胎盘中孕育，依靠原始的权力意志和历史变革的破坏力量得以进行。埃森曼独一无二的建筑作品是史无前例的；有组织的空间却是由一系列的“偶然”（kairós希腊语）作用产生的，所以这种打断连续运动的暂时性的胜利是从来——事实上也不可能被保留下来的。就像自治的游牧民族创造了对定居文化奇迹般的冲击一样，埃森曼的实践活动已经集聚力量成为一场拥有精神力量的运动。而且游牧民族和埃森曼都是通过侵入、瓦解，并且将暂时性的力量释放到自由动作和重新组合中去完成这个过程的。

对于埃森曼，我要说的是这些运动和剥蚀是在三个独立又相互关联的层面上展开的：理性的历史观，散漫的文本以及物质的形式感。就算不考虑埃森曼在同行中受到惊人的持续不断的怀疑（多数是无趣的和不自主的），仍然不存在埃森曼的专属领地，没有有效的“政治”权力范围，只有持续

It is difficult to say which is the more impressive career accomplishment: to have generated an endlessly renewed trail of agitative hypotheses over a 30-year period, or to have eschewed nearly all the comforts of consolidation—and the inevitable complacencies—afforded by conventional, repeatable “successes” such as the production of “great” buildings or the development of a signature style. In both these respects, Peter Eisenman differs not only from other architects of his own generation (it would, after all, be charitable to say that the work of his fellow “New York Five” architects has now degenerated into nothing better than mannerism), but from nearly all other architects working today.

When Eisenman’s work began in the early sixties, it was, and remains to this day, a primarily *tactical* enterprise: its force from the outset was drafted from that of the enemy—classicism—but was also turned aggressively against it. The Eisenman parti has always been to deploy mobile entities such as *historical circumstances* (holocaust, Hiroshima), *situations* (death of God, transformations of domesticity and its mores) and *idea-moments* (generative grammar, structuralism, conceptualism, anti-humanism) against the ethos of established orders and places, reversing the age-old bourgeois victory of values of domain over values of time. Eisenman’s task has been to develop a practice that, to borrow an expression from Foucault and Nietzsche, would come *from outside*—a new type of modernist adversarial practice to be launched from a placeless but volatile “steppe,” home of disembodied fluxes, raw will to power, and the destabilizing forces of historical change. There is not now, nor has there ever been, a fixable Eisenmanian alternative architecture; tactical space after all is made up of a series of seized “occasions” (Greek *kairós*), so that the momentary triumphs that punctuate its unfolding campaign are never—indeed cannot be—stored. Like the autonomous, fluid nomad civilizations who made legendary assaults on sedentary cultures, Eisenman’s practice is assembled and articulated *in movement* and in the spirit of movement. Both operate through invasion, disruption, and the release of temporarily trapped forces into free motion and recombination.

In the case of Eisenman, I will argue, these movements and abrasions unfold on three distinct yet interconnected levels: the intellectual–historical, the discursive–textual, and the material–formal. Yet despite an amazing and persistent paranoia among colleagues (primarily the dull and unfree), there exists no Eisenmanian fiefdom, no domain of

被迫地对“无节制”以及客观意志力量的收敛，这两者都来源于无组织的外界，并指向了具有静止形式的刻板世界。埃森曼的影响就像海滩上波浪的冲蚀作用：构成我们今天丰富多元的战后历史的思想成果和理性潮流，它们都被有规律地，巧妙地用来表达建筑和“人”。

埃森曼从来没有自称是哲学家。虽然他写文章带有一种严谨和规范确是事实，然而他的文章，就像他的建筑作品一样，是集各种杂乱材质的碰撞之大成；拥有激烈的外观，是由塑造我们这个世界形状的力量作用产生的。使得这些想法发生作用的具体方式（说它是一种原始的创造力也不为过），无疑引出了罗伯特·史密森和其他一些美国极少主义者的作品，纵然埃森曼的大部分作品包括早期的方案草图和完成的作品在他的整个实践过程中都是那么的有组织性而又抽象，但是我们仍然很难说清他的建筑是在哪里产生的，或者说他的建筑是否是最早的“正在”发生的建筑。

对于埃森曼的偶尔缺乏条理我们很容易误解他，然而这种说法也是不无道理的。对于埃森曼（以及这个患有理性缺失症和历史健忘症的时代）来说，最重要的是，他是近代历史上第一位迎接了未来派艺术家的挑战从而考虑了所有的文化因素的建筑师，这些可塑的、历史的同时又充满智慧的文化因素被他整合成一个单一的、连续的并且是相互关联的领域。在埃森曼之前的建筑界往往狭隘地认为建筑最好是具有文化特性并赋予理性的；然而今天所有的文化和各种精巧的理智都能够成为建筑，至少暂时能够成为建筑。这个新的领域的可塑性，属于我们现代社会最伟大的文化发展之一，这种可塑性完全是一个新的空间类型，因为它被赋予了理性的、文本的因而也是无限伸展的维度。这些想法的起源部分地来自于尼采的“权力意志”理论。尼采第一个认为形式是其背后不可见的斗争力量的具体显现。在尼采的哲学体系中，历史成为了赋予形式的力量之历史，它就是说它是一个

concentrated "political" power, only the continuous forced convergence of "wild," impersonal idea-forces both drawn from the amorphous outside and directed at the stolid world of quiescent form. The Eisenman-effect operates like the abrasions of a wave on a beach: the parade of ideas and intellectual currents that make up our collective post-war history are made to render, through rhythmic, directed encounters, what to a humanistic tradition was once solid—both Architecture and "Man"—a shifting fluid as well.

Eisenman has never claimed to be a philosopher. It is true that he writes with seriousness and discipline, yet his texts, like his architecture, are more than anything else promiscuous material fields of collision; aggravated surfaces onto which are drawn the raw, active forces that give shape to the objects of our world. The concrete way in which ideas are here assembled (it would not be out of line to ascribe to it a barbaric creativity) elicits, to be sure, the work of Robert Smithson and certain of the American minimalists, though most of all, Eisenman's own early drawings and built work, which together are so textual and abstract that across the continuum of his practice it remains hard to say where his architecture takes place, or whether it is even primarily architecture that *is* taking place.

It has been easy to fault him for an occasional lack of rigor, yet that does not mean that such claims do not seriously miss the point. For what is important in Eisenman (and in this era of intellectual poverty and historical amnesia it merits being pointed out again) is that he is the first architect in recent history fully to take up the Futurists' challenge to conceive of all of culture—plastic as well as historical, intellectual—as a single, continuous and connected field. In the parochial, pre-Eisenmanian architectural world, it could be said that architecture was at best cultivated and intelligent; whereas today, all culture and elaborated intelligence can—at least potentially—become architecture. The ductile nature of this new field—a new type of space entirely, because it is endowed with intellectual, textual and therefore infinitely extendable dimensions—belongs to one of the greatest cultural developments of our modernity. The origins of this program can be found in Nietzsche's concept of "will to power." Nietzsche was the first to proclaim that form was but the concrete *appearance (Schein)* of invisible conflicting

基本的美学现象而不再是一个道德现象，并且在今天这种理论的政治上的引申含义仍然在发挥其功用。那些在精神世界被创造和谈论的东西也许和那些具体的事物有着一系列的物质联系，而这些事物却是在另一个完全不同的甚至和他们没有任何联系的领域里，这种领域就是具体的物质的环境；这两个具有完全不同性质的相似事物似乎不仅相互影响，而且事实上还可能由同一类力量产生的，至今这还是认识论所持的观点。那么，从福科的术语说，那些杂乱的物体、杂乱的实践，是如何影响那些具体的不杂乱的物体呢？事实上他们之间已经相互相错在一起了。虽然答案显然是太复杂以至于不太可能在这里完全展开，但是对问题的回答确需要明显的特征或者元素，通过它们语言（文化与被表达的精神物质）内部表达哲学的属性可以与之相连，并在具体的或者建成的环境中与其表述行为的属性相连接。[1]

就像许多战后的语言哲学家包括富科所主张的那样，谈到世界上的某种事物，就是预先假定地做某事。简单地说，这意味着语言上和头脑中的事实是存在的，并且凭借它们的能力来改变物质的状态：它们安置，充满，并以相同间隔将世界重新划分。一个连续的被调整好的事物影响的概念是一个主义的成就，如果不是战后的存在论，那么一定是战后的美学主义，这一概念将各种不同的现象联系在一个丰富而又和谐的类型中。语言在刚出现的推测中首次成为完全的，手势的建构，从而用它的能力来引起并指导社会和物质集体的力量；客体的世界，包括制度和建筑正在逐渐地被看作——至少是被富科、德勒兹和瓜塔里以及撒克逊人哲学传统的法国继承人是相似的高密度形式、基本的主题环境。

1　表述行为的方式之概念是由英国语言哲学家约翰·奥斯汀发展起来的。它起初的表达方式是想和非事实的表达方式相区分，也就是那些简单的声明或者事实的描述，而不是事实，不过是说法罢了。他最早给假定的物体命名，但是他的整个职业生涯的后期却是逃避了那些正式的区别，并将那些现行的表述行为的功能延伸到所有的演说中。在这一扩展的领域中，在更加细微差别的层面上，他引入了内部表述行之哲学名词，用其来指那些语言学之外的完全的事实的转变（在剧院叫喊“着火”或者在婚礼上说“我愿意”（yelling a fire in a theater，saying“I do” in a marrariage ceremony）），并引入了言语表达效果之哲学名词，以其来表达那些只是在讲述者和聆听者位置中变化的事实（说服、恐吓、打扰等等）。

forces working below and across it. History, in the Nietzschean cosmos, became the history of *shaping forces*—that is, a fundamentally aesthetic phenomenon, and no longer a moral one—and this idea is one whose political implications have still today only begun to be worked out. That what is created and *said* in a "mental space" might be materially continuous with what is given shape in a domain that is entirely distinct and removed from it in nature and modality, that is, in a concrete, physical milieu; and that these two parallel but disparate types of phenomena might not only affect one another but in fact be engendered by the very same genus of forces, remains a radical epistemological claim. For how, to use Foucault's terminology, do discursive objects—discursive practices—impose their effects upon, indeed form a tissue with, concrete or non-discursive domains? Though the answer to this problem is clearly too complex to develop fully here, it is enough to say that its solution entails a resonant feature or element through which an *illocutionary* property of language (culture and expressed mental objects) connects to, and communicates with a *performative* property within the concrete or built environment.[1]

To say something in the world, as many post-war language philosophers besides Foucault claimed, is pre-eminently *to do* something. What this means in a nutshell is that linguistic and intellectual acts exist and operate by dint of their capacity actually to *change material conditions:* they program, suffuse, and in each instance, redistribute the physical world. The concept of a continuous and modulated tissue of effects that connects disparate phenomena (such as language, ideas and matter) together in a type of manifold or consistency, is a principle achievement, if not of post-war ontology, then certainly of post-war aesthetics. Language, in this emerging conjuncture, became for the first time fully and gesturally tectonic in its capacity to provoke and direct the forces of social and material assembly; the worlds of objects, institutions and buildings were increasingly seen—at least by Foucault, Deleuze and Guattari, the French inheritors of this Anglo-Saxon philosophical tradition—as hyper-dense forms of these same, fundamentally *programmatic,* milieus.

[1] The concept of the performative utterance was developed by British language philosopher John Austin. Its original formulation was meant to distinguish it from utterances which were not acts—that is, simple statements or matter-of-fact descriptions which were not actual *doings*—but only *sayings.* He originally named these latter objects *constatives,* but his entire late career was committed to withdrawing the formal distinction and extending the active, performative function to virtually all speech acts. In this extended domain, and at a level of higher nuance, he introduced the terms *illocutionary* to describe complete acts of transformation in an extra-linguistic domain (yelling 'fire' in a theater, saying 'I do' in a marriage ceremony), and *perlocutionary,* to describe acts that merely induce changes of state in the interlocutor or hearer (persuading, frightening or boring, etc.).

在建筑中，这些发展在埃森曼的建筑中得到了最充分的表达，在埃森曼的建筑中就像在富科的神经错乱的，妄想的制度语境中一样，埃森曼的示意图、文章和建筑是在一种和谐、果断之杂乱而又完好的连续体中相互关联并且相互影响。这里，所有的文化指的是物质文化，而如果用生物学家的话说，历史在总体上就成为一个鲜活的“兴奋的媒体”，这一媒体和它所包含的事物有着紧密的联系，它被一个交流波动的传播系统穿过并连接。

埃森曼最早的思想渊源当然不是来自于欧洲大陆的传统领域，而是来自英国和美国，并且很多是来自建筑教育和形式美学的狭小领域。20 世纪 60 年代的埃森曼是威特科尔和罗（而非尼采和富科）的追随者，20 世纪 70 年代的时候，他则追随主流的结构主义和乔姆斯基的转换语言学。对于逻辑和数学引导的分布规则的探索是埃森曼最初的兴趣所在，特别是当这些嵌入的结构通过一种严谨的方式被拿到表面的时候，还有哪些更能引起他的兴趣呢？但是尽管他的表达方式不是很直接，我要说的是，最重要的是埃森曼对朱塞佩·泰拉尼的作品长期的执迷。因为就像大多数历史学家所认为的那样，泰拉尼的工作并不是理性主义的，也不是新帕拉第奥静态结构语法，事实上却是永远富有变化的，是真正持久的影响波动，这种波动不断转换或者移动从一种状态到另一种状态，很像在布鲁塞尔机容器中的化学反应。[2] 这些新的活动类型决定向平静发出了挑战，柏拉图式的表达秩序与其他系统的抗衡产生了。事实上，埃森曼的工作一直是在企图（无意识的？）发现或者发展来自古典机制的波动。[3]

一个人并不需要深入研究他的强制的甚至是傲慢的工作过程，因为在埃森曼早期的系列作品中，他已经创作了有关

2 化学反应，是一个液体容器，在这个容器里面化学药品生成了稳定持续的蒸汽。化学药品之间的接触反应产生了连贯的色波，反应结果中色波的图案和形式有规则的间隔。关于自动接触反应系统与建筑的关系，请参看我在《彼得·埃森曼和弗兰克·盖里》（Rizzoli 出版社 1991）一书中，所写的文章“天造之物：埃森曼的辛辛那提作品”，还有文章“Maxwell 的魔力和埃森曼的创造：信息时代的竞赛”（A+U，1993 年 9 月）。

3 埃森曼至少是有意识地将古典阅读系统的不足区分出来，这一事实是无可争辩的，尽管没有明确地将它们区分。可以参考他的研究例子，“从物体到关系”，*Casabella*，334 页，1970 年 1 月。

In architecture these developments found expression most fully in Eisenman where—just as in the delirious, paranoid, institutional milieus of Foucault—drawing (diagram), text and building actually came to connect with and interpenetrate one another in a promiscuous and unbroken continuum of determination and resonance. Here, all culture is *material* culture, while history, to speak like a biologist, becomes a living "excitable medium" in total intimate contact with all of its objects, shot through with, and correlated by, a propagative system of communicational waves. Every disturbance in the continuum is instantly converted into movement, registered and transmitted like an irrigating flow throughout the system.

Eisenman's earliest intellectual roots did not, of course, grow out of the traditions of continental Europe, but from those of England and America, and all too often from the narrow milieus of academic architecture and formalist aesthetics. The Eisenman of the sixties was a follower of Wittkower and Rowe (not Nietzsche and Foucault), and in the seventies, of mainstream structuralism and Chomsky's generative grammar. The search for logical or mathematically driven distributional rules appeared to be his primary interest, especially insofar as these embedded structures could be brought to the surface by rigorous operations, and there rhetorically hyper-developed at the deliberate expense of a founding "humanist" creator-subject. But of far greater importance, I would argue, even if its expression remained indirect, was Eisenman's career-long fascination with the work of Giuseppe Terragni. For Terragni's work was not, despite what most historians have argued, a rationalist, neo-Palladian grammar of static structures, but in fact a container of perpetual movement, a veritable standing wave that switched or migrated from state to state not unlike the chemical fluctuations in a Brusselator tank chemical clock.[2] This newly identified type of activity defied the calm, Platonic play of expressed orders of which these other systems were built. Indeed, Eisenman's work has always been a search (unconscious?) to find, or develop, this wave from within the classical machine.[3]

One does not need to search far to see this forcible—even hubristic—process at work, for in the early *House* projects Eisenman had already laid down the choreographic lexicon from which his later work would never fully depart. Each of these ten or so projects may be said at the outset to develop

[2] The chemical clock is a container of liquid into which a steady stream of chemicals are fed. The catalytic effects that the chemicals have on one another provoke coherent waves of color, pattern and form to appear in the solution at regular intervals. On the relation of these autocatalytic systems to architecture, see my essays "The Genius of Matter: Eisenman's Cincinnati Project," in *Peter Eisenman and Frank Gehry*, (Rizzoli, 1991), and "Maxwell's Demons and Eisenman's Conventions: Challenge Match for the 'Information' Age," (*A + U*, September 1993).

舞蹈的辞典，而这些在他后来的作品从来没有完全舍弃过。这十个住宅作品的每一个或者其他的作品可以说是从一个边界固定的立方体开始发展的。当然说立方体的边界是固定的，并不是说它们不是相互连续的或是未经雕琢的。事实上，它们和分裂以及削减紧密地联系在一起，疯狂地就像自动钢琴的游戏程序在表演一样。当然最重要的是在这些实验性的结构中，乐器或者共鸣体以及音乐符号系统（活页乐谱和游戏程序）彼此之间都是完全同步的。这样才有了美好的甚至是神奇的信息传达。这种系统类型的结构和古代土著居民的网格方式是非常相象的，土著人梦想轨迹和歌之版图（songlines）就像富有动力的地图，能够真正清晰地表明澳大利亚陆地的每一个物理特征。当然没有一个部落或者个人真正理解其他部落的语言，也包括其他相邻部落的歌之版图；然而通过人们自身的方式和语调，一个有着特定特征和路径的陆地就会清晰地呈现在一个人的面前，就像一个宣传手册是通过每一页而形成的。

埃森曼住宅的偶然性有反漫步（anti-promenade）这种类型的文化力量，或者用一个词来说是闲庭信步，他的住宅这种偶然性基于与古典主义建筑的联系，并且它正是从古典主义建筑中产生的。这些房子产生的晕眩被说成只是中产阶级神经质图谱和超然性的象征，它们能够通过“空间”从客观世界分裂出来得以产生。然而我则认为这些房子应该被看成是一种深思熟虑的意识形态的分离，从静态、古典空间（大不列颠的殖民地经济或者更多的一般的欧洲城市）中分离，以及在一种液体中的浸入，一种无限复制的弹性脉冲十字线液体；一个“位置”独特地通过“事件”来展开的系统，这些“事件”是特定的材料特征的永久的呼唤。就像在澳大利亚人口稀少的内陆一样，埃森曼的住宅中，被歌唱的“歌曲”以及风景是不能在本质上分开的，主要是因为他们都是嵌入在一种相似的深度时间中。当然对于土著人来说，深度时间是指对于宇宙起源的无限的猜测。对埃森曼来说，深度时间是指句子的构法，对错误的遥远起源的放弃。埃森曼的游走（Walkabout）住宅从殖民时期的一维空间中解放出来，也通过记忆和回归——并且将它转化成一种制造和进行的可

within an essentially boundary-fixed cube. Of course to say that the boundaries are fixed does not mean that they are either continuous or inviolate. They are, in fact, maniacally articulated with disruptions and deletions, crazily perforated like the program cards that drive a player piano. What is important of course is that in these experimental structures the "instrument" or resonating body, and the notational system (sheet music or program cards) are entirely coextensive with one another. There is here a very beautiful and almost mystically efficient compression of information. The structure of this type of system resembles the webways of ancestral Aboriginal dreaming tracks or songlines that articulate, like a dynamical map, virtually every physical feature of the Australian continent. No single clan or individual, of course, actually "understands" the language of any but their own, and their immediately adjacent clan's, songlines; yet by means of deeply embedded patterns and intonations (a kind of deep structure of melodic contours and phrases available to intuition though not—yet—to analysis) a continent of specified details and trajectories appears to open transparently before one like a hyper-book ever further called into being with each turn of a page.

The encounter with the Eisenman House, at least in relation to classically based architectures from which it broke, has the cultural force of this type of anti-promenade, or, in a word, of the *walkabout*. The vertigo that these houses are said to provoke is but a bourgeois symptom of the neurotic preoccupation with maps and the transcendence they are able to induce by dissociating "space" from the object-world. Rather, I propose, the houses should be seen as a deliberate ideological break from a static, time-hating space (the economy of the colonial British, or more generically, European, city), and an immersion into the fluid criss-cross of infinitely multiplied trajectorial pulses; a system where "location" is established uniquely by "events"—the perpetual "calling out" of designated material features. In the Eisenman House, as in the Australian outback, the "song" and the landscape that is sung, are materially inseparable from one another (it is impossible to say which engenders which), primarily because both are embedded in a similar kind of *deep time*. In the Aboriginal case, of course, deep time refers to the infinite conjuring

[3] That Eisenman at least consciously identified the insufficiency of these classical systems of reading, even if unable to get definitively beyond them, is irrefutable. See for example his study, "From Object to Relationship," in *Casabella*, no. 344, January 1970.

更新进程。简言之，埃森曼住宅从来没有意味着提供一个家庭居所，而是将建筑向动态敞开，将其归还给已经丢失的田园风光。

住宅中的结构的分布也是遵循着这种逻辑。这里，创造过程是一个从表面边缘或者实体内部穿行的“词汇特征”的稳定的腐蚀性的传播，因此室内表面的打破就被认为是方形波干涉的模式。[4] 从每个住宅的起源来看（也就是它的概念和示意图的发展），它实际上是从非常简单的单元（正方形）开始发展的，将这些单元以一个灵活的或者移动的中心进行移动或者旋转。比如说一个简单的九宫格就可以同时向东面和南面发出相同的节奏，2号住宅便是如此，先创造一个4m×4m的网格，接着在其上面加了一个新的扩展的九宫格。（我们知道这是一件复杂的事情，因为在东南角有一个“附加”的方形。）

于是产生了一个多旋律的类型，它的逻辑结构对埃森曼的作品有十分重要的影响，也就值得我们仔细研究一下。最早的网格模式的发展像水晶结构一样：它以相同的间隔尺寸不断地重复原始结构的比例（如果不是明确的单元），仿佛它是一个简单的树枝状的延伸。当然，最初的中心已经向东南方向平移了（从坐标 $x=1.5$，$y=1.5$，到 $x=2$，$y=2$），形成了一个内嵌式的钩或者说是对惰性持久波的引导。[5] 现在新的更大的九宫格不仅简单表达了在结构频率上的增加，而且也是在振幅上的增加。有两个截然不同的但是已经结合在一起的过程：成长加上繁殖，音调上的改变以及在强度或者音量上的改变；这两者都是同时存在的，一个存在于另一个之中。这结果是一系列的神经元系统，是一个通过该系统可逆的过程，它产生了在每个表面变化时期的振动现象。就像在最开始的时候被移动以达到延伸面、增加室内频率、重新校正振幅的作用，同样面和体也可以采用这种处理方法。手法是相当简单的：一个面和一个体是可以沿着一个钩切割的，

4 方形波是方形的能量脉冲，这种脉冲在两个函数之间振荡，但是在振动的过程中它的振幅保持不变。

5 一个持久波（A standing wave）是一种已知的固定的波。它是由反射波的叠合产生的，当一种波通过振动的媒介传播的时候，在它上面与其相随的波发生了重合。相随的波是在媒介的边缘进行传播的，在媒介的边缘一些或者全部的波的能量都反射回它来源的地方，于是干涉波的模式就产生了，因此在传播的表面就产生了波峰和波谷。为了强调我的观点，能量或者信息（energy, or information）已经无数次地讲到，但是模式仍旧没有改变，即使不是完全一致至少也大体相同。

(re-enactment?) of the origins of the universe; in Eisenman, a flight into pure syntax, a renunciation of the false and distant origin. The Eisenman Walkabout House frees the origin from the one-dimensional colonial time of a fixing—a remembering and a return—and transforms it into an ever-renewable process of an engendering or *a proceeding*. The Eisenman House, in short, was never meant to furnish a home, but rather to open architecture to movement, to re-turn it toward the lost pastorality of the *nomos*.

The distribution of structure in the Houses follows this particular logic as well. Here, the impression is of a steady, corrosive propagation of "lexical features" traveling from the edge surface or envelope inward, such that the breaking up of internal surfaces may be read as square-wave interference patterns.[4] From the point of view of each House's genesis, however, (that is, its conceptual and graphic development) it actually proceeds by very simple units (squares) shifted or rotated on elastic or mobile centers. A simple nine-square grid, for example, might emit a compound beat simultaneously to the east and to the south, as in House II, producing a sixteen-square grid over which is subsequently laid a new, expanded nine-square. (We know it is a compound wave, because an "extra" square is produced at the south-east corner of the new "el".)

What has been engendered here is a type of polyrhythm whose logic is sufficiently important to all of Eisenman's work that it is worth examining a moment. The first propagation of the grid-pattern is crystal-like: it repeats the proportions (if not explicitly the units) of the original structure at the same granularity as if it were a kind of simple dendritic extension. And yes, the germinal center has shifted 45 degrees to the south-east (from $x = 1.5$, $y = 1.5$ to $x = 2$, $y = 2$), forming a kind of embedded jig or guide for the resultant standing wave.[5] Now the new, larger nine-square represents not simply an increase in the *frequency* of the structure but in its *amplitude* as well. There are two distinct, but now wedded processes: growth plus multiplication, or change in tonality, plus change in intensity or volume; and these both exist simultaneously, one inside the other. The effect is a series of nervous, reversible phase shifts that travel through the system, generating vibratory phenomena of varying period across

[4] Square waves are rectangle-shaped energy pulses that oscillate regularly between two values but whose amplitude remains constant between jumps.

[5] A standing wave is sometimes known as a stationary wave. It is created by the superimposition of a reflected wave on top of an incident wave that is propagating through a vibrating medium. The incident wave is "processed" at the medium's boundaries where either some or all of the wave's energy (and therefore its structure) is returned back toward its source so that an interference pattern is created, distributing nodes and antinodes across the transmitting surface. The energy or information, to underscore my main theme, propagates endlessly while the pattern remains, if not the same, at least highly stable.

当它掠过或者发生冲撞的时候可以“分解”几何体，并且通过延展和复制来吸收被迫的干扰和干涉波。尽管这些是冷酷的机械的手法，但是它们产生了永不停歇的稳定性之流动模式。这种合成的系统类型与那熟悉的古典类型的区别在于，这里秩序是一种找寻和创造，绝非一种平衡，也不是相近。埃森曼的空间很明显已经成为一种规则系统，运动而且充满活力。

在埃森曼的每个作品中，结构总是来源于一个原始的模式，而这个模式与平衡往往发生碰撞。所带来的冲击波四处游走，达到一定的极限后转回来从而形成一个自我干扰的波。结构上的信息或者模数沿着这个波以断断续续的步伐行进，然后获得一个暂时的歇息，接着开始下一个振动。每一个振动都可以被看成是空间用来表示基础的底或者正弦波曲线。但是振动模数的通道也会使空间倾斜，澄清也会混淆沿着这条路径的所有虚设，它是根据另一个增值中心的逻辑而放置的物体，还是只不过是曾经的或者将来的波的一个记忆幻影，这个波已经或者即将通过神经元。倾斜的过程将断断续续的残余与包含他们的持久动力波相连。这样的一个系统与在克拉尼数值曲线看到的沙滩形状没有任何差别，克拉尼数值曲线是一个精确的规则模式，这种模式用来使金属盘子与作用于它们的持久波产生共鸣。这样，我们就可以自信地推出连续性甚至存在于静态之中。

可以说埃森曼所有的作品都是寄居在特定的现代主义的悖论中：连续统一体的理论。这就关系到能量（光子、电子等等）在波中的传播方式，虽然能量曾经被安置、组织、固定在一定的位置上，它也只能作为一个粒子来表达自己。埃森曼是在特拉尼那里首次遭遇这个悖论：一种平衡的不确定性，在这其中，粒子和波彼此同时存在，并早于薛定谔（Erwin Schrodinger）和布罗格利的平衡所描述解析的“波功能的崩溃”，并与这一理论有着鲜明的区别。布罗格利在1923年提出了“物质波”的设想，同时在1925年和1927年戴维

every surface. And just as lines are moved in the initial stages in order to extend surfaces, to multiply internal frequencies, and to recalibrate amplitudes, surfaces and volumes too are set into motion. Again the operation is relatively simple: a surface or volume is permitted to slide along a jig, to "shed" geometry as it sweeps and frays, and to absorb the resultant disturbances and interferences by stretching and multiplying. Yes, these are cool, mechanical operations, but they produce fluid patterns of restless stabilization. The difference between this type of compositional system and the familiar classical ones is that here order is sought, and produced, far from, rather than close to, equilibrium. Eisenman-space has demonstrably always been approaching the algorithmic, the active, and the living.

In an Eisenman work, structure always emanates from an initial pattern that is knocked away from equilibrium. The disturbance then travels, reaches a limit, then turns back toward itself to form a self-interfering wave. The structural information, or modulus, proceeds in discrete steps along the wave, coming to a momentary rest at the next, and then at each subsequent beat. Each of these beats may be conceived of spatially as marking the floor of a basin or the trough of a sine wave. But the passage of this flickering modulus also *rakes the space,* articulating and disarticulating all that lies along its path: be it the real substance deposited according to the logic of another propagative center, or simply the ghost memories of a previous or future wave that has, or soon will have, passed through the synapse. The raking process links the discrete residues to the dynamic standing wave that subtends them. Such a system is no different from the sand shapes seen in Chladni figures—the elaborate regular patterns formed on resonating metal plates when a standing wave has been applied to them. Thus the continuous, one is lead confidently to conclude, exists even within the (homeo)static.

The entire Eisenman project, one could say, is lodged within the specific Modernist paradox: the theory of *continua.* This concerns the manner according to which energy (photons, electrons, etc.) may travel in waves, yet once located, arrested, and fixed in position, can express itself only as a particle. It was in Terragni that Eisenman first confronted the mechanics of this paradox—a kind of quantum indeterminacy *where the particle and the wave coexist*

森在两个独立的试验中证实了布罗格利的假设，同时期泰拉尼已经开始有自己的作品诞生。[6] 埃森曼总是试图用特定的语言模型分析在文字上阐明泰拉尼的成就，这些语言模型的分析困惑了同时也激发了和他同时代人的作品，但是通过语言模型的分析，泰拉尼的成果还是不能被表达出来。[7] 然而埃森曼的草图和作品却总是拥有一个额外的部分，这部分寂静无声甚至一点也不明显，它却在分析的范围外移动。这里，在这些超出了简约派分析多余的、不明显的部分中，我们发现了完全盛开的埃森曼的影响和埃森曼之波。

在事实的每个层面上，埃森曼对建筑文化的冲击已经表现出连续性和运动性，而这两者从前是完全分开的而且是惰性的。总是在断断续续、散乱的环境中介入的连续体会释放交流分裂的过程。但是有一种特殊的反古典主义；它没有任何地方比埃森曼的特殊的脚本用法更明显（或者暗中）清晰。在排版印刷界，罗马字体——分散的垂直字母——模仿石头夸张的秩序和纪念性——在埃森曼的手里的时候立刻就被单一的流动的线条奚落并集合起来，这些线条用一种独特的、连续的——几乎是夸张的创意表达出来。这里草书的形式捕捉到一种力量，直观上使得罗马形式在微观戏剧舞台屈服了它的规则，在这一舞台中将埃森曼所有的可塑的、图表的工作都投放到了一个新的清晰的缓解中。尽管有多语主义，语言学家的埃森曼想像自己不停地创造了多语主义，还有形式主义的埃森曼构想自己去编造的多数几何顺序之多振（polyresonance），还会存在另一个也许是更突出的埃森曼，尽管更多的甚至是对他自己来说都是被隐藏起来的：那就是埃森曼的运动，他的草书形式（cursive form），他的连续的研究领域以及广泛传播的影响力。

6 布罗格利在 1929 年因这项成果获得诺贝尔奖。

7 见《彼得·埃森曼》，《朱塞佩·泰拉尼》（未出版）。

within one another prior to the analytical "wave function collapse" described by the equations of Erwin Schrödinger and Louis de Broglie, that splits them definitively apart. De Broglie posited the concept of "matter waves" in 1923, while C.J. Davisson confirmed the hypothesis in two separate experiments in 1925 and 1927, the same years in which Terragni had begun to produce his first significant work.[6] Eisenman always sought to articulate textually the intuited paradox in Terragni's work with the particular language model of analysis that obsessed and inspired the work of most of his generation, but through which it simply could not be expressed.[7] Yet Eisenman's drawings and works nonetheless always possessed an *excessive* part that moved—silently and even unconsciously—beyond the limits of the analytical paradigm. It is here, in this excessive and unconscious space beyond the reach of reductionist analytics, that one finds the full blooming of the Eisenman effect and the Eisenman wave.

On virtually every level, Eisenman's impact on architectural culture has been to render continuous and active what was previously separate and inert. It is always the introduction of a continuum into a discrete and disjunctive milieu that unleashes the processes of communicative disruption. But here is an anticlassicism of a very specific kind; one that is nowhere more obviously—or furtively—apparent than in Eisenman's idiosyncratic use of script. In the typographical world, the roman forms—discrete, upright letters that mime the bombastic orders and monumentality of stone—are, in Eisenman's hand at once ridiculed and mobilized by the single, fluid line that renders the same letters in a unique, continuous—almost exaggerated—cursive stroke. Here, the cursive form seizes power, visibly forcing the roman form to submit to its rule in a microdrama that throws all of Eisenman's plastic and graphic work into newly clear relief. For beyond the polysemantism that the linguistic Eisenman imagined himself to be producing, beyond the polyresonance of multiple geometric orders that the formalist Eisenman conceived himself to be orchestrating, there lies another, perhaps more salient, Eisenman, though for that all the more hidden, even to himself: the Eisenman of movement, of the cursive form, of the continuous field, and of the propagating wave.

[6] De Broglie was awarded a Nobel Prize for this work in 1929.

[7] Peter Eisenman, *Giuseppe Terragni*, (unpublished).

就像光子一样，埃森曼始终是两个紧密相连的不可调和的状态的创造者：当他谈论和思考他所做的事情的时候，他属于经典粒子世界，但是当他的笔在纸上绘图或者他的想法在各个文化领域穿梭的时候，他形成了强大的冲击波。也许正是因为这样事情才会很好，因为在当代，现代人和埃森曼的作品是从思想界的混乱分离出来，从而脱颖而出的，对于今天的新一代人来说，正是已经建成的建筑作品和草图构思将用他们全部的勇气来歌唱勤勉的、无声的精髓，包括他们的可塑性的以及视觉上的额外部分，这些精髓都是关于新的领域，在新的领域中埃森曼的特殊的现代主义夸张的烙印从来没有办法清晰地阐述出来，但是在新的领域中，标志着埃森曼30年职业生涯的许多尝试和形式，经久不衰地占有一席之地。

Like the photon itself, Eisenman has always been a creature of two intimately linked but irreconcilable phases: when he speaks and thinks about what he does he belongs to the classical particle world, but when drawing pen across paper, and moving ideas across the cultural spectrum, he forms a formidable wave. And yet it is perhaps well that this is so; because for the new generations emerging today, systematically removed from the intellectual turbulence out of which both the Modern and the Eisenmanian projects emerged, it is the built objects and the drawn artifacts that will continue to sing, in all their gritty, assiduous and mute refinement, in all their plastic and visual *excess,* about the new world to which Eisenman's particular brand of Modernist rhetoric itself could never explicitly speak, but to which the multiple risks and forms that mark his 30-year career unfailingly give place.

GRIDDINGS SCALINGS TRACINGS FOLDINGS

House I

Design/Completion 1967/1968
Princeton, New Jersey
Mr and Mrs Bernard M. Barenholz
3,000 square feet
Wood frame
Exterior: painted wood panels
Interior: painted wood and brick panels

住宅 I

设计/竣工　　1967/1968
普林斯顿，新泽西州
伯纳德·M·巴伦豪兹夫妇
3000平方英尺
木结构
室外：涂漆的木板
室内：涂漆的木头和砖板

This house symbolizes the new family structure of a professional couple that must occasionally live apart because of their separate work schedules.

The design relates the miniature scale of their toys to the normal scale of the individual. This creates a series of transitions from overly large to excessively small, where the individual must stoop to look at the toys. The new house seems to be a large toy next to the existing house.

该住宅象征了职业夫妇家庭结构的特征，那就是他们由于不同的工作作息时间必须偶尔分开居住。

该设计将他们玩具之缩小的比例与正常人的比例联系起来。这就创造了一系列的转换场所，从过大到极小，在转换场所中，个人必须弯腰来观看玩具。新建成的住宅相对旁边的已经存在的房子来说，就像一个玩具。

1

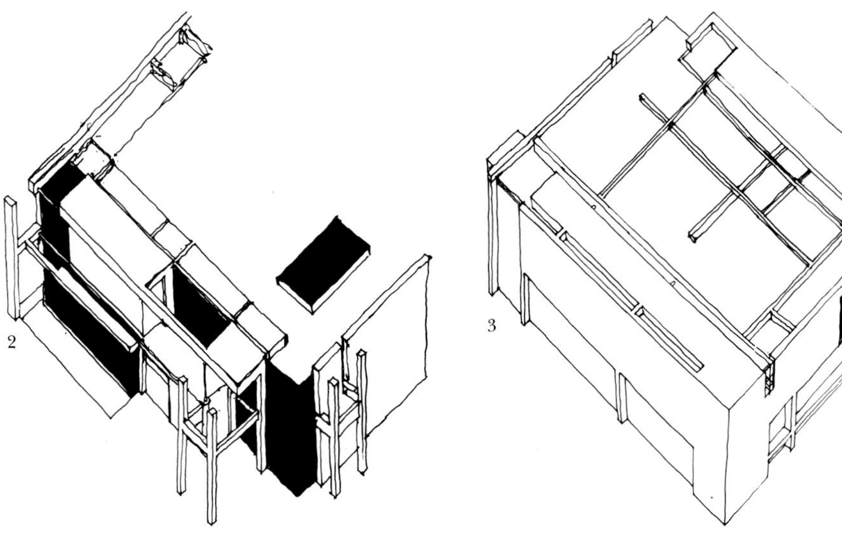

2　3

1 View from the south-east
2 Axonometric study sketch: view from the north-east
3 Axonometric study sketch: view from the south-west
4–6 Ground level study sketches
7–9 Upper level study sketches

1 从东南方向看
2 轴测研究草图：从东北方向看
3 首层研究草图：从西南方向看
4－6 首层平面研究草图
7－9 二层研究草图

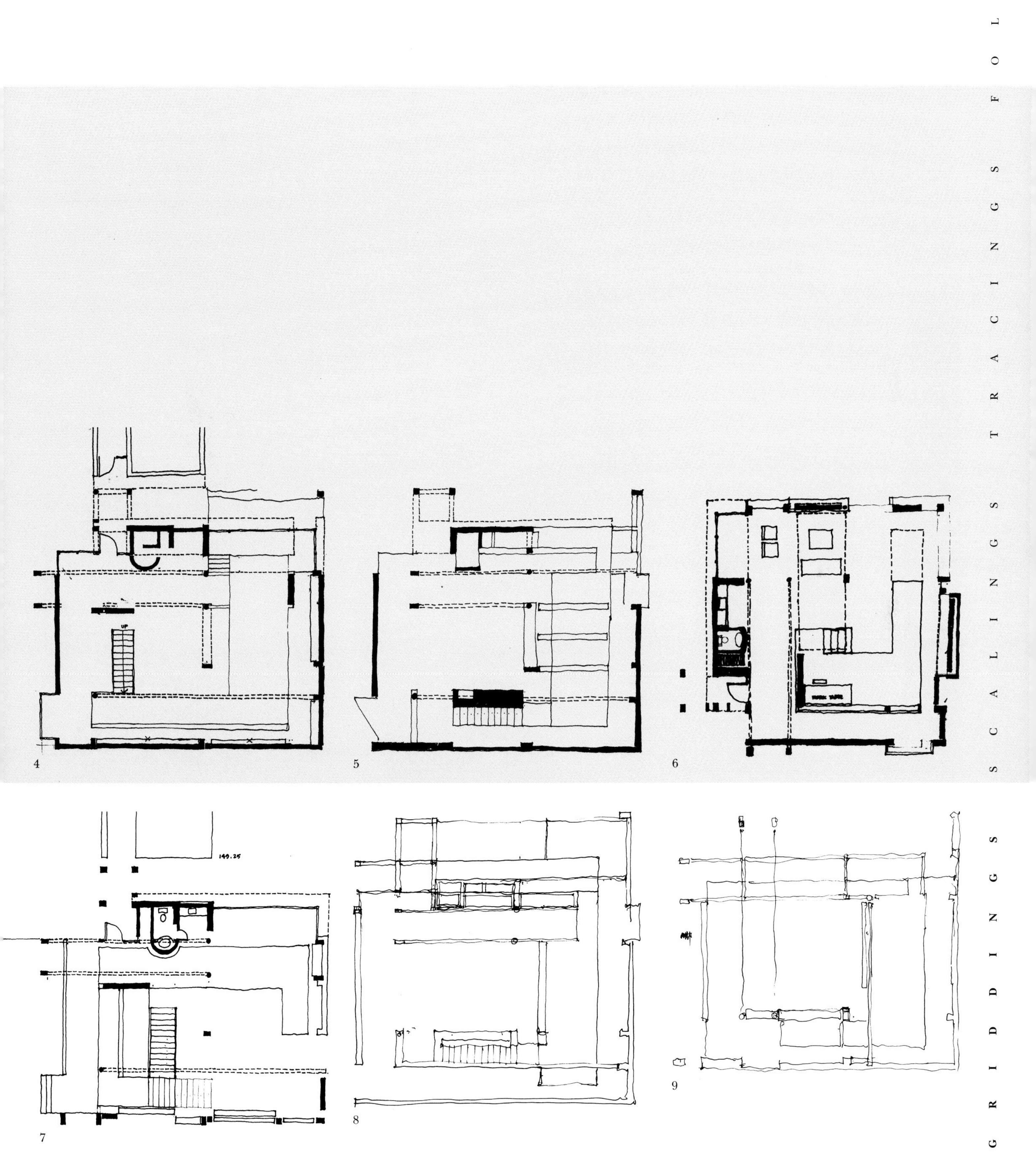

10

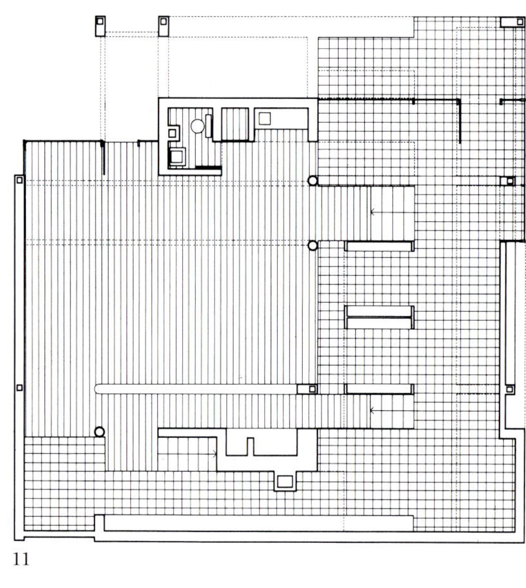

11

10 View from the south-east
11 Ground level plan
12 Axonometric view from the north-east
13 Interior ground level, view from the north
14 Interior ground level, view from the north-west
15 Upper level plan

10 从东南方向看
11 首层平面
12 东北方向轴测图
13 首层室内实景，从北部看
14 首层室内实景，从西北方向看
15 二层平面

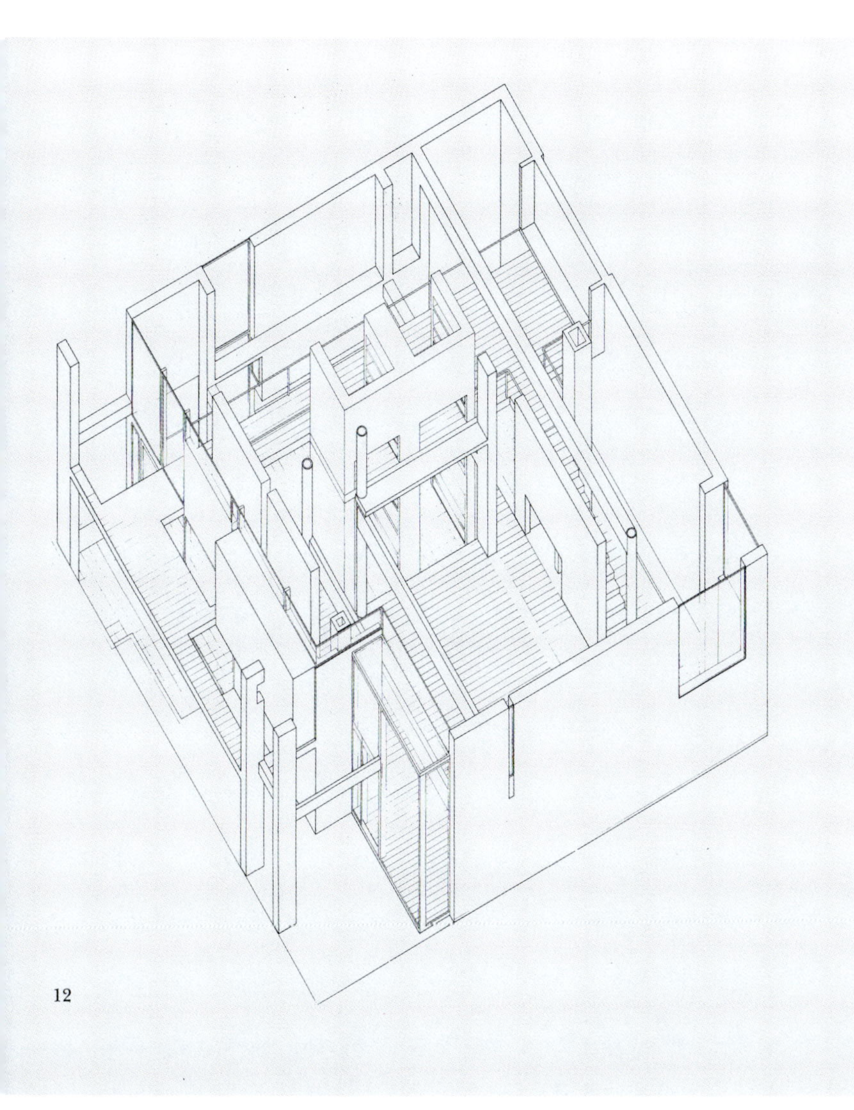

12

13

14

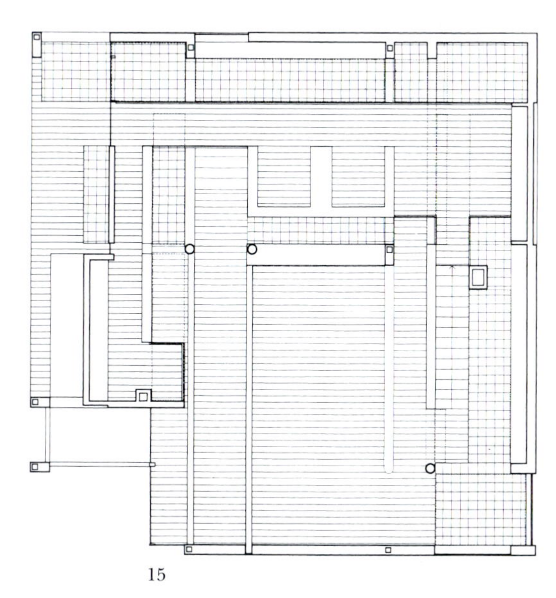

15

16

17

18

19

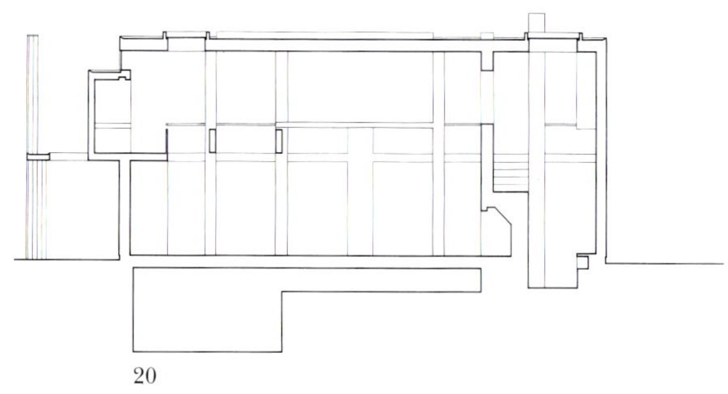
20

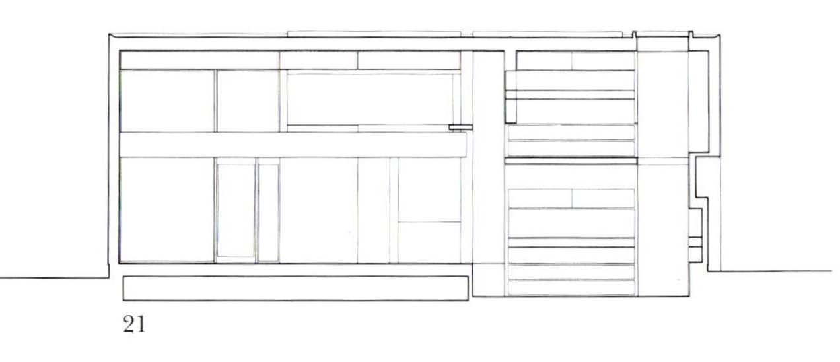
21

16 View from the south-east
17 View from the north
18 Upper level interior, view from the south-west
19 Upper level interior, view from the north-east
20 Section, view from the north
21 Section, view from the west
22 North elevation
23 South elevation
24 West elevation
25 East elevation
26 Ground level interior, view from the north
27 Upper level interior, view from the west

16 从东南方向看
17 从北部看
18 二层室内，从西南方向看
19 二层室内，从东北方向看
20 剖面：从北部看
21 剖面：从西部看
22 北立面
23 南立面
24 西立面
25 东立面
26 首层室内实景，从北部看
27 首层室内实景，从西部看

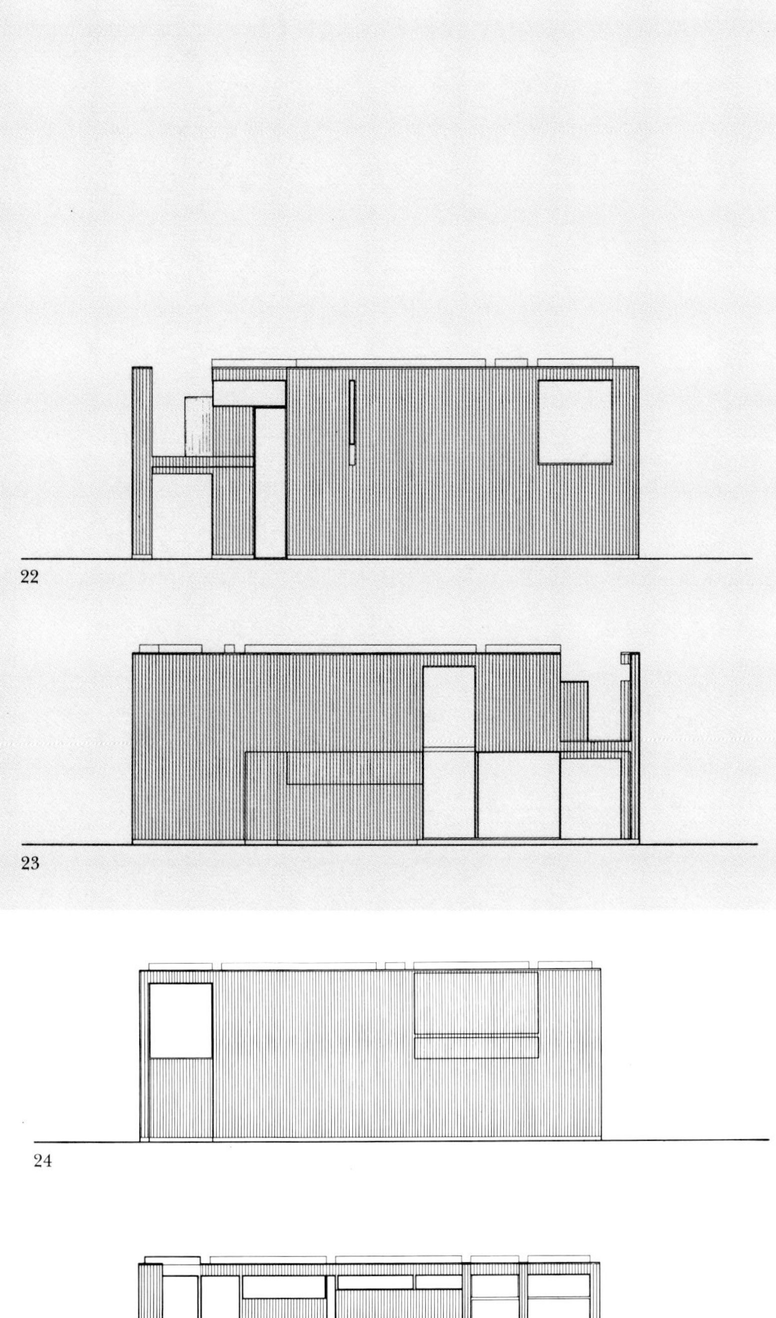

22

23

24

26

27

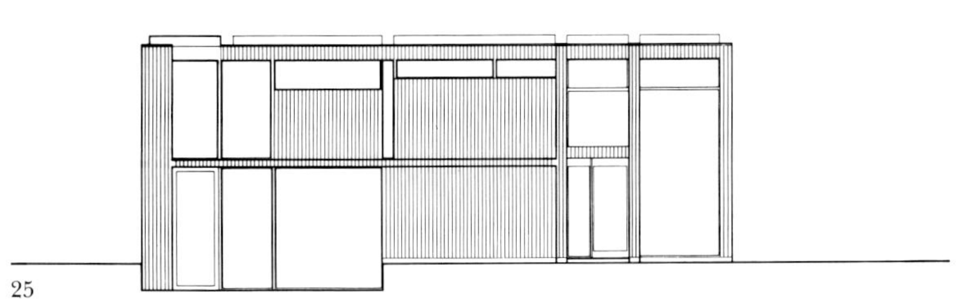

25

House II

Design/Completion 1969/1970
Hardwick, Vermont
Mr and Mrs Richard Falk
2,000 square feet
Wood frame
Exterior: painted wood panels
Interior: painted wall board

住宅Ⅱ

设计/竣工　　1969/1970
哈德威克，佛蒙特州
理查德·福克夫妇
2000平方英尺
木结构
室外：涂漆的木板
室内：涂漆的墙板

The house is situated on the highest point of a 100-acre site with panoramic views on three sides which extend for 20 miles.

The design simulates the presence of trees and hedges, which are non-existent on the barren hilltop, through a sequence of columns and walls. These architectural elements frame and focus the view and ensure a transition from extroverted summer activities to the introverted security of the winter fireplace.

此住宅位于一块100英亩地段的最高点，周围三面拥有延伸20英里的开阔视野。

该设计通过一系列的柱子和墙面模仿了树木和篱笆的形式，而在这个贫瘠的山顶上，树木和篱笆都是不存在的。建筑元素起到了框景的作用，并提供了一种转换，从夏天室外活动到冬天壁炉的室内安全性。

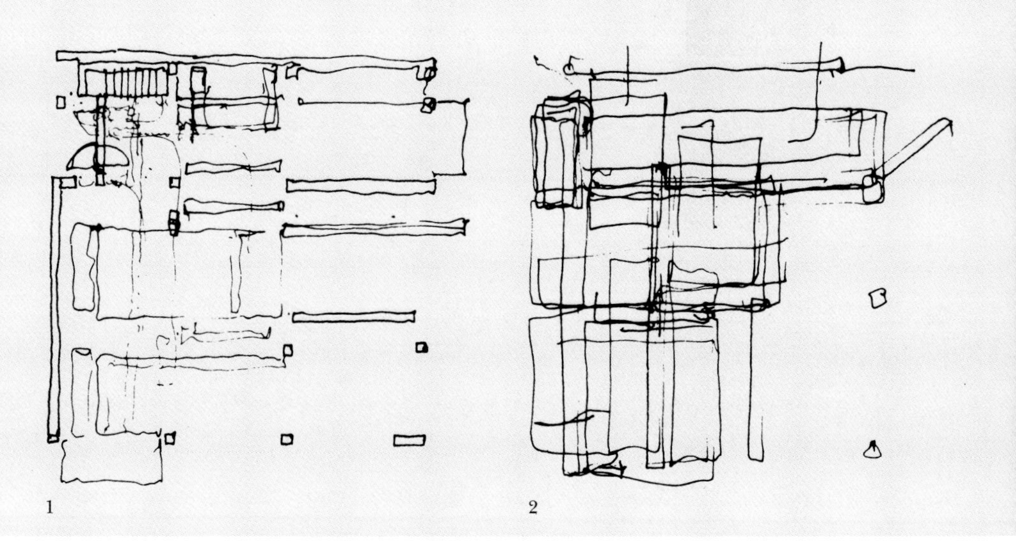

1　　2

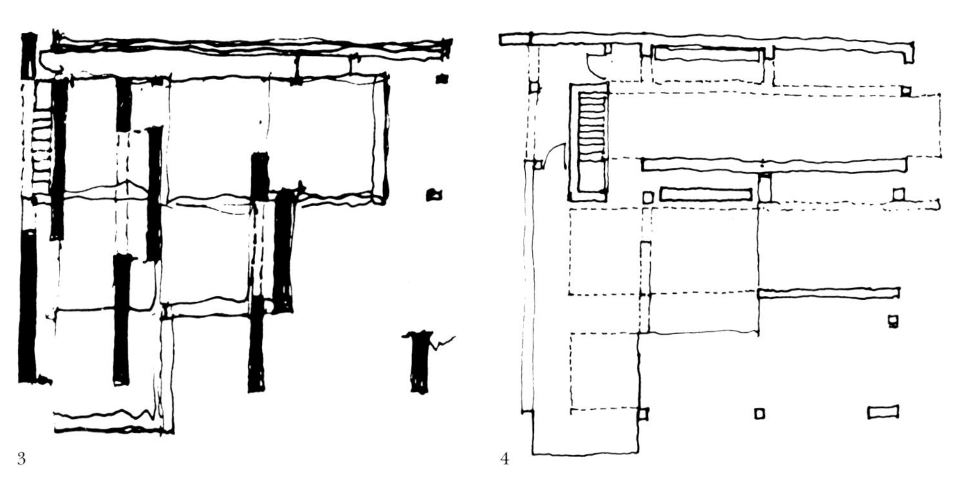

3　　4

1–4 Plan study sketches
5 View from the south
6–8 Plan study sketches

1-4 平面研究草图
5 从正南方向看
6-8 平面研究草图

5

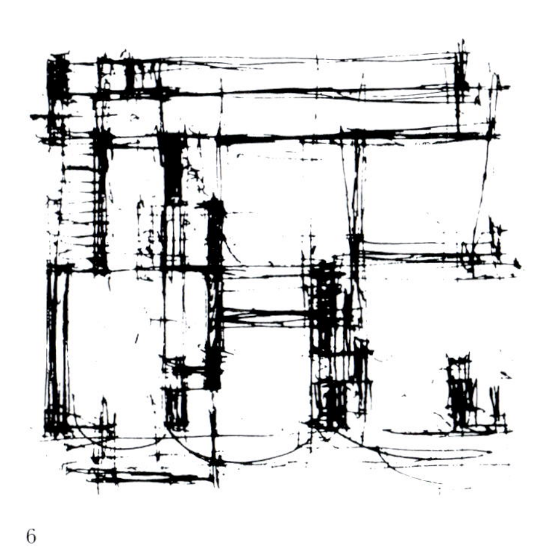

6

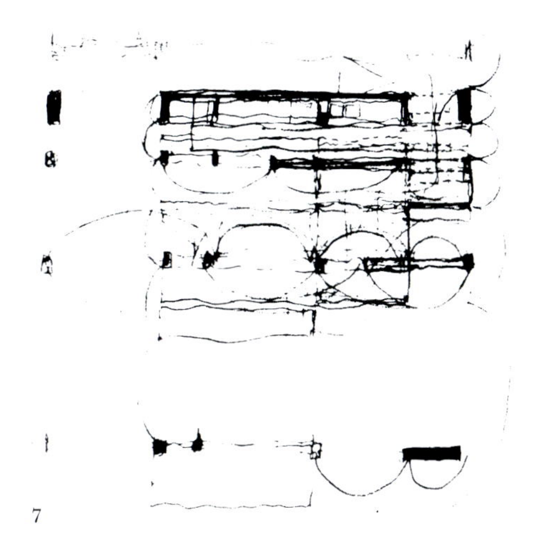

7

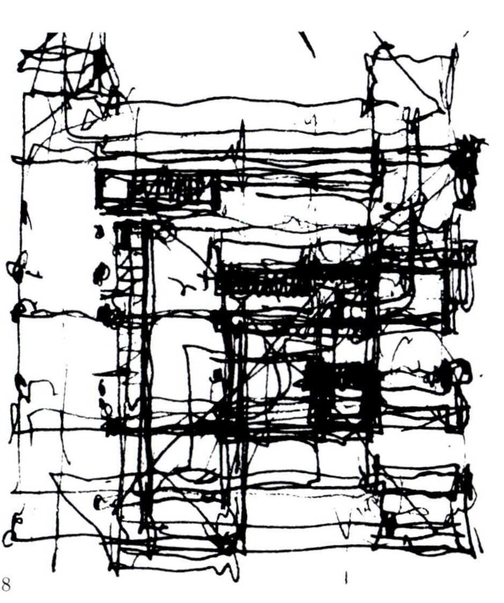

8

9

11

12

13

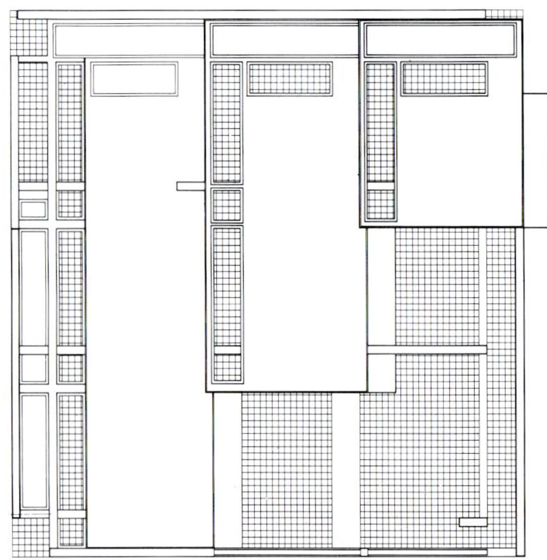

10

9 View from the north
10 Roof plan
11 View from the south-west
12 View from the north-west
13 View from the north-east
14 Upper level plan
15 Ground level plan
16 View from the south-east
17 Detail view from the north-east

9 从正北方向看
10 屋顶平面
11 从西南方向看
12 从西北方向看
13 从东北方向看
14 二层平面
15 一层平面
16 从东南方向看
17 从东北方向看细部

16

17

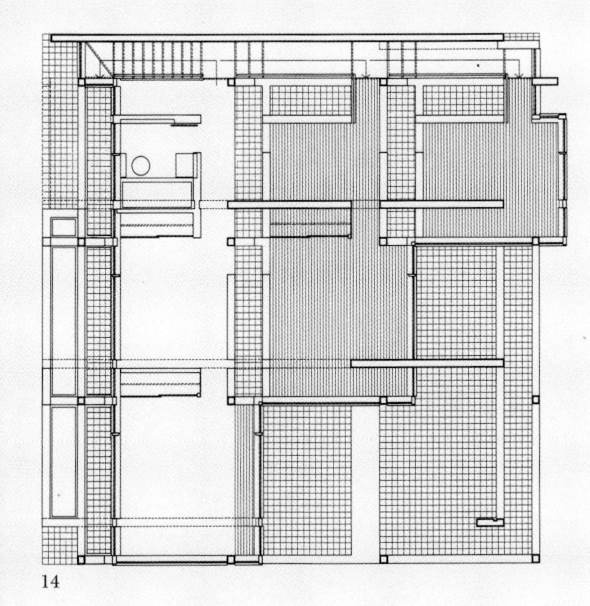

14

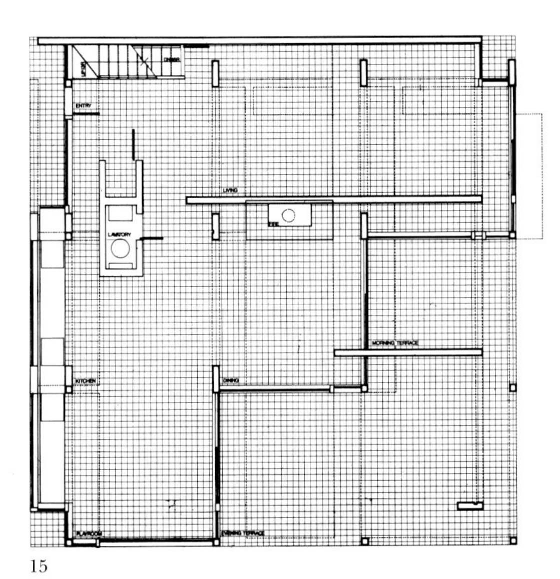

15

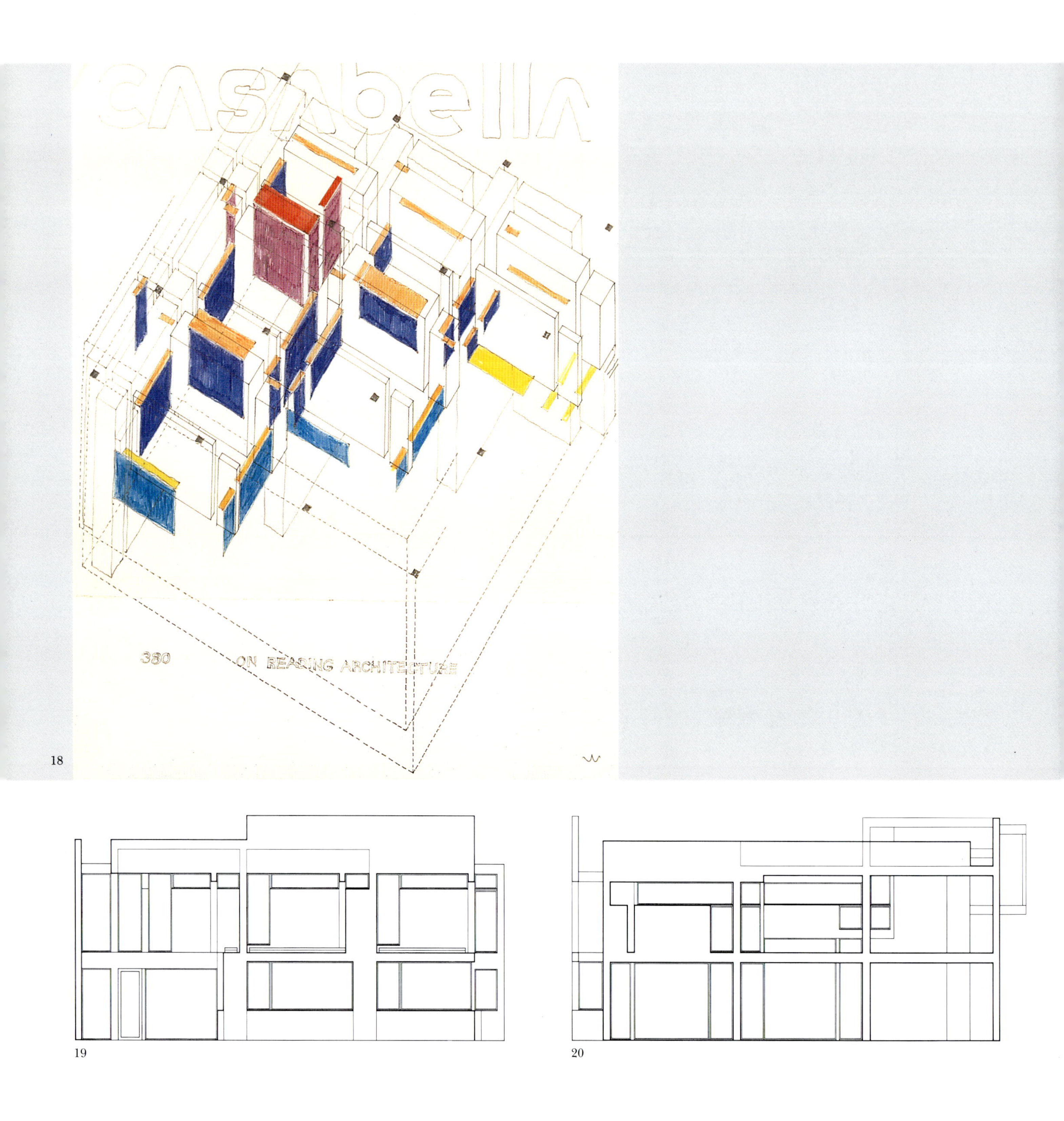

18

19

20

18 Magazine cover visual with building axonometric
19 East elevation
20 South elevation
21 Axonometric
22 Building section
23 Conceptual model, view from the south-east
24 Conceptual model, view from the north

18 带有建筑轴测图的杂志封面
19 东立面
20 南立面
21 轴测图
22 建筑剖面
23 概念模型，从东南方向看
24 概念模型，从正北方向看

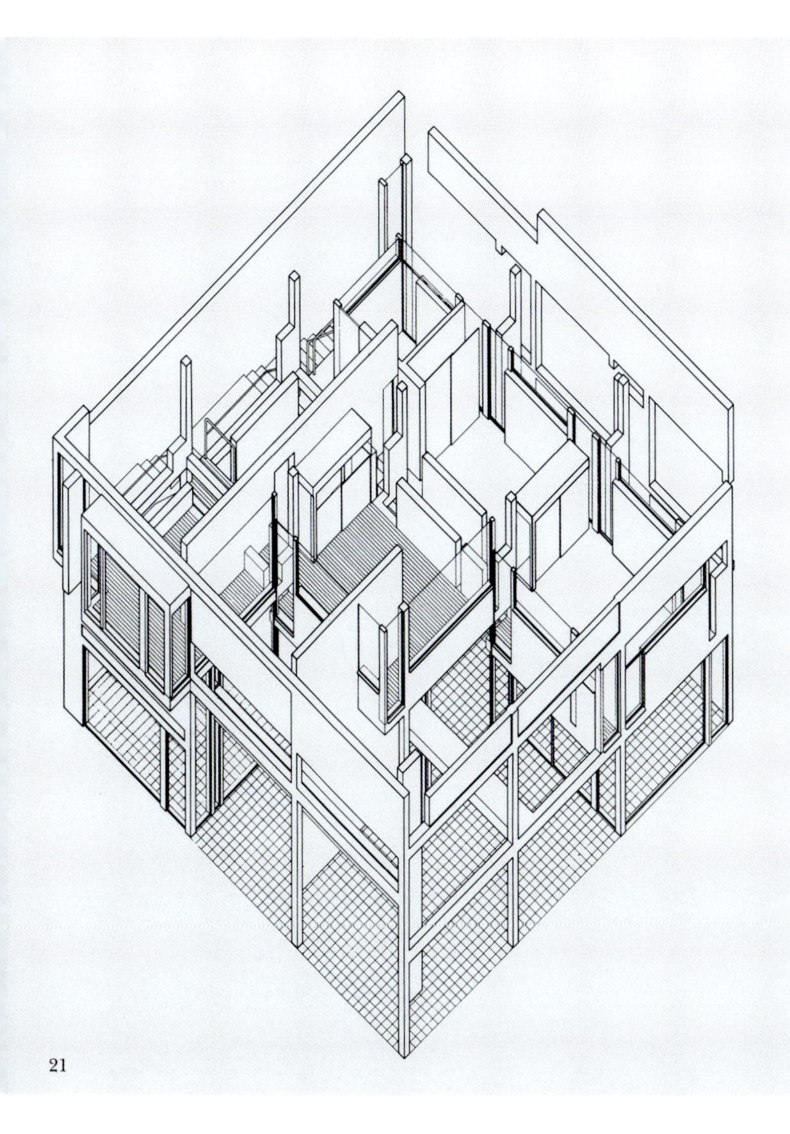

21

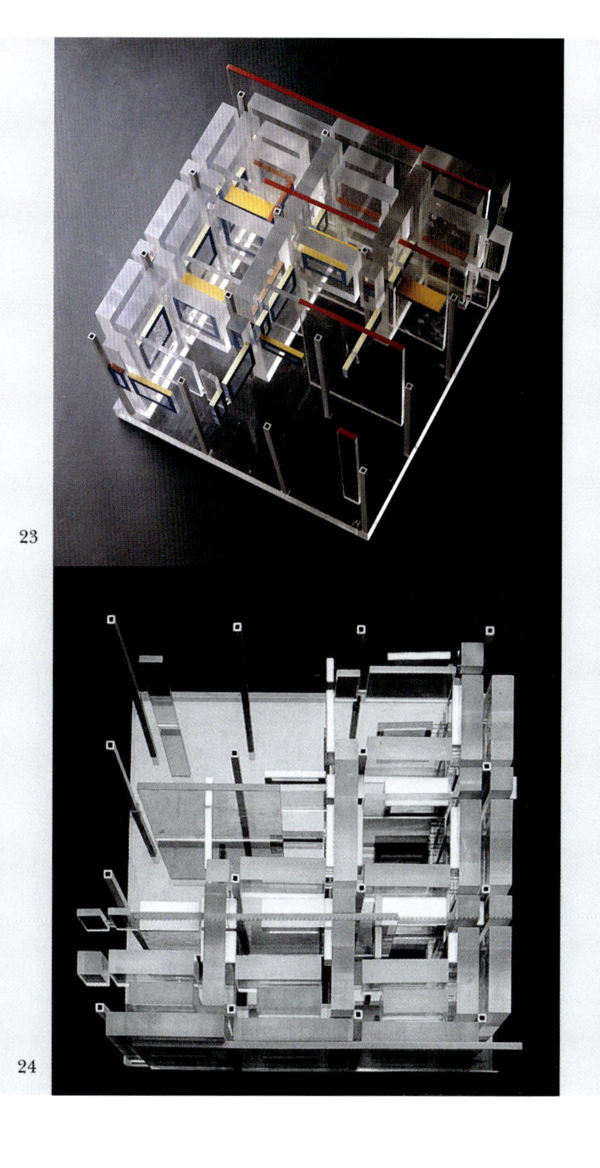

23

24

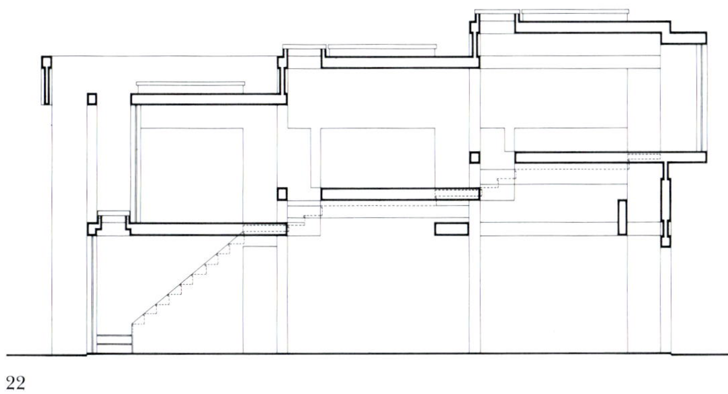

22

House III

Design/Completion 1969/1971
Lakeville, Connecticut
Mr and Mrs Robert Miller
3,500 square feet
Wood frame
Exterior: painted wood panels
Interior: painted wall board

住宅Ⅲ

设计/竣工 1969/1971
莱克维尔，康涅狄格州
罗伯特·米勒夫妇
3500平方英尺
木结构
室外：涂漆的木板
室内：涂漆的墙板

House III is the third in a series of investigatory projects that search for the form–meaning relationship in architecture. The owner enters the house as an intruder in an attempt to regain possession and, consequentially, destroys the unity and completeness of the architectural structure. The interior void of the structure acts as both background and foil, as a conscious stimulant for the activity of the owner.

No longer concerned with imposing some preconceived idea of good taste, the house works dialectically to stimulate the owner to a new kind of participation.

住宅Ⅲ是一系列探讨建筑中形式意义之关系的研究性项目的第三个住宅。业主被看成是一个入侵者带着他的目的进入住宅，他企图收回所有权并破坏建筑结构的统一性和完整性。室内空间的结构起到背景和衬托的作用，同时在意识方面是对业主活动的一个刺激。

不再考虑将一些有良好空间体验的设想加入到住宅中，该住宅以辨证的方式促使业主参与到一个新的类型中。

1

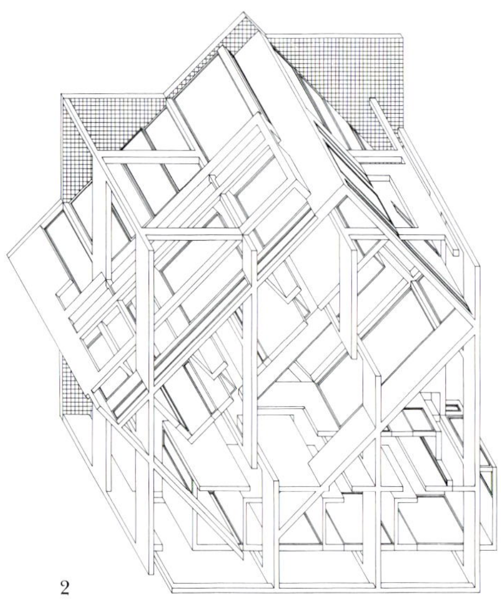

2

1 View from the east
2 Axonometric
3 View from the east
4 First level plan
5 Second level plan

1 从正东方向看
2 轴测图
3 从东部看
4 首层平面
5 二层平面

3

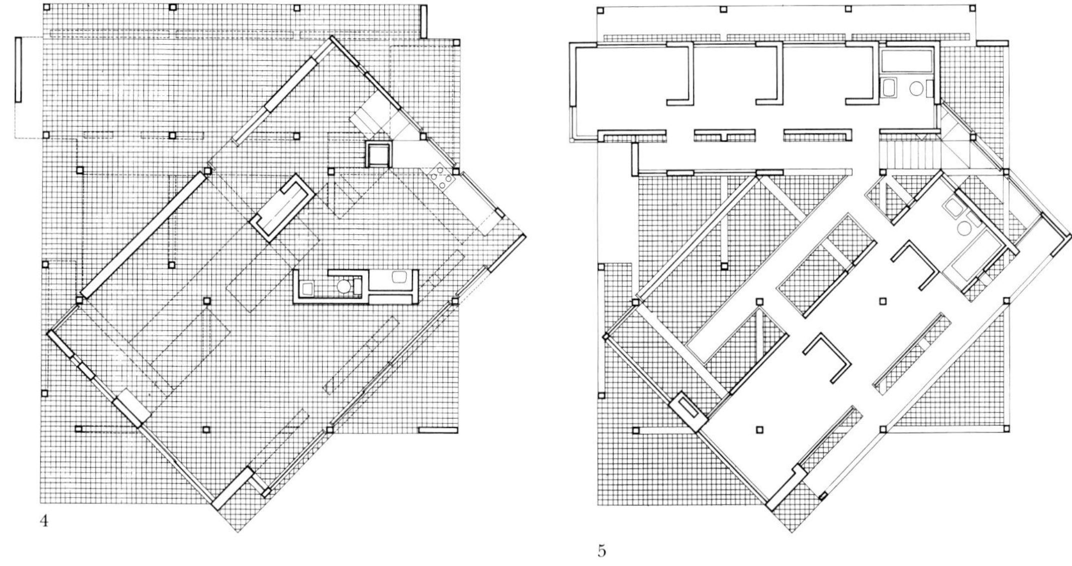

4

5

6

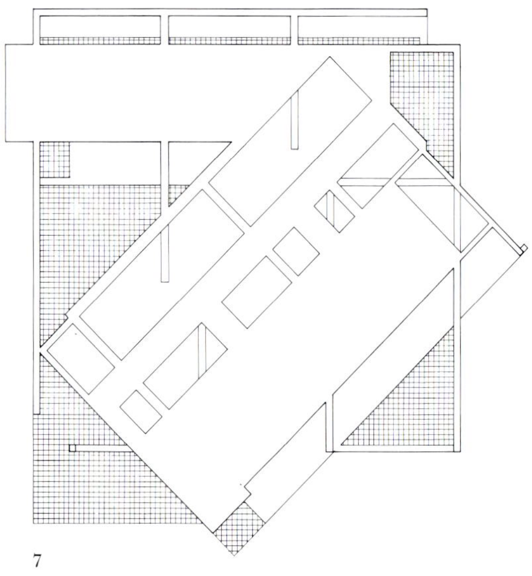

7

6 Interior view
7 Roof level plan
8 South-east elevation
9 North-west elevation
10 North-east elevation
11 South-west elevation
12–14 Interior views

6 室内实景
7 屋顶平面
8 东南立面
9 西北立面
10 东北立面
11 西南立面
12–14 室内实景

12

13

14

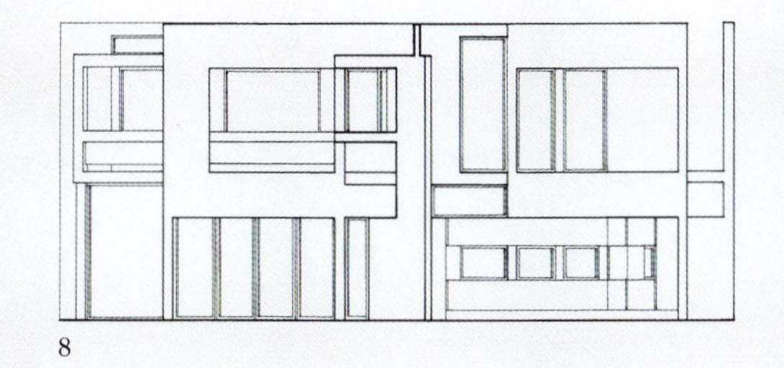

8

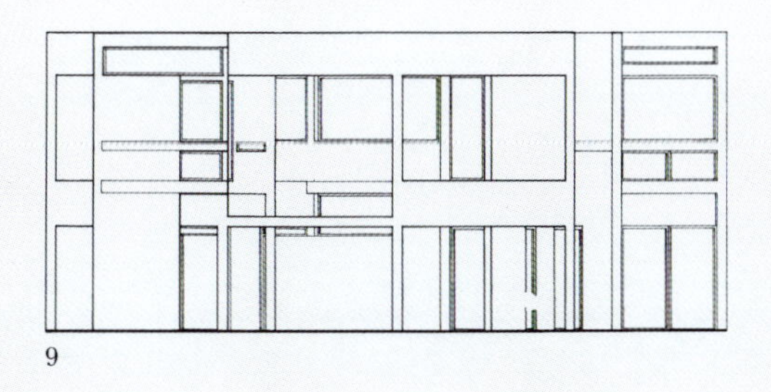

9

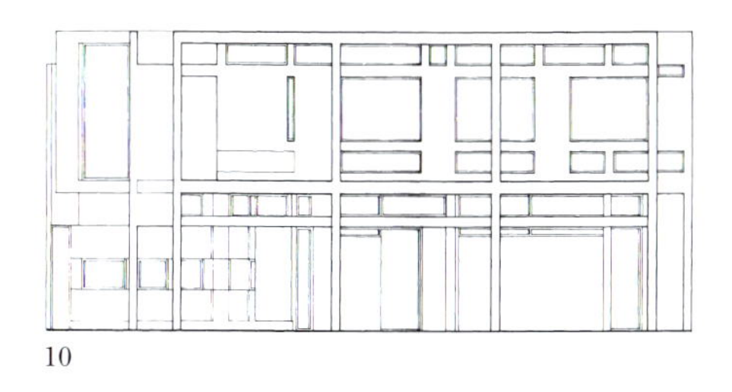

10

11

House IV

Design 1971
Falls Village, Connecticut
3,000 square feet

住宅Ⅳ

设计　　1971
福尔斯村，康涅狄格州
3000平方英尺

In House IV, a physical environment has been produced which is semantically and culturally diminished or more neutral. To do this, the conceptual structure has been overstressed to give it primacy over the perceptual or traditional structure of understanding an architecture.

This house is an attempt to produce a physical environment which could be generated by a limited set of formational and transformational rules. Spatial relationships are in the syntactic domain of architecture and, since our present knowledge of the nature of these relationships is rather imprecise, it is difficult to code an architecture or produce a precise set of transformational rules.

在住宅Ⅳ中，产生了一个自然环境，这个环境在语义和文化上的含义已经消失了，而且更加趋于中性。为了达到这一目的，概念上的结构被着重强调，从而使其地位在先，或者位于传统的支撑建筑的结构之前。

该住宅试图创造一种自然环境，并试图将此环境在有限的结构和转换的规则下产生。空间关系存在于建筑的符号关系学领域中，由于我们现有的知识对这种内在关系的了解不够精确，所以很难破译建筑的密码或者制造出精确的转换规则。

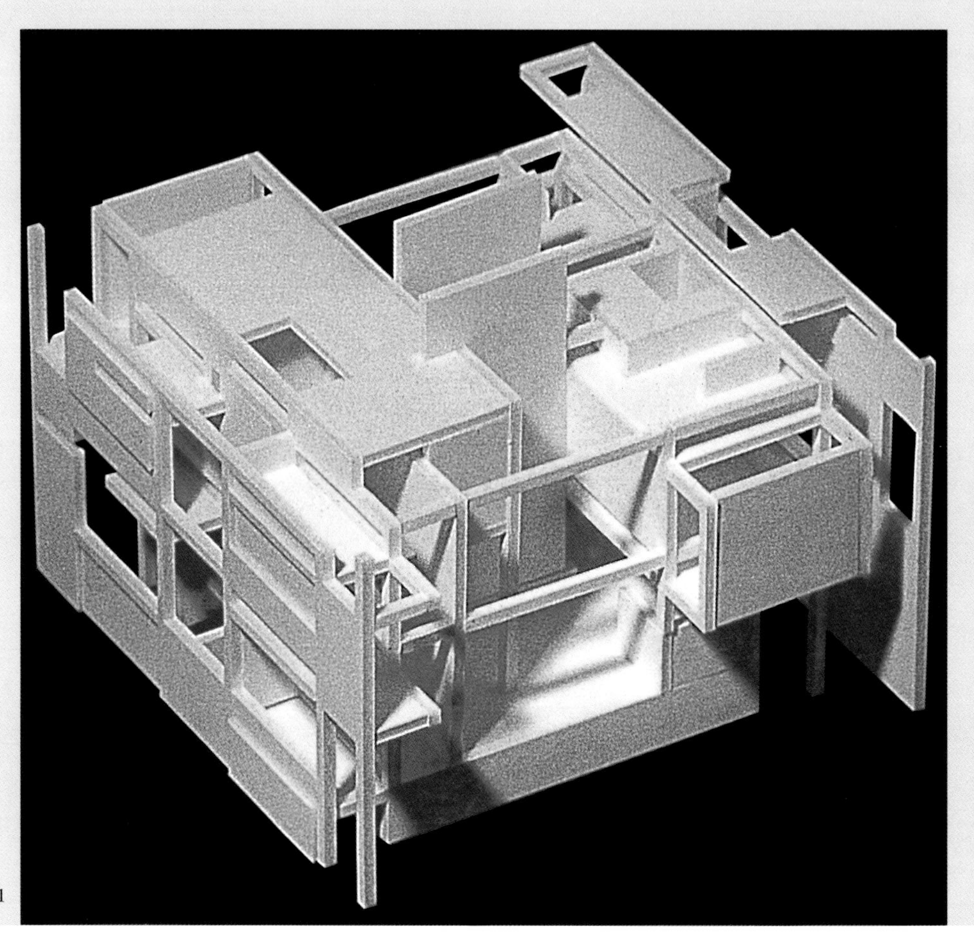
1

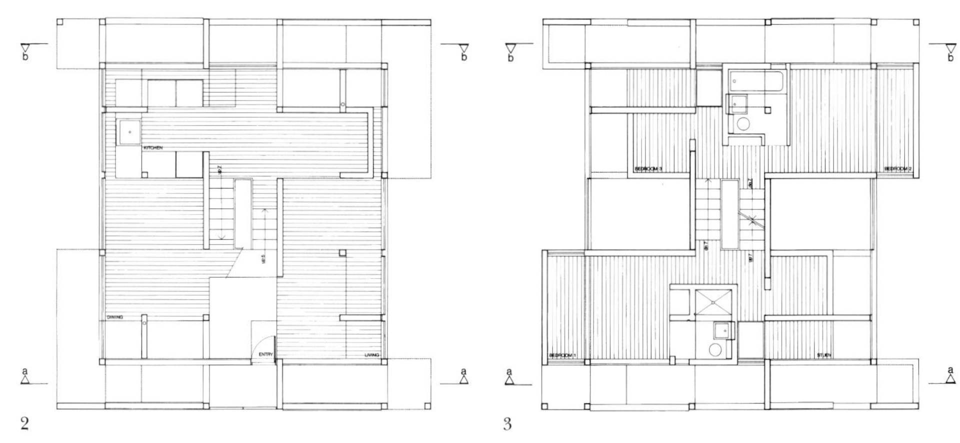
2　　3

1 Study model, view from the north-east
2 Ground level plan
3 Upper level plan
4–7 Axonometric drawings

1 研究模型，从东北方向看
2 首层平面
3 二层平面
4－7 轴测图

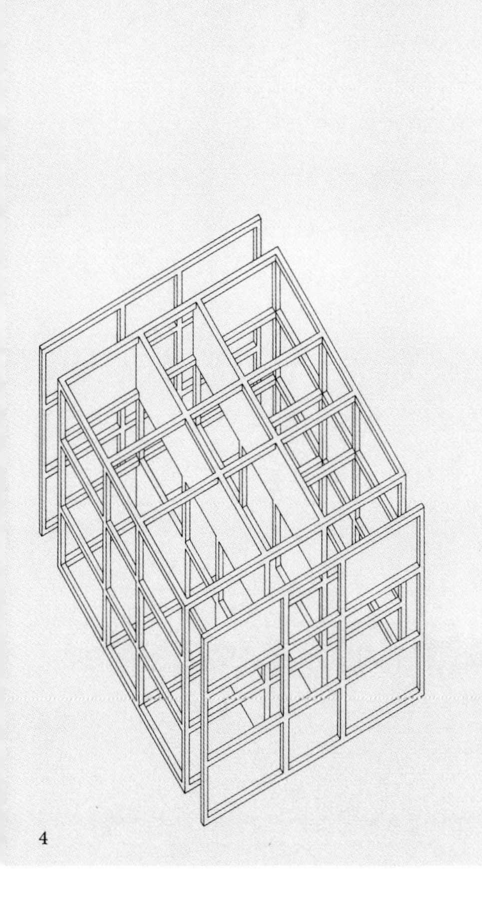

4

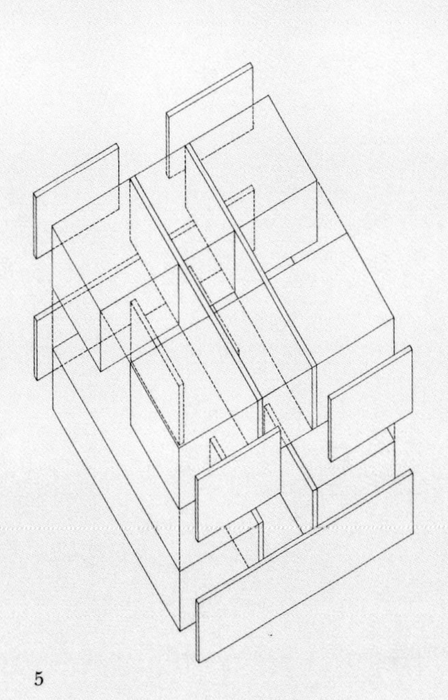

5

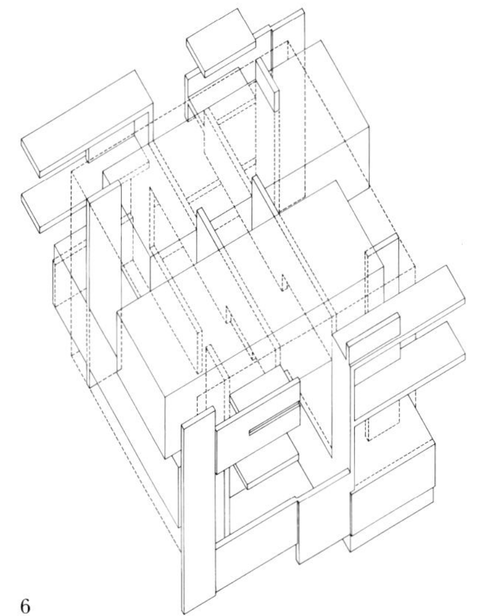

6

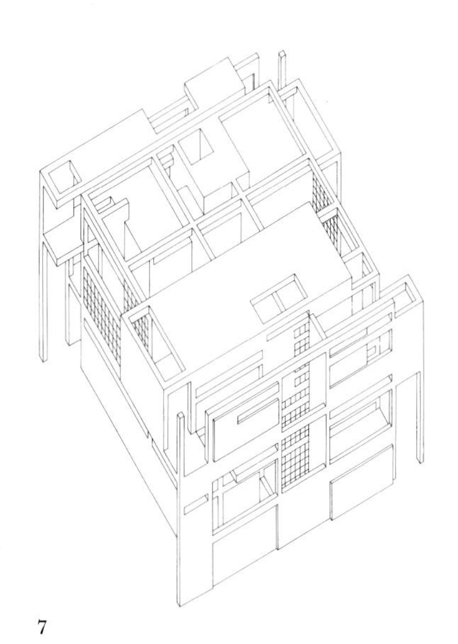

7

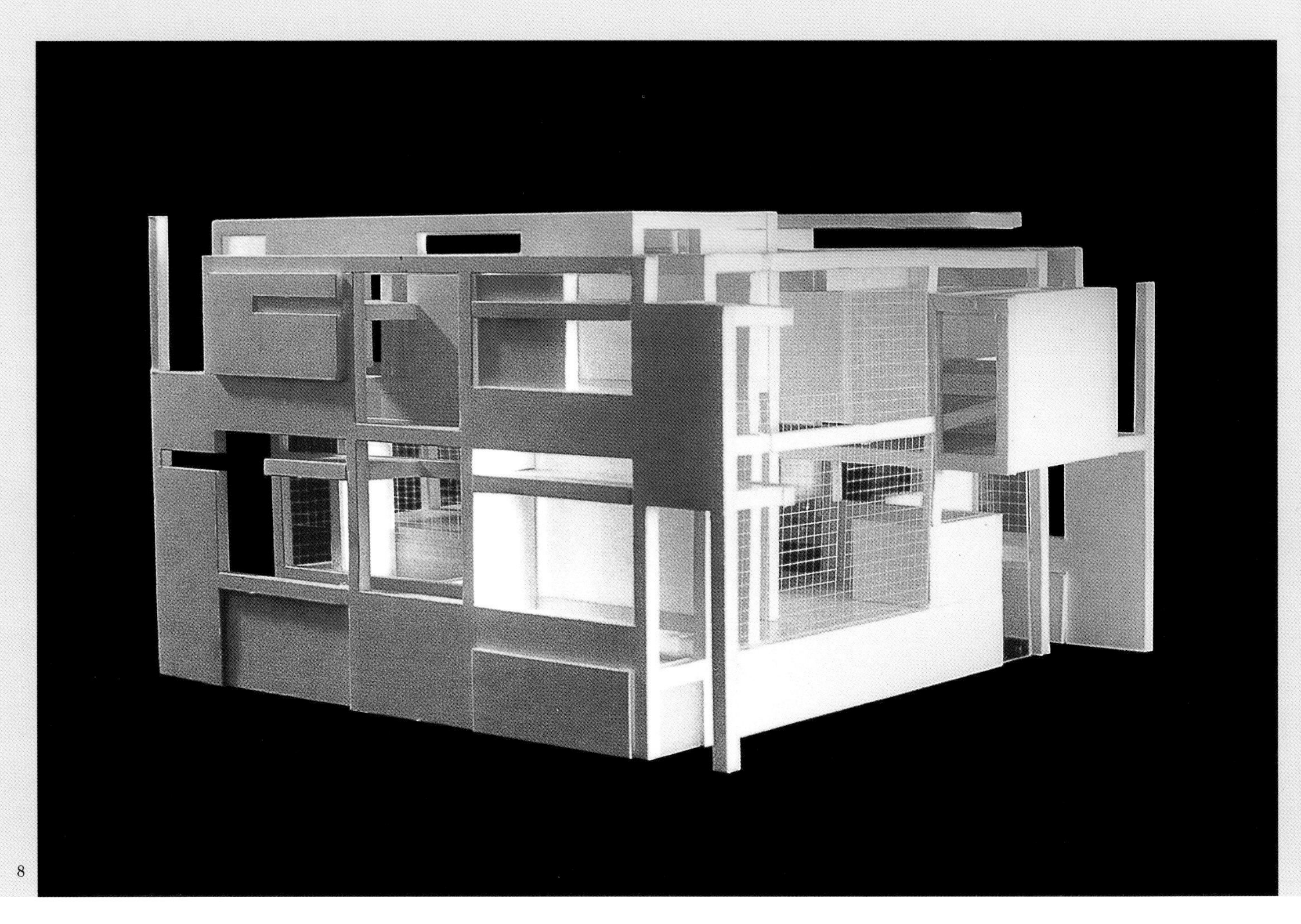

8

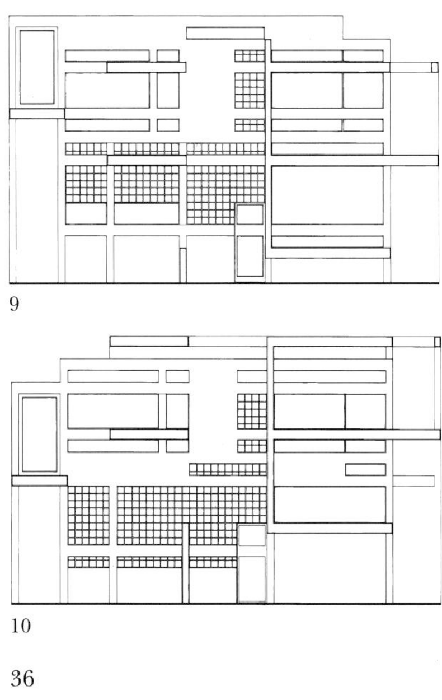

9

10

8 Study model, view from the north-west
9 Section AA
10 Section BB
11 North elevation
12 East elevation
13 South elevation
14 West elevation
15 Study model
16 Study model, view from the north-east
17 Study model

8 研究模型，从西北方向看
9 AA 剖面
10 BB 剖面
11 北立面
12 东立面
13 南立面
14 西立面
15 研究模型
16 研究模型，从东北方向看
17 研究模型

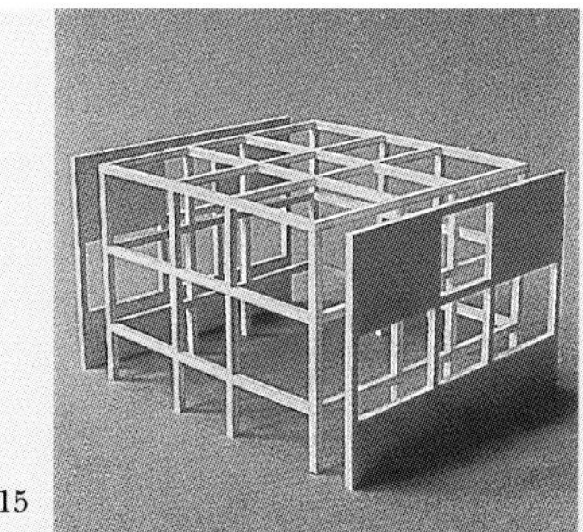
15

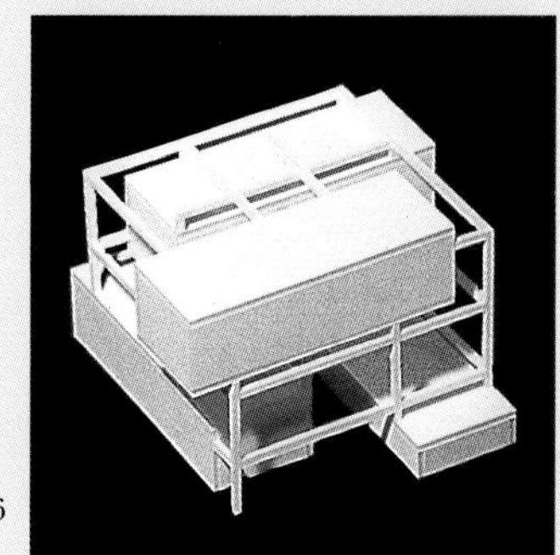
16

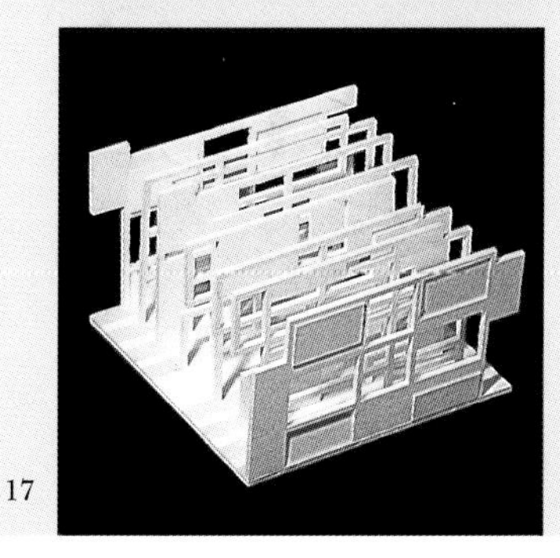
17

11

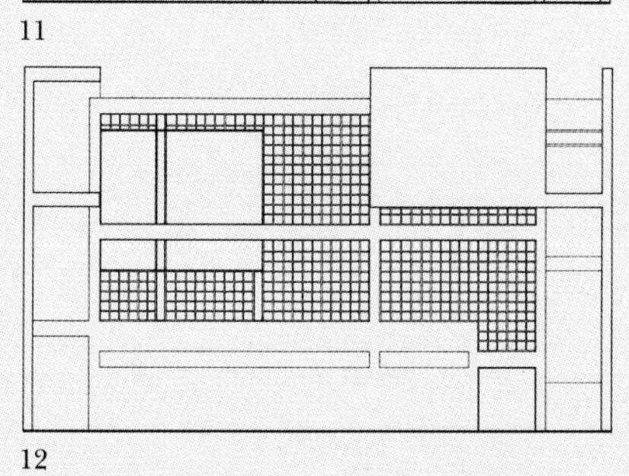
12

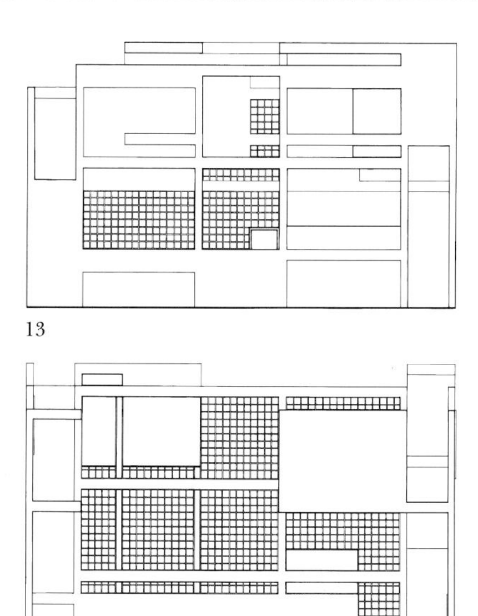
13

14

House VI

Design/Completion 1972/1975
Cornwall, Connecticut
Mr and Mrs Richard Frank
2,000 square feet
Wood frame
Exterior: painted wood panels
Interior: painted wall board

住宅Ⅵ

设计/竣工　　1972/1975
康沃尔，康涅狄格州
理查德·弗兰克夫妇
2000平方英尺
木结构
室外：涂漆的木板
室内：涂漆的墙板

This weekend house on a small rural site in north-western Connecticut provides the owners—a photographer and his wife—with a sensuous and playful environment, full of continuously changing light, shadows, color, and textures. The house is a studio landscape, providing an abstract background for the photography of still life and people. In doing so, the house and its occupants become part of a series of daily "living portraits."

坐落在康涅狄格州西北郊区的这个周末住宅，为居住者——摄影师夫妇提供了一个风景优美而有趣的环境，环境中充满了连续的光线、阴影、颜色、质感的变化。房子是一间风景工作室，为静谧的生活和人们的摄影提供了抽象的背景。为达到这一目的，建筑和它的居住者已经成为一系列日常"活雕塑"的组成部分。

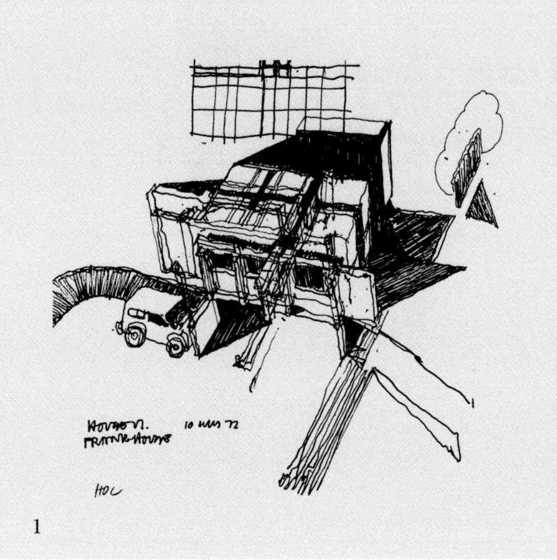

1

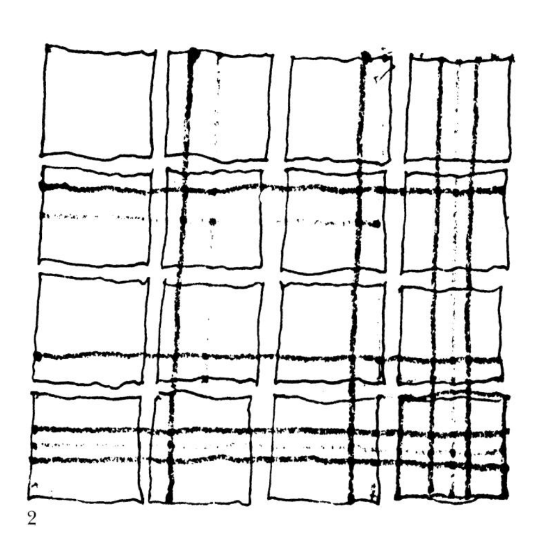

2

1 Study sketch
2 Plan study sketch
3 View from the east
4–5 Plan study sketches

1 研究草图
2 平面研究草图
3 从东部看
4-5 平面研究草图

3

4

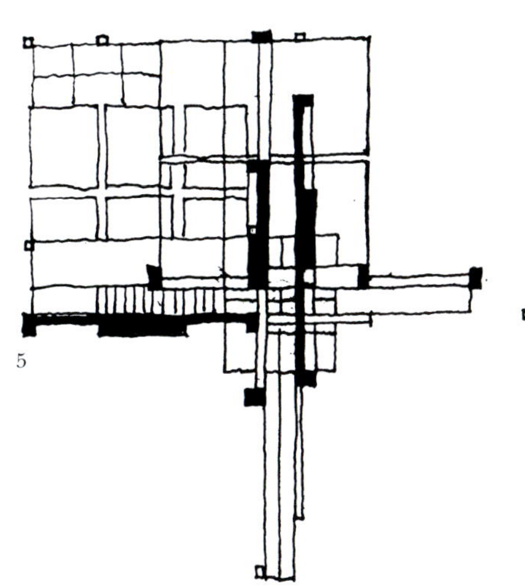

5

6

8

9

10

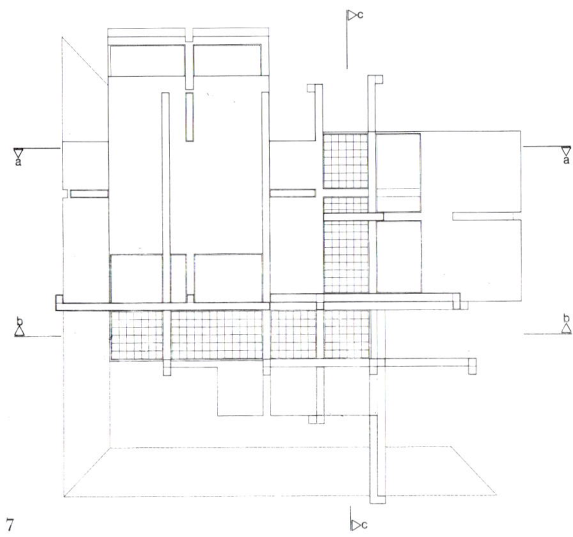

7

6 View from the west
7 Roof plan
8 Partial west elevation, view from the north-west
9 Detail view from the north-east
10 Entry, view from the north
11 First level plan
12 Second level plan
13 Living room, view from the west
14 Living room, view from the south-east

6 从西部看
7 屋顶平面
8 部分西立面，从西北方向看
9 从东北方向看细部
10 入口，从正北方向看
11 首层平面
12 二层平面
13 起居室，从西部看
14 起居室，从东南方向看

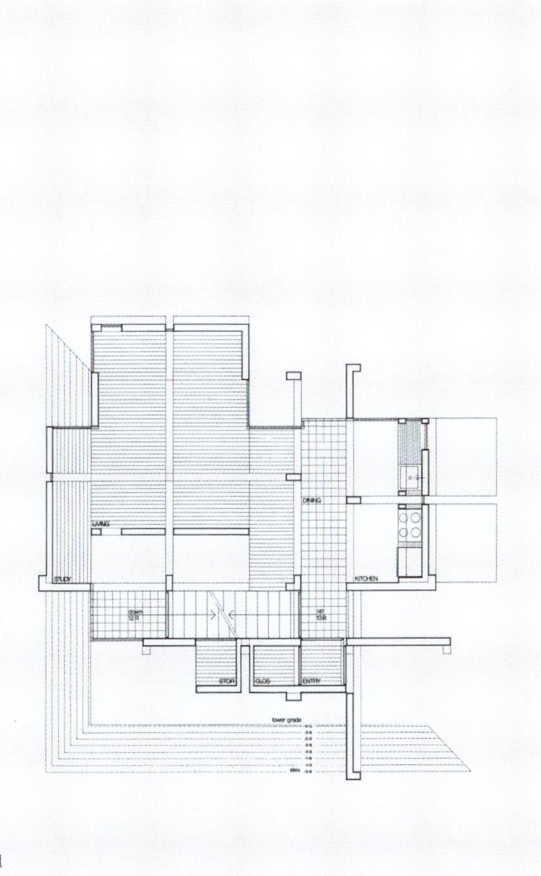

11

13

14

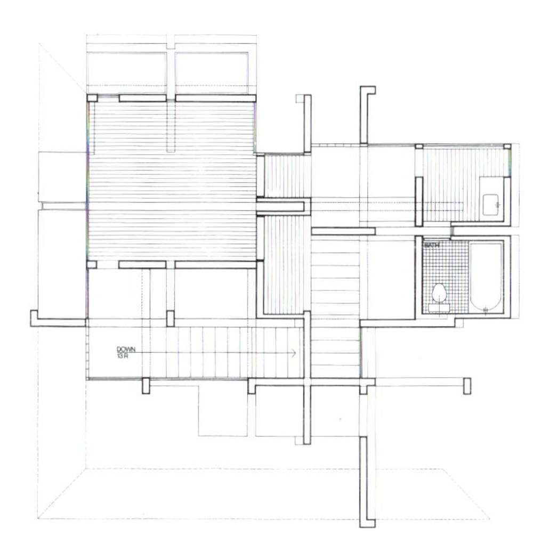

12

15

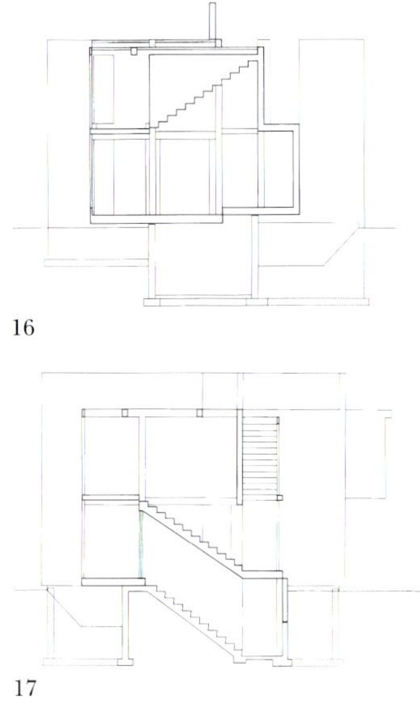

16

17

15 Interior stair, view from second level
16 Section CC
17 Section BB
18 West elevation
19 South elevation
20 East elevation
21 North elevation
22 Interior stair, view from the north-west
23 Interior stair, view from the north
24 Interior stair, view from the south

15 室内，从二层看
16 CC 剖面
17 BB 剖面
18 西立面
19 南立面
20 东立面
21 北立面
22 室内楼梯，从西北方向看
23 室内楼梯，从北部看
24 室内楼梯，从南部看

22

23

18

19

24

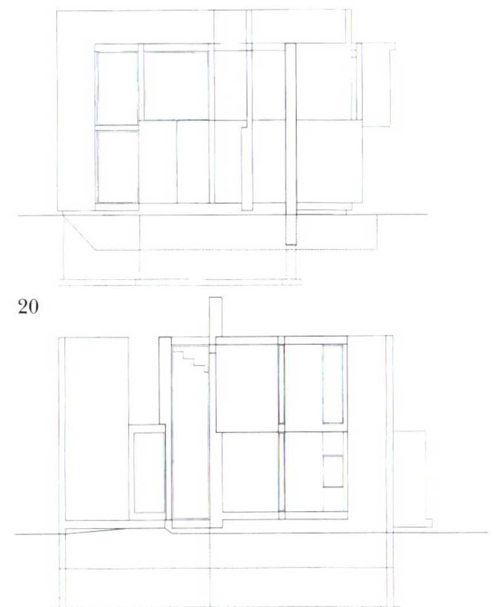

20

21

House X

Design 1975
Bloomfield Hills, Michigan
Mr and Mrs Arnold Aronoff
3,000 square feet

住宅X

设计　　1975
布卢姆菲尔德希尔斯，密歇根州
阿诺德·阿罗诺夫夫妇
3000平方英尺

This private residence is situated on a large, wooded, sloping site, adjacent to a country club. It is surrounded by a swimming pool, a tennis court and a summer house. The design uses the slope of the land in such a way that the natural landscape runs through the house, splitting it into four quadrants and reducing its scale.

这栋私人住宅位于一个大的树木繁茂的斜坡上，与一个乡村俱乐部相邻。它周围被一个游泳池、一个网球场和一栋避暑别墅所包围。设计合理地利用地形斜坡，让自然的风景穿过房子，将其分为四个部分，同时降低了建筑的尺度。

1

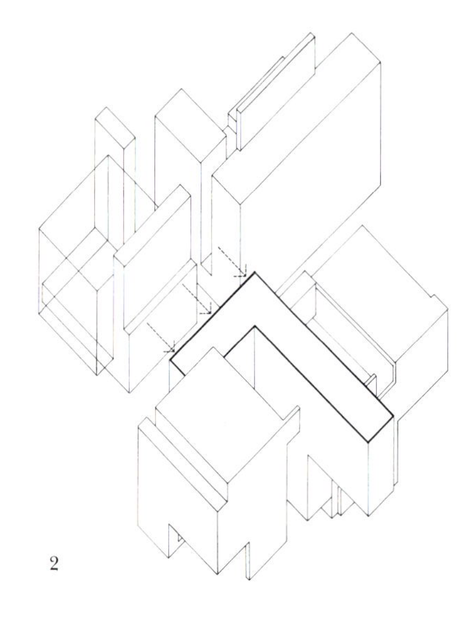

2

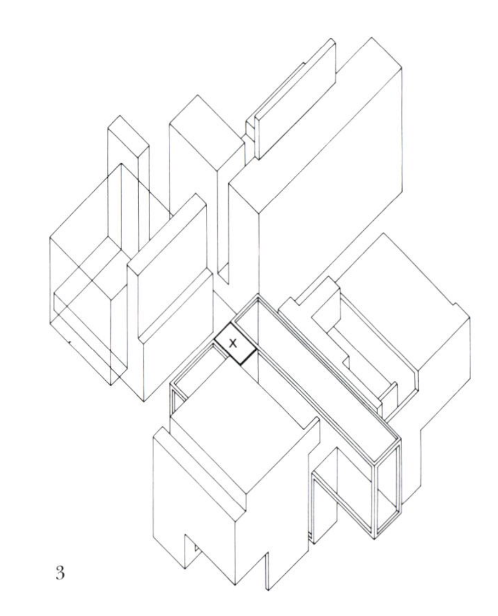

3

1 Model (Scheme G), view from the north-east
2 Axonometric diagram showing east–west arm of glass el pushing into north-east quadrant
3 Axonometric diagram showing perimeter frame of glass el
4 Axonometric diagram showing center as point of intersection
5 Axonometric diagram showing introduction of square gridding in glass el
6 Axonometric diagram showing glass el pulled away from north-east quadrant and pushing into north-west quandrant
7 Plan diagram of two north quadrants showing glass el pulled

1 模型（方案G），从东北方向看
2 东西方向轴测图，玻璃房子插入建筑东北角部分
3 表现玻璃房子边界框架的轴测图
4 表达交叉中心的轴测图
5 表现玻璃房子中的方格之轴测图
6 轴测图，表达玻璃房子被从建筑东北角拉出来，插入西北角
7 北边两部分的平面，表达被拉出的玻璃和盒子

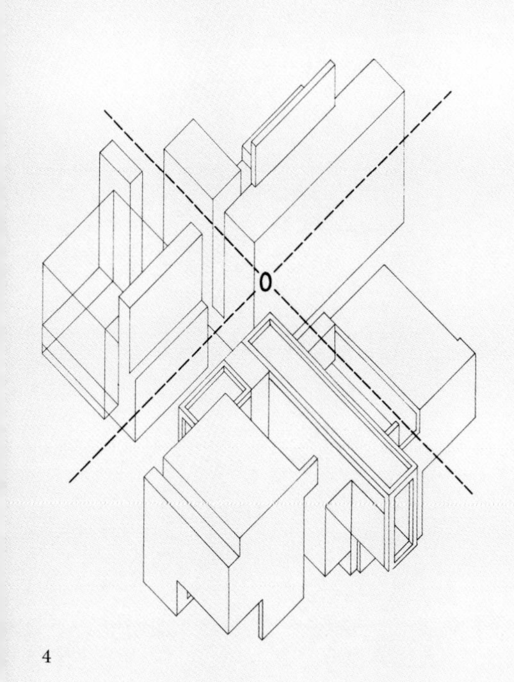

4

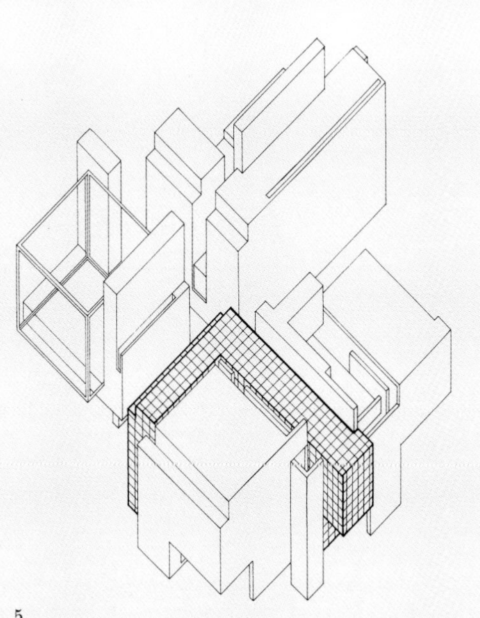
5

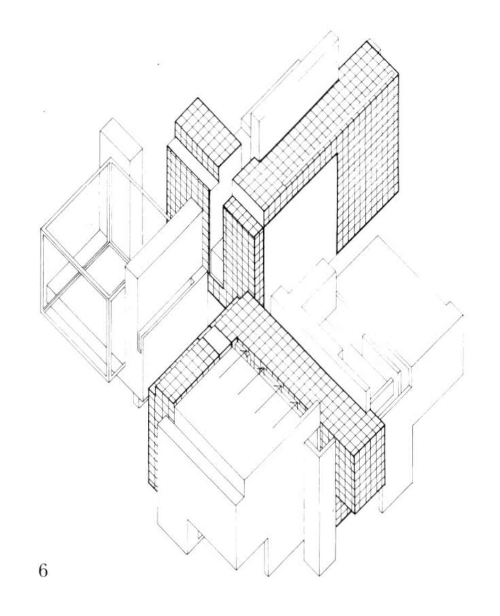
6

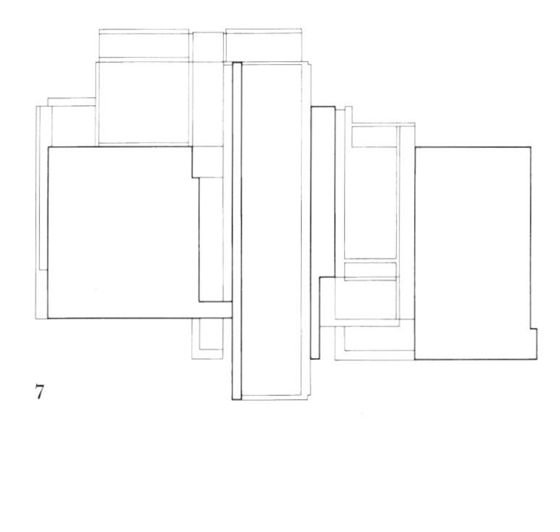
7

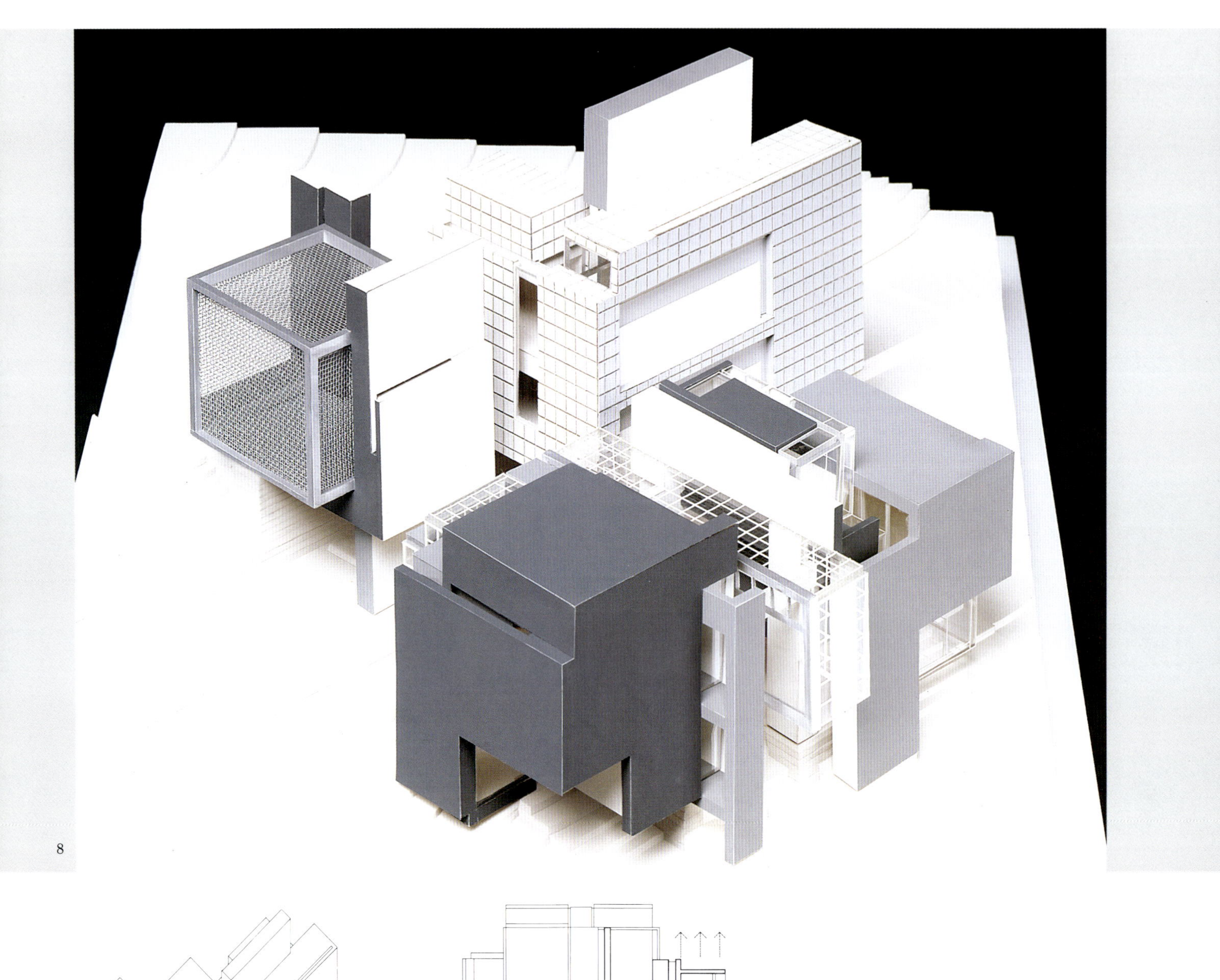

8

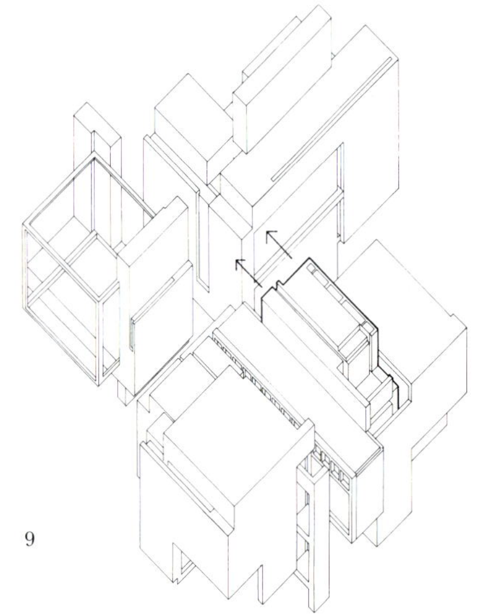

9

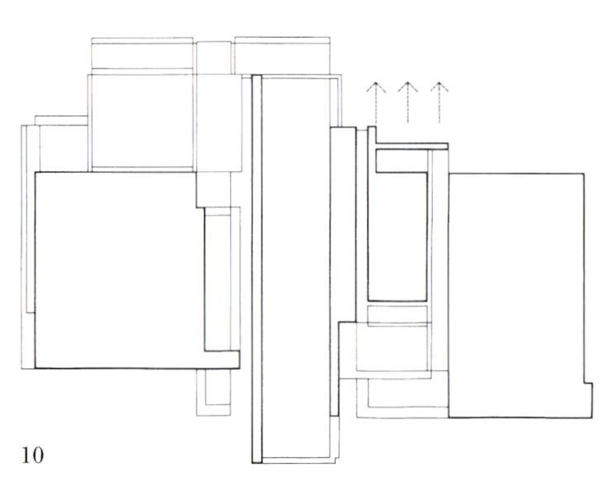

10

8 Axonometric model (Scheme H), view from the north-east
9 Axonometric diagram showing upper portion of north-west quadrant seemingly pulled away from the north face of the quadrant
10 Plan diagram of two north quadrants showing upper portion of north-west quadrant seemingly pulled away from the north face of the quadrant
11 Scheme G, first level plan
12 Scheme G, second level plan
13 Scheme G, third level plan
14 West elevation collage
15 East elevation collage

8 轴测模型（方案H,），从东北方向看
9 表现建筑西北部的上部的轴测图，看起来好像是被从北部表面拉出来的一部分
10 建筑北部两个部分的平面，西北部分的上部看起来好像是从北部表面拉出来的一部分
11 方案G，首层平面
12 方案G，二层平面
13 方案G，三层平面
14 西立面构成
15 东立面构成

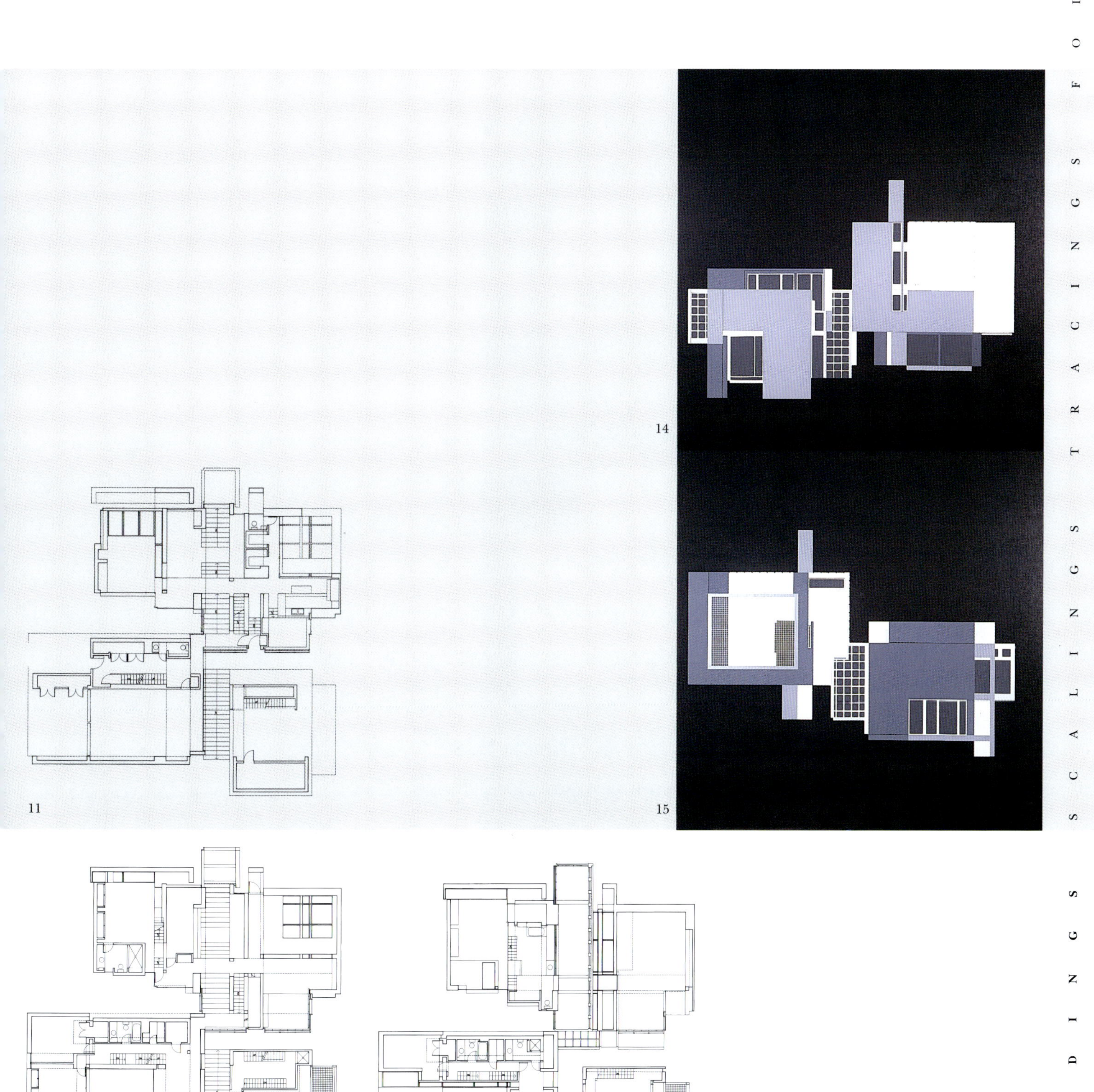

14

11

15

12

13

16

17

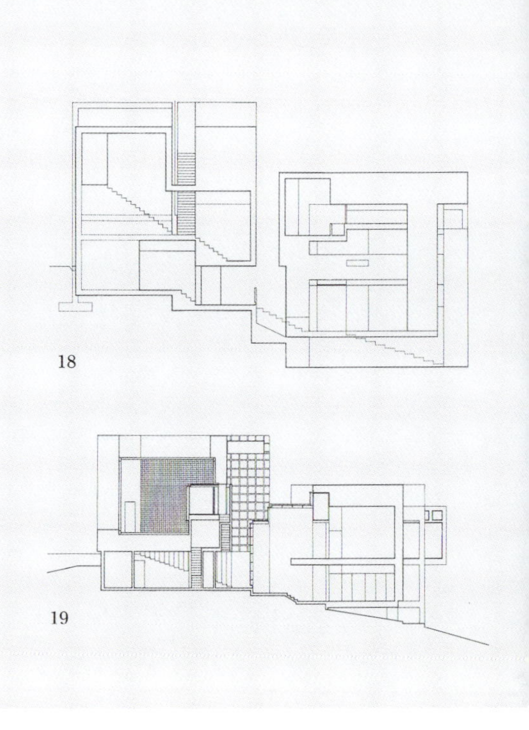

18

19

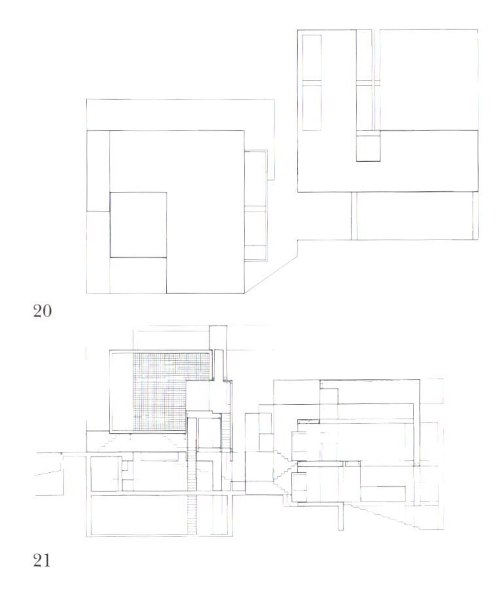

20

21

16 North elevation collage
17 South elevation collage
18 Scheme F, Section BB
19 Scheme G, Section CC
20 Scheme F, west elevation
21 Scheme F, Section AA
22 Axonometric model (Scheme H), view from the north-east
23 Scheme G, east elevation
24 Scheme F, north elevation
25 Axonometric collage
26 Axonometric model (Scheme H), view from the north
27 Axonometric model (Scheme H), view from the east of interior elevation

16 北立面构成
17 南立面构成
18 方案 F，BB 剖面
19 方案 G，CC 剖面
20 方案 F，西立面
21 方案 F，AA 剖面
22 轴测模型（方案 H），从东北方向看
23 方案 G，东立面
24 方案 H，北立面
25 轴测构成
26 轴测模型（方案 H），从北向看
27 轴测模型（方案 H），从内部东立面方向看

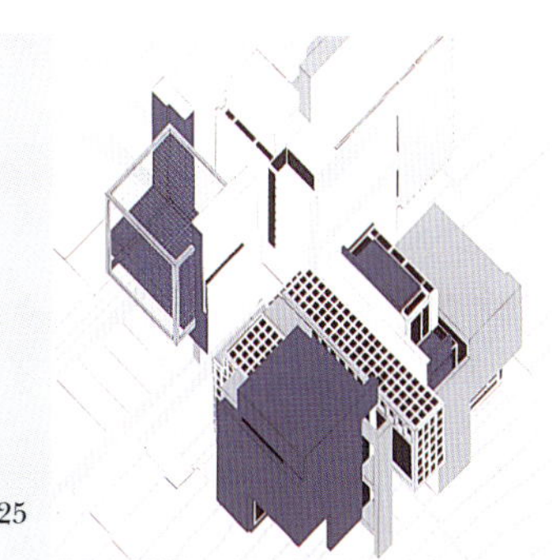
25

22

26

27

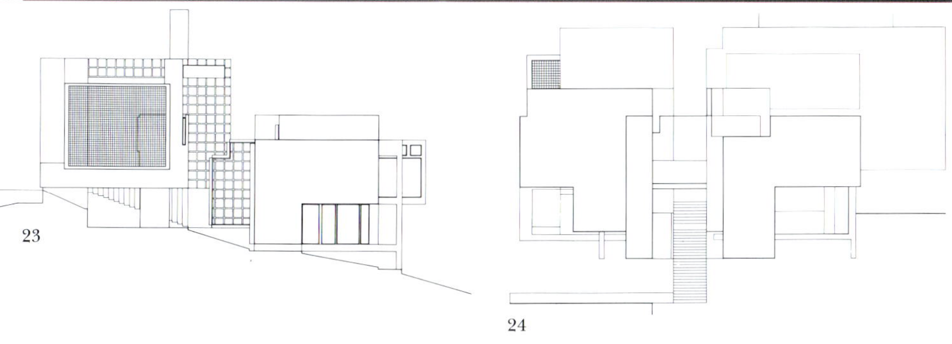
23

24

Cannaregio Town Square

Design 1978
Venice, Italy
Municipal Government of Venice

坎纳莱焦城镇广场

设计　1978
威尼斯，意大利
威尼斯市政府

This project is derived from an architecture that invents its own site and program. Rather than reproducing an existing Venice, it constructs another, fictitious Venice. The grid of Le Corbusier's Venice Hospital is continued as a structure over the site. This grid marks a series of voids which act as metaphors for man's displacement from his position as the centered instrument of measure. Architecture becomes the measure of itself.

The objects in this landscape are variations of House 11a, shown at different scales. The small object is too small to provide shelter, but raises the question of whether it is a house or a model of a house. The middle object contains the smaller object inside it, while the large object is twice the size of the middle object.

1

这个项目起源于一个建筑，该建筑创造了它自己的位置和项目意图。与其是重新创造已经存在的威尼斯，不如说它构筑了另一个虚构的威尼斯。柯布西耶的威尼斯医院的网格作为一种结构被延伸到该场地。网格标识了一系列的空间，这些空间起到一种隐喻作用，为人们从自身到中心衡量尺度的转换。建筑成为自身的衡量尺度。

场地中放置的物体都是住宅 11a 的变异体，以不同的比例被展示。小的物体因为太小只是一个遮蔽体，但它却提出了究竟它是建筑还是建筑模型的质疑。中等大小的物体内包含了小的物体，大的物体是中等物体的两倍大。

1 Presentation model
2 Site plan including Cannaregio West and Le Corbusier's Hospital
3 Presentation model including Cannaregio
4 Presentation model
5 Site plan
6 Plan
7 Site plan

1 展示模型
2 总图，包括西坎纳莱焦和柯布西耶的医院
3 包括坎纳莱焦镇的展示模型
4 展示模型
5 总平面
6 平面
7 总平面

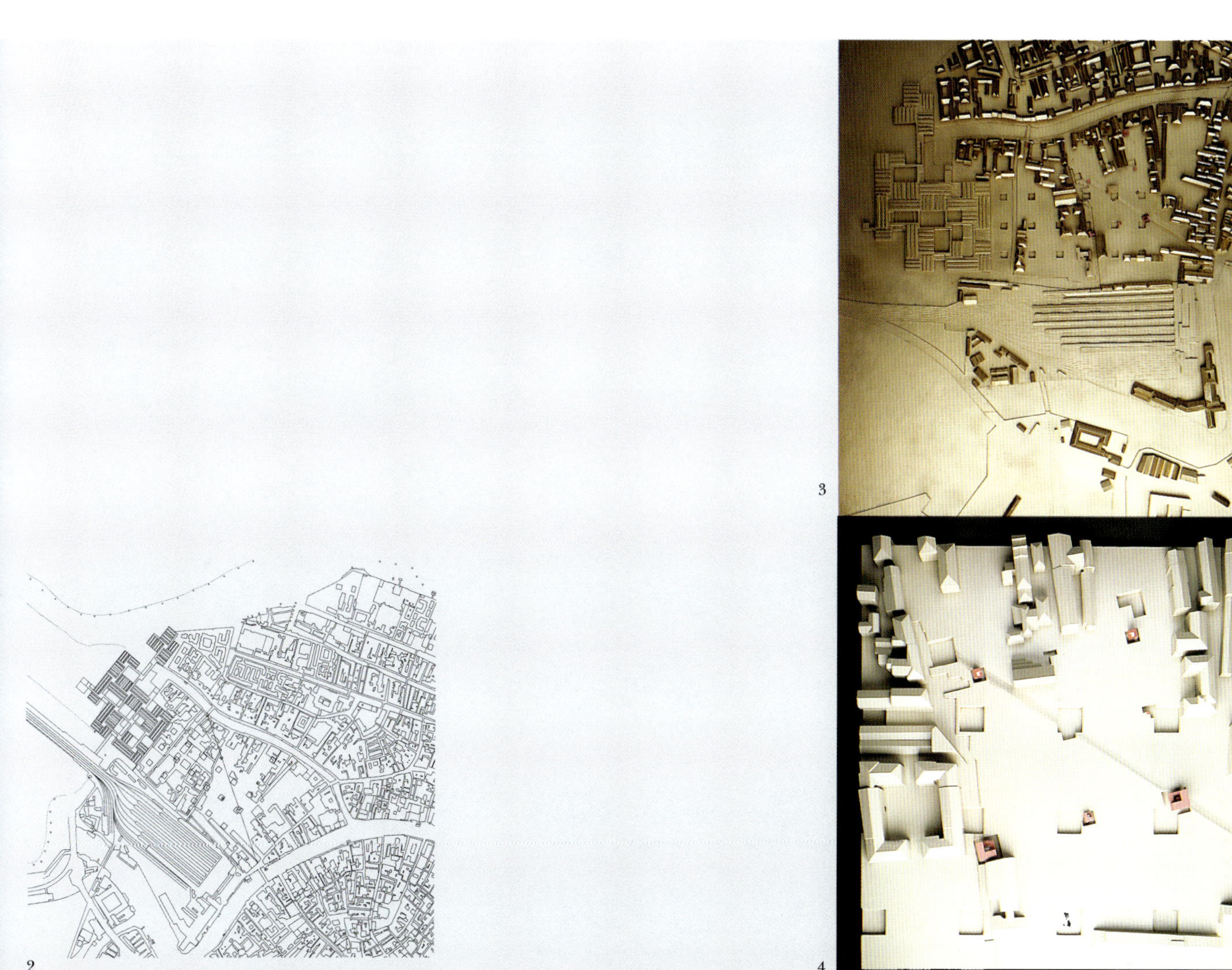

3

2

4

5

6

7

8

8 Presentation model
9–12 Site sections
13 Site plan
14 Section model of El structure

8 展示模型
9–12 地形剖面
13 总平面
14 房子结构的剖面模型

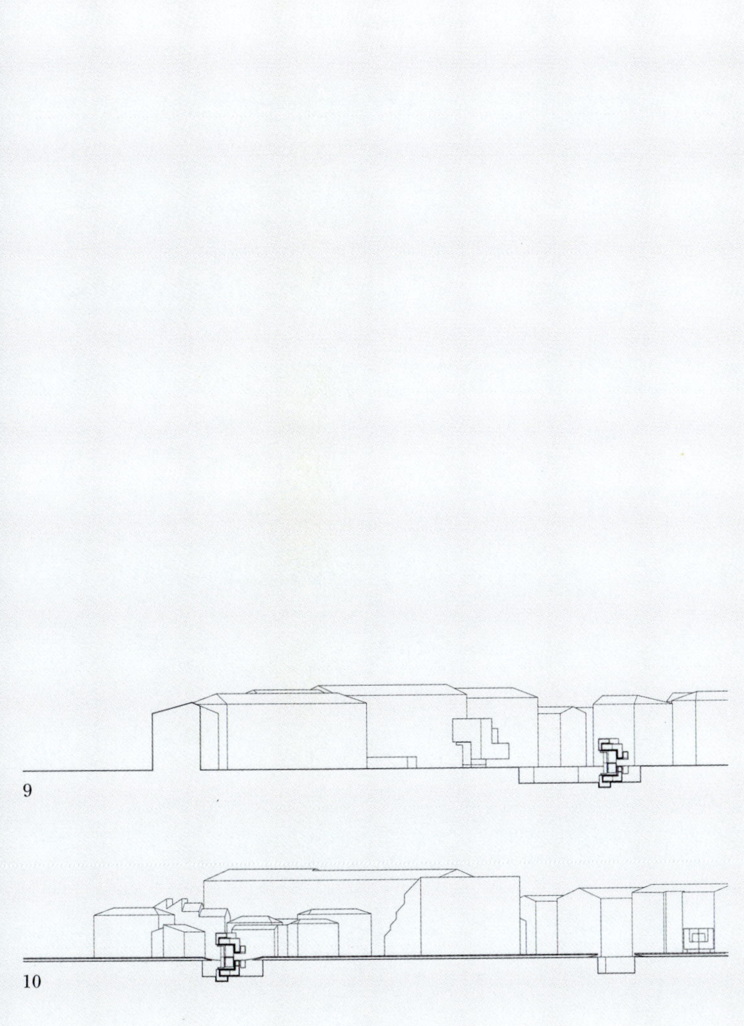

9

10

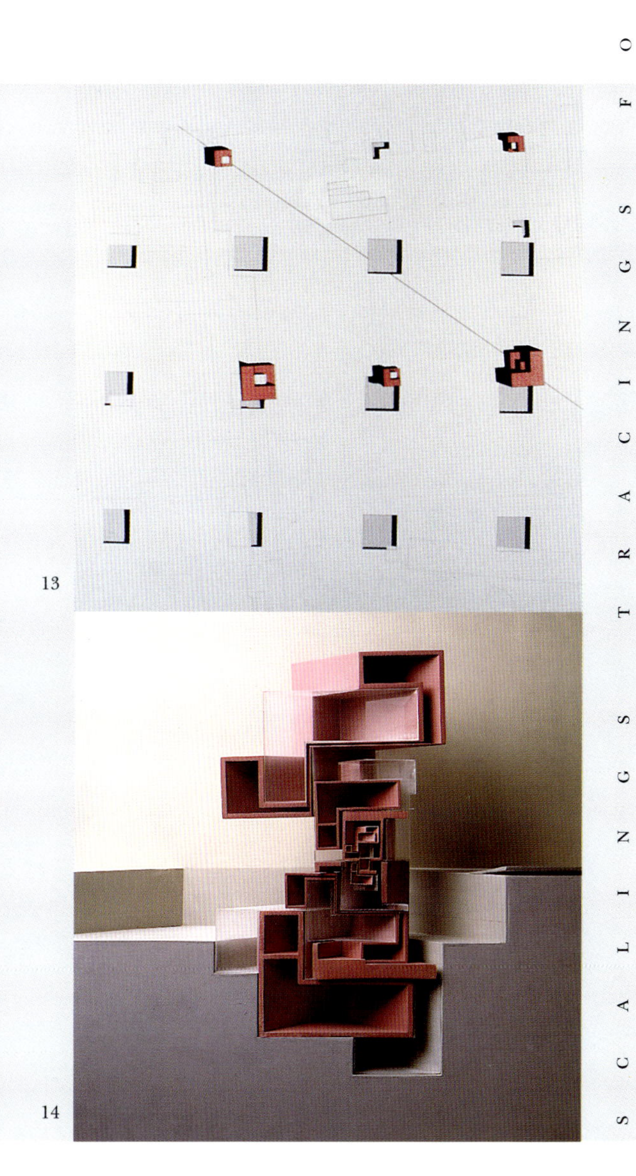

13

14

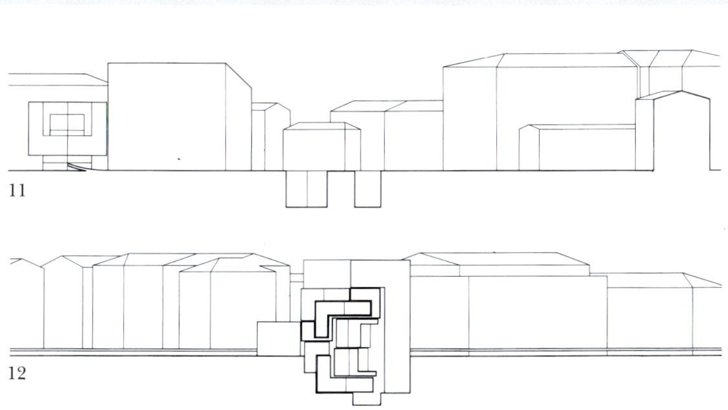

11

12

House El Even Odd

Design 1980
Palo Alto, California

E1 奇偶住宅

设计　1980
帕洛阿尔托，加利福尼亚州

House El Even Odd begins with an el-shaped axonometric object as its initial condition of reality. Two axonometric transformations then take place, allowing it to appear simultaneously as a three-dimensional object, an axonometric projection, and a plan. The object is then turned upside-down and placed below ground, so that the element that seems to be a plan is actually a roof. A smaller el-shaped volume which fits within the cut-out of the larger one is suspended in space, allowing two possible readings. A third and smaller volume, concentric to the first, suggests the same two readings. The three nesting els together ask, which is the actual size, and which is the model of the actual size?

E1 奇偶住宅起源于一个直角的轴测体。于是就产生了两个轴测方向的转换，这使得它可以同时作为一个三维物体出现、一个轴测投影以及一个平面的形式出现。物体被倒置并放在底层，于是看起来是平面的部分实际上是屋顶。在大的方体中的小方体被悬挂在空间中，包含两个可能的物体。第三个更小的物体与第一个物体同心，暗示了两个相同的事物。这三个嵌套着的物体不禁让人疑惑：哪一个是模型尺寸，哪一个又是实际的大小呢？

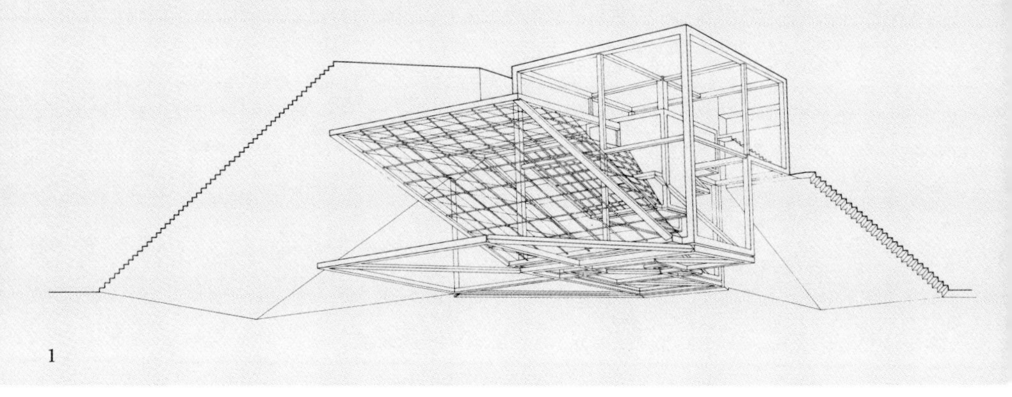

1

1 Axonometric
2 First level plan
3 Second level plan
4 Presentation model
5 Third level plan
6 Fourth level plan

1 轴测图
2 首层平面
3 二层平面
4 展示模型
5 三层平面
6 四层平面

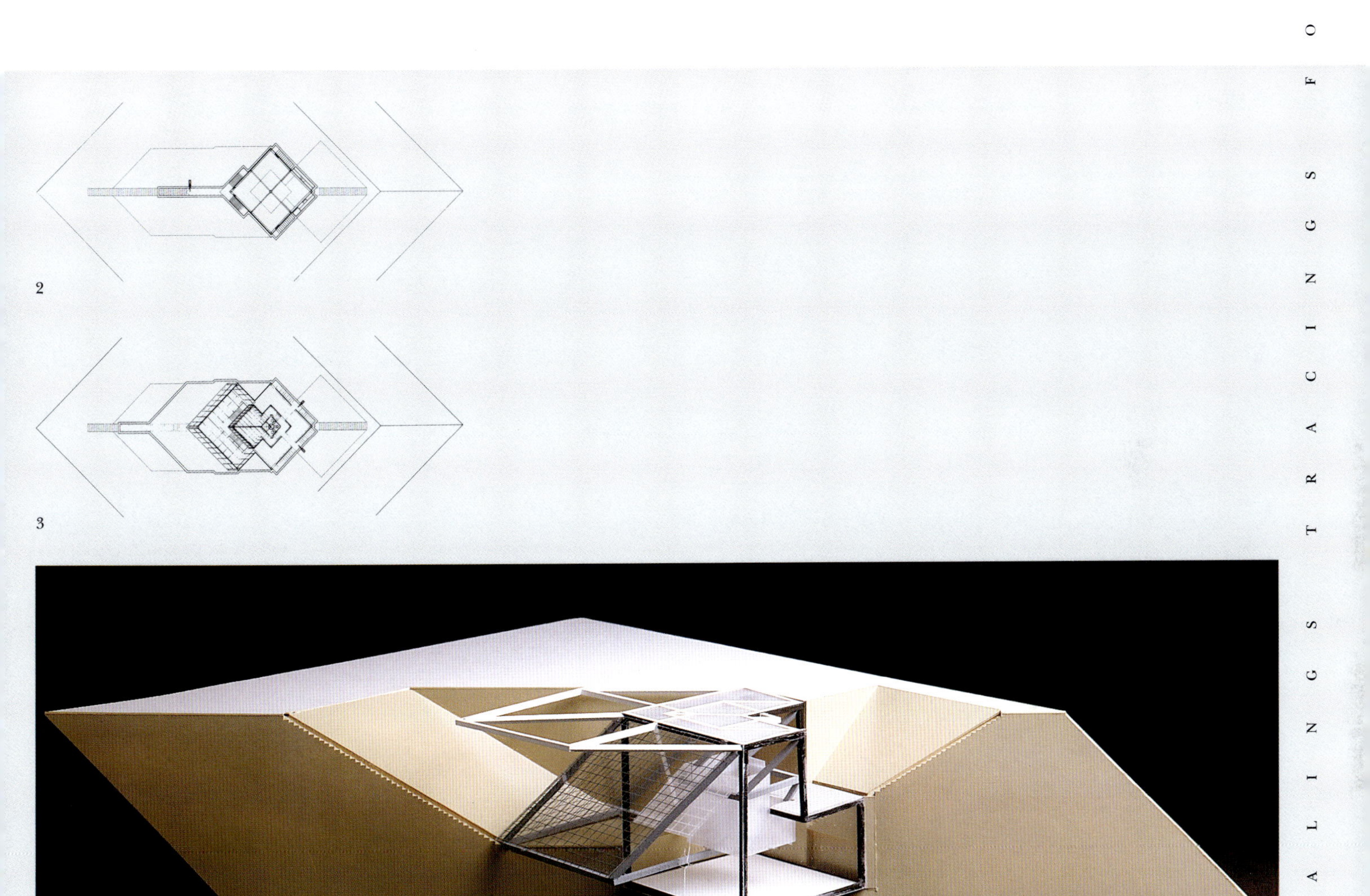

2

3

4

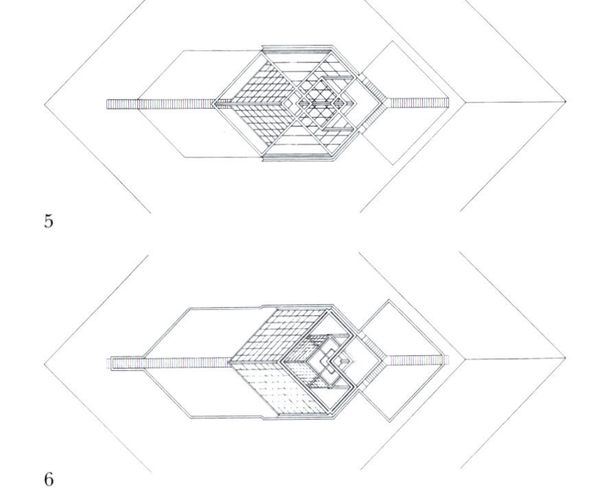

5

6

7

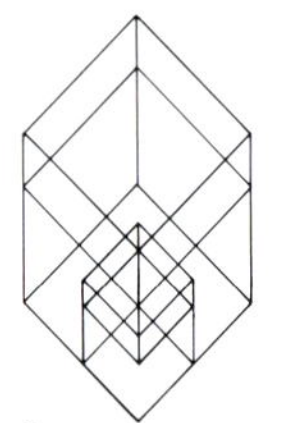

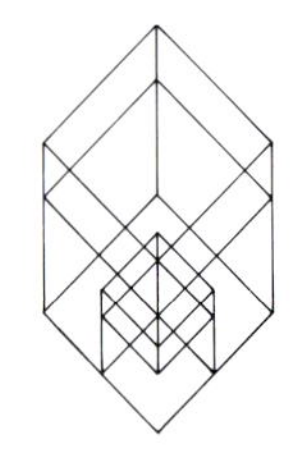

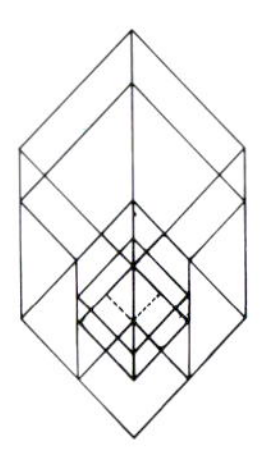

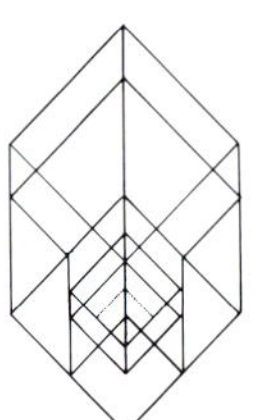

8

7 Presentation model
8 Concept diagram, oblique elevation
9 Concept diagram, front elevation
10 Concept diagram, plan
11 Concept diagram, perspective
12–14 Presentation models

7 展示模型
8 概念示意图，斜立面
9 概念示意图，前立面
10 概念示意图，平面
11 概念示意图，透视图
12－14 展示模型

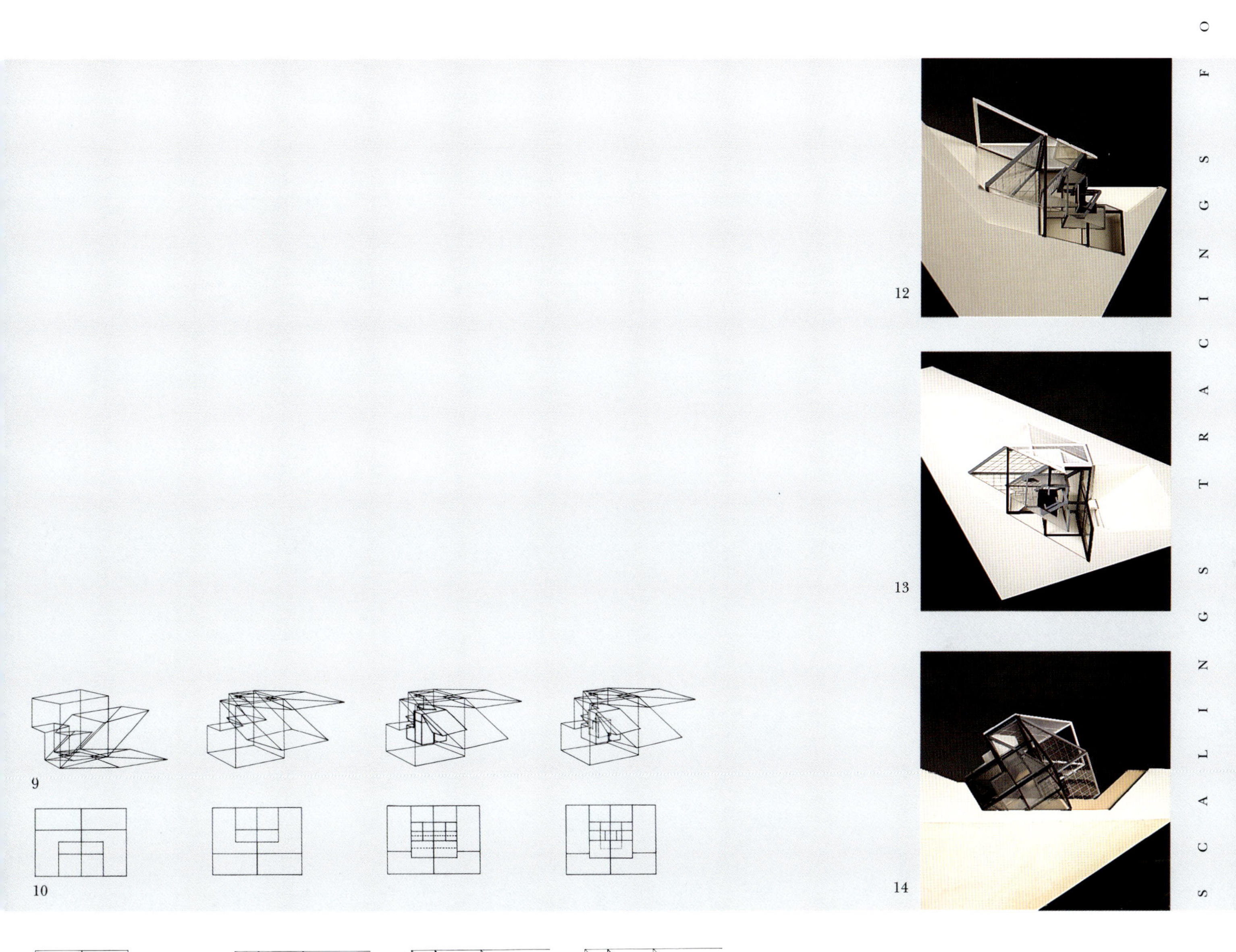

11

Madison Components Plant

Design/Completion 1981/1982
Madison, Indiana
Cummins Engine Company
326,000 square feet
Steel and concrete block

麦迪逊元件厂

设计/竣工 1981/1982
麦迪逊，印第安纳州
康明斯机械公司
326000 平方英尺
钢筋混凝土结构

This industrial building was designed to house a turbo-charger and diesel engine components manufacturing process. The focus of the design was to create a well ordered and smoothly functioning interior layout: a working environment which ensured worker safety and enhanced productivity.

The manufacturing plant was designed as a single-story rectangular building with a dramatic, angled skylight running its length. In the master plan, the existing plant and new administration center are surrounded by the industrial space to allow manufacturing activities to be viewed in all varieties of light without glare off the machinery.

这一工业建筑的设计是为了容纳涡轮充电器和柴油机器元件的生产过程。设计的重点在于创造一个井井有条、功能流畅的室内布局：也就是在保证工人工作安全的同时提高生产效率的工作环境。

生产车间被设计成单层的方形建筑，沿着它的长边有生动的与之成一定角度的天光洒下。在总体规划中，原有的车间和新的行政中心被工业空间包围，从而使得生产活动在所有的光线变化情况下都可以被看到，而没有眩光。

1

1 Presentation model, view from above
2 Presentation model, view from the south-east

1 展示模型，从上方看
2 展示模型，从东南方向看

2

3

3–5 Interior perspectives
6–7 Interior views

3-5 室内透视图
6-7 室内实景

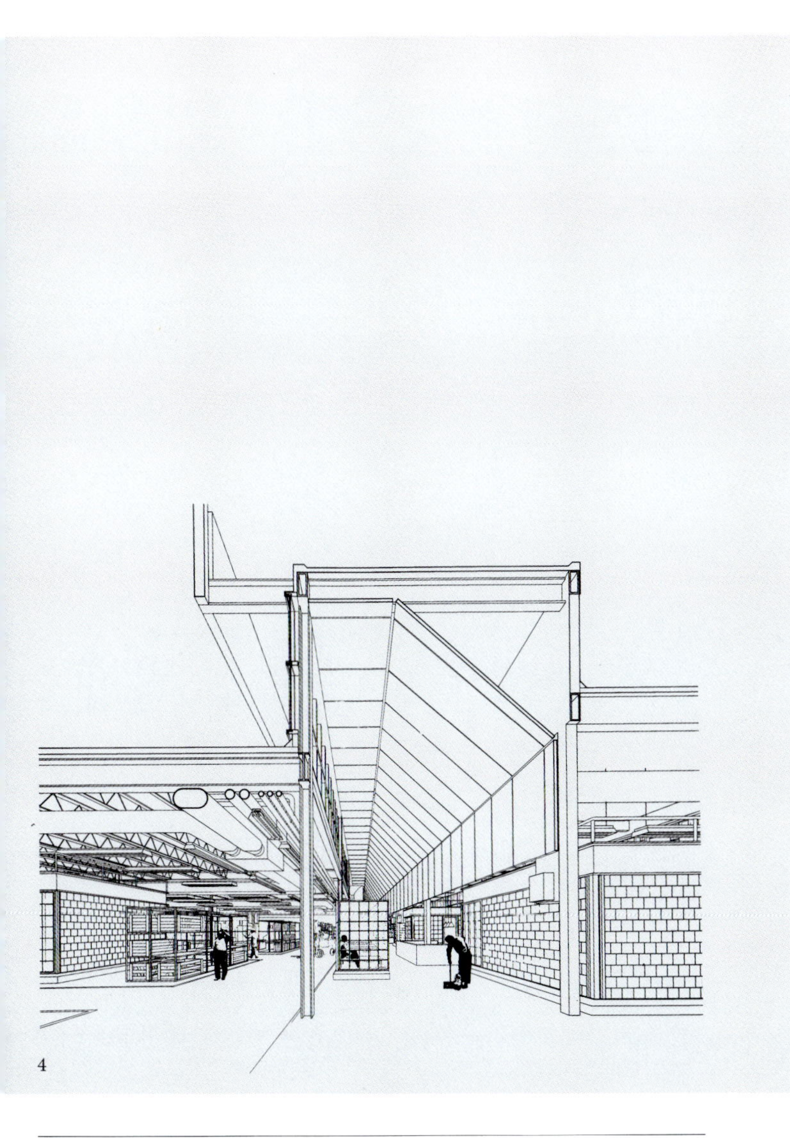

4

6

7

5

IBA Social Housing

Design/Completion 1981/1985
Berlin, Germany
Hauert Noack, GmbH & Company
50,000 square feet
Concrete frame
Stucco and metal panels

IBA 社会住宅

设计/竣工 1981/1985
柏林，德国
Hauert Noack GmbH 公司
50000 平方英尺
混凝土框架结构
拉毛灰泥和金属板

This apartment block is intended not only to help meet the pressing need for housing in Berlin, but also to commemorate the events that have taken place around the site.

This project is designed for social (low-income) housing on a corner site in Berlin, located on the block adjacent to the former Berlin Wall and Checkpoint Charlie. It is the first phase of a two- or three-phase project which will eventually cover the entire block.

The design, in addition to meeting the very restrictive functional and financial requirements for social housing in Berlin, responds in a unique way to two general architectural problems: context and symbolism.

该公寓街区不光是为了解决柏林居住迫切的问题，同时还是为了纪念在其周围所发生的事件。

该项目位于柏林一街道的转角，被设计成社会住宅（提供给低收入者），其所处的街区与从前的柏林墙和查理检查站相邻。它是一个2～3期工程的第一期工程，整个工程最终将覆盖整个街区。

为了兼顾柏林社会住宅功能的限制和财政要求，该设计以一种独到的方式对建筑的两个普遍问题进行了回应：文脉和象征性。

1

1 View from the south-west
2 View from the south
3 Axonometric, view from the south-west
4 Concept sketch

1 从西南方向看
2 从正南方向看
3 轴测图，从西南方向看
4 概念草图

2

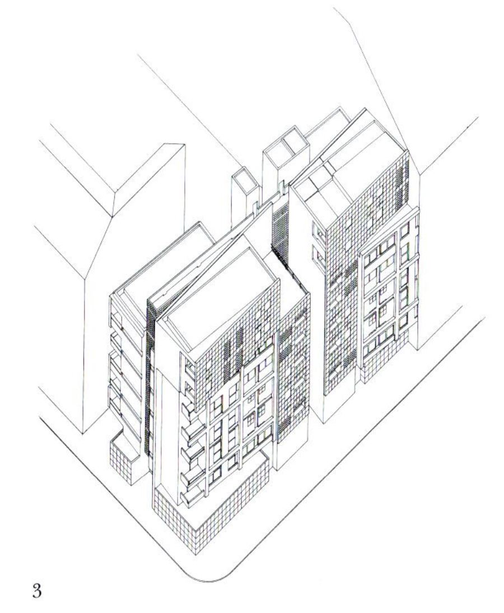

3

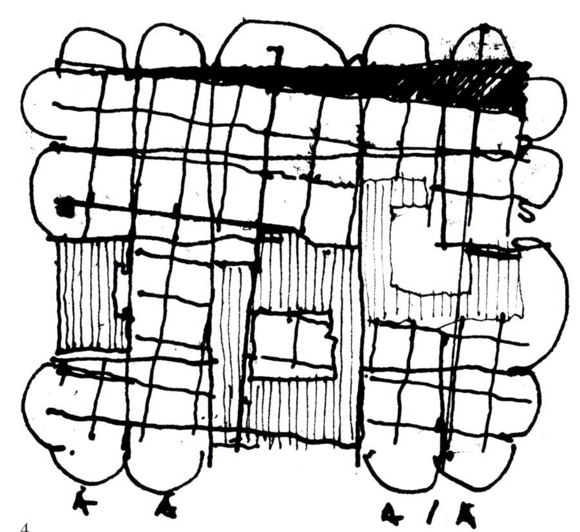

4

5

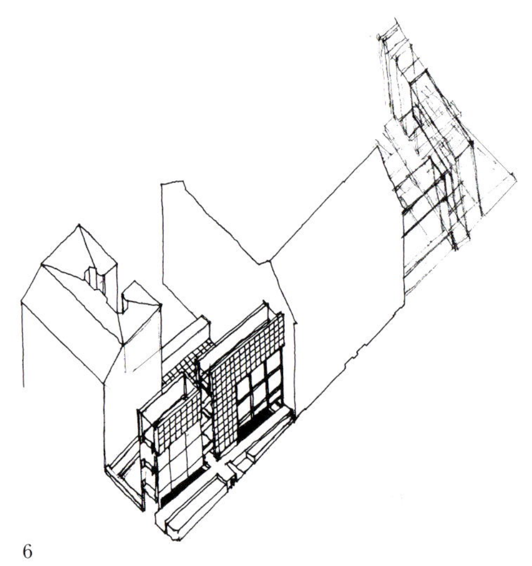
6

5 View from the south
6 Facade study
7 Block plan
8 Second level plan
9 Basement level plan

5 从正南方向看
6 立面研究
7 首层平面
8 二层平面
9 地下层平面

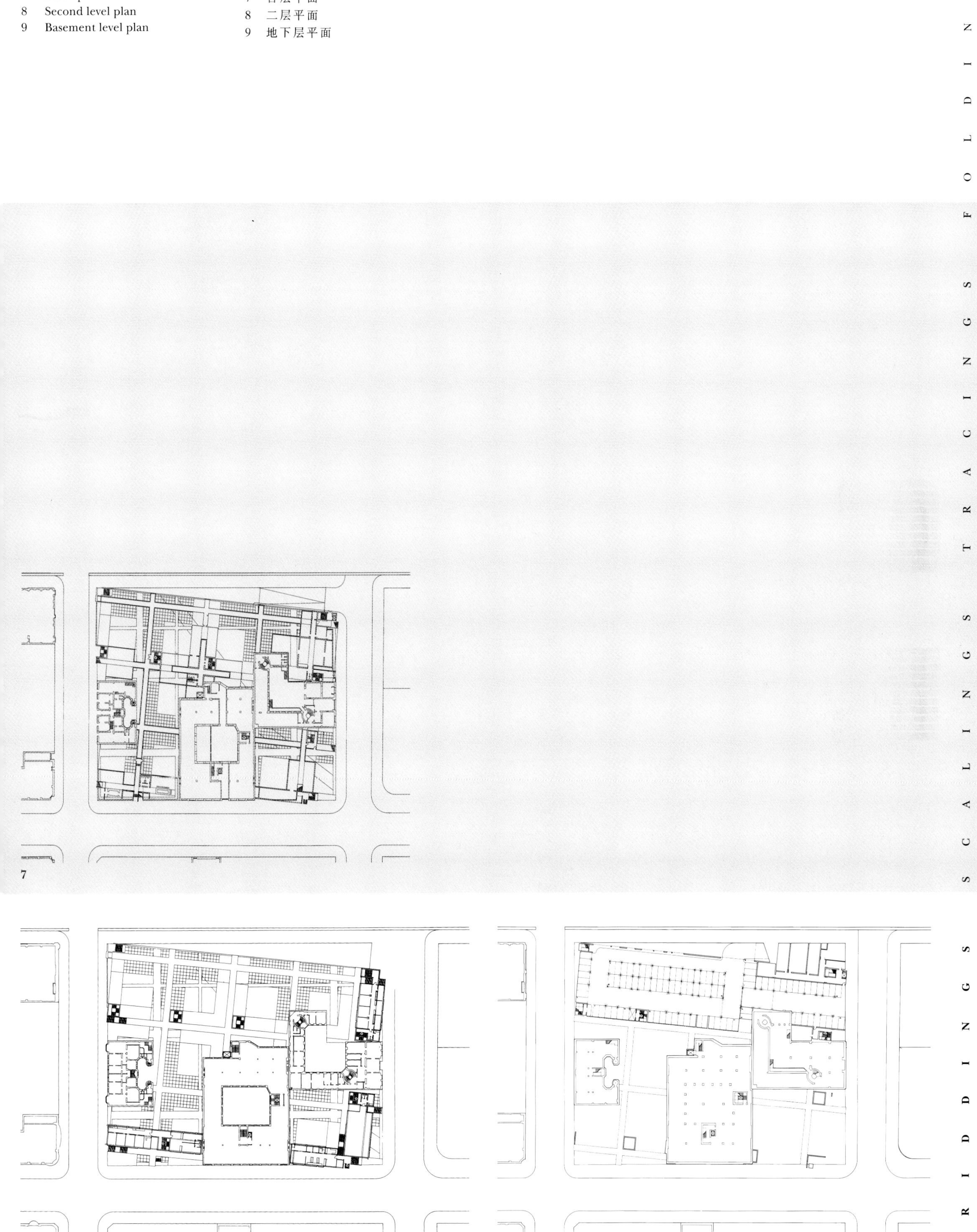

7

8

9

10

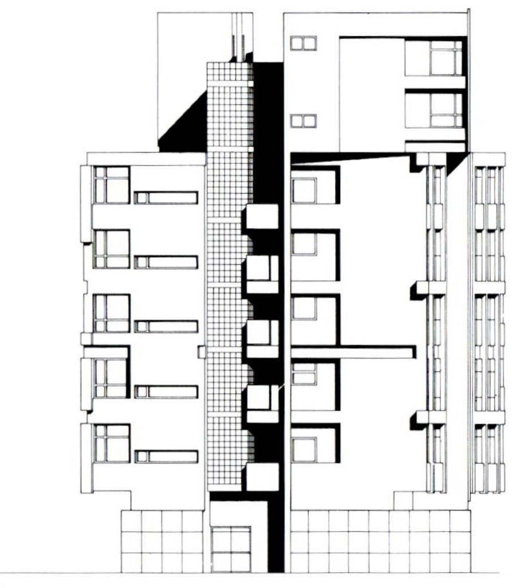
11

10 South elevation, view from the south-east
11 West elevation
12 View from the south-west
13 North elevation
14 South elevation

10 南立面，从西南方向看
11 西立面
12 从西南方向看
13 北立面
14 南立面

2

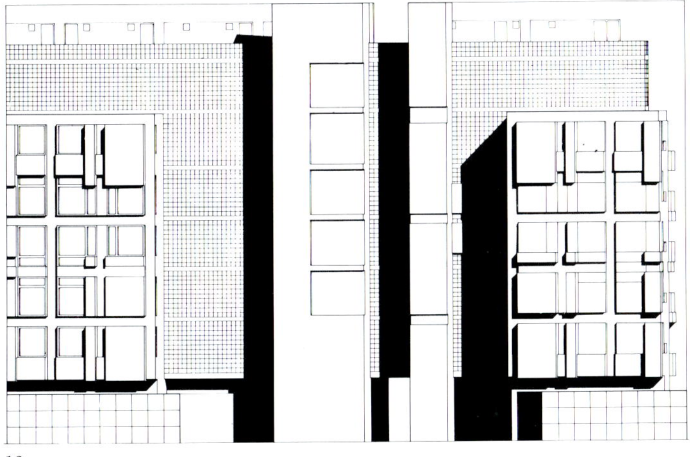

13

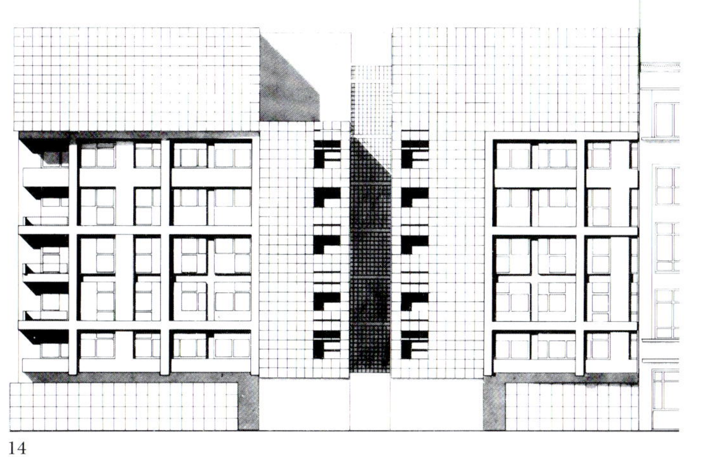

14

15

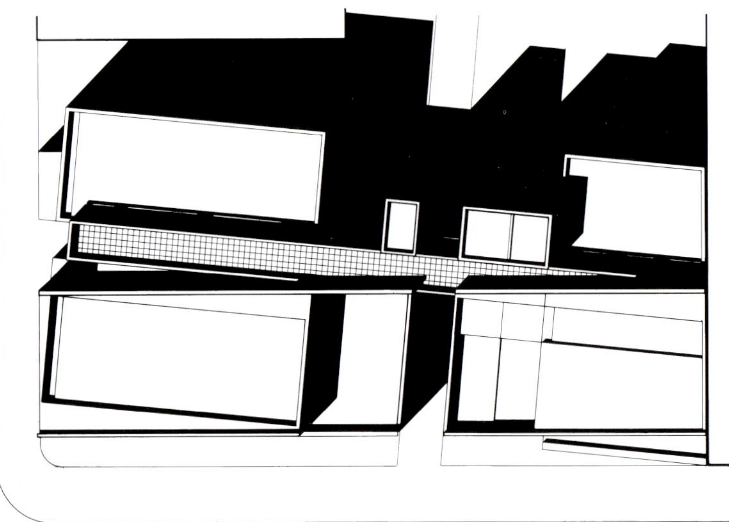

16

15 Typical apartment
16 Roof level plan
17 Block elevation, view from the Kochstrasse
18 Block elevation, view from the Zimmerstrasse
19 Detail view from the north
20–21 Detail views from the south

15 标准公寓类型
16 屋顶平面
17 街区立面，从科赫大街方向看
18 街区立面，从齐默大街方向看
19 从北向看细部
20 20～21，从南向看细部

19

20

21

17

18

Travelers Financial Center

Design/Completion 1983/1986
Hempstead, New York
Fair Oaks Development/Schottenstein Properties
235,000 square feet
Steel frame
Glass and aluminum curtain wall

旅行者金融中心

设计/竣工　1983/1986
亨普斯特德，纽约
费尔奥克斯开发/Schottenstein 房地产公司
235000 平方英尺
钢结构
玻璃和铝壁板

The design for this 10-story office building on Long Island consists of eight floors of office space, with retail facilities on the ground floor and a lower level containing a private dining area and building services.

The building demonstrates a plasticity of form and surface not ordinarily associated with curtain-wall office buildings. This "glass box" is effectively broken into several different readings by a number of shifts in the plans and elevations. The two geometries of the site are encapsulated in the small-scale interplay of the wall, surface and grid in the ceilings, floor and walls of the main lobby level.

这一坐落在长岛上的 10 层办公楼建筑是由 8 层办公空间和底层的零售设施以及地下层的私营餐饮、服务空间组成。

该建筑证实了形式和表面的可塑性，而不同于以往的幕墙办公建筑。这个“玻璃盒子”被许多平面和剖面的变化有效地分成若干不同的部分。场地上的两个几何形状被压缩到下边的墙的交界处、表面、顶棚的网格、地面以及门厅的主要墙面中。

1

2

1 View from the south-west
2 South elevation
3 Axonometric of lobby ceiling, view from below
4 North elevation
5–6 Detail views from the south
7 View from the west

1 从西南方向看
2 南立面
3 门厅顶棚轴测图，从下往上看
4 北立面
5-6 从南向看细部
7 从西向看

5

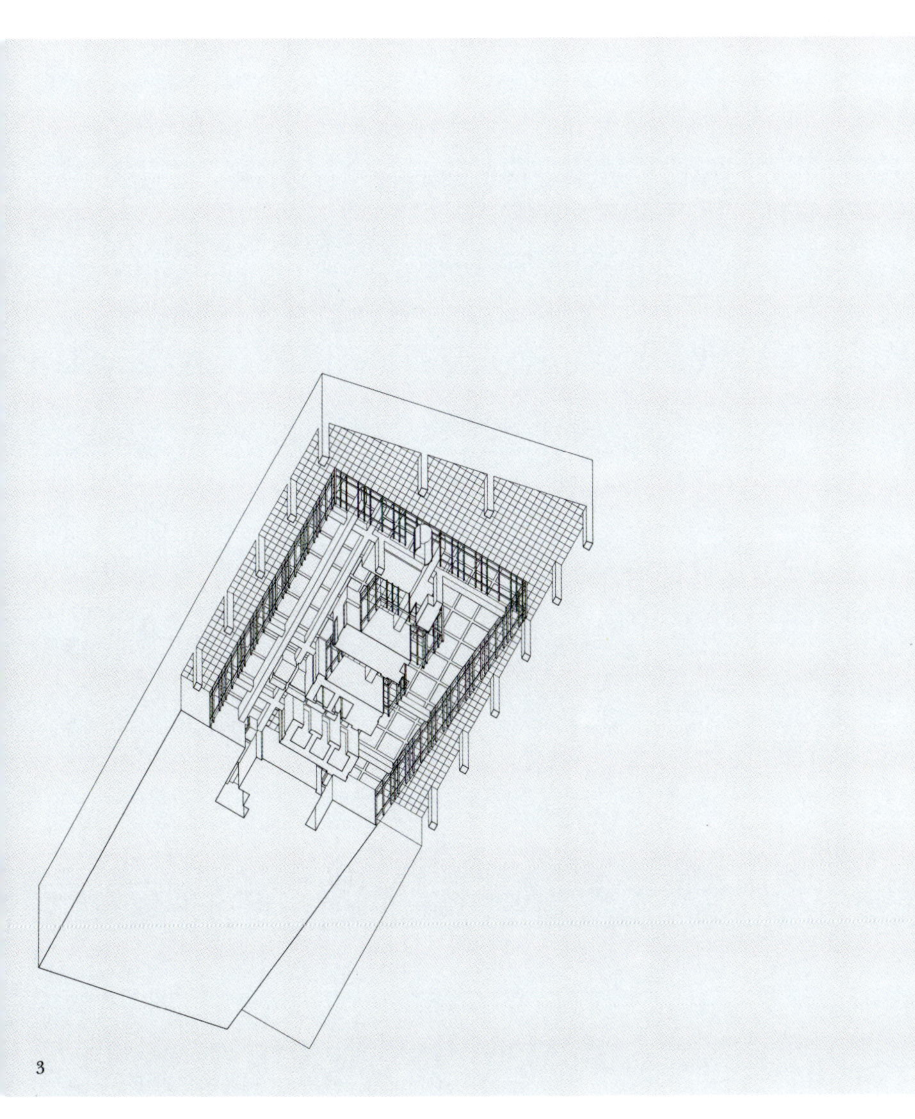

3

6

7

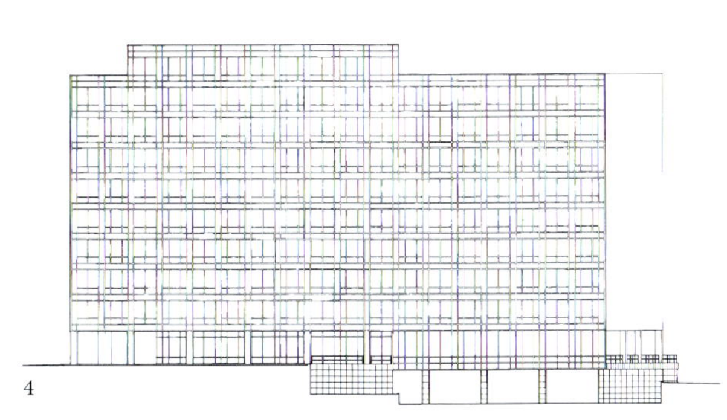

4

8

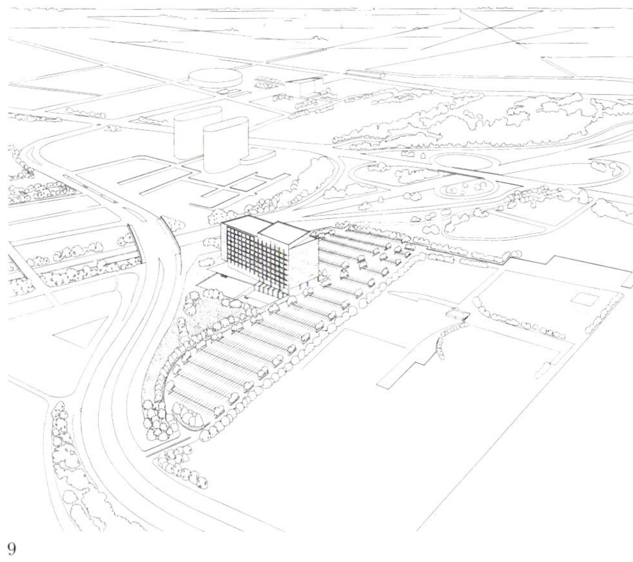

9

8 View from the south-east
9 Site perspective
10 Lower level plan
11 First level plan
12 Typical level plan
13 Ground level lobby
14 Elevator lobby

8 从西南方向看
9 总图环境鸟瞰
10 地下层平面
11 一层平面
12 标准层平面
13 首层门厅
14 电梯间

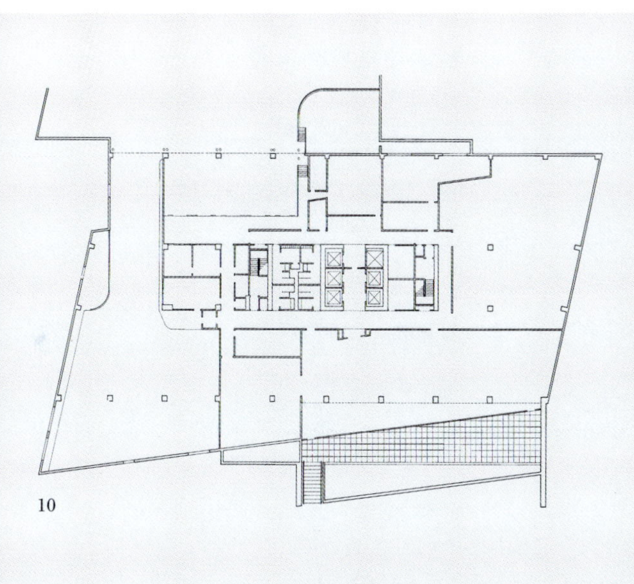

10

13

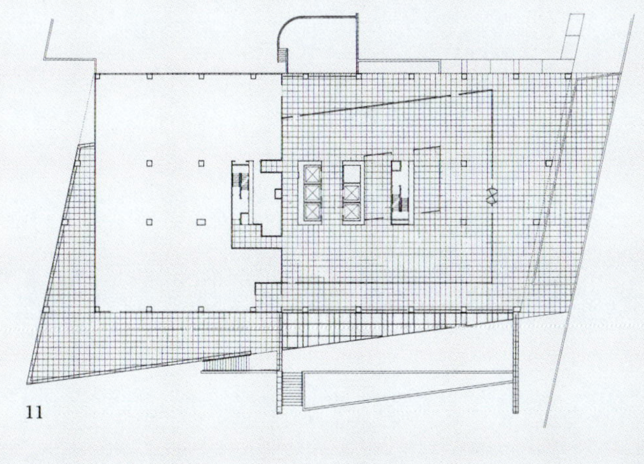

11

14

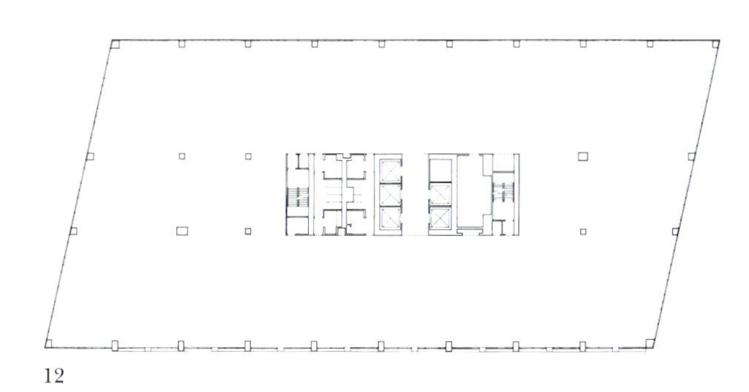

12

Firehouse for Engine Company 233 and Ladder Company 176

Design/Completion 1983/1985
Brooklyn, New York
City of New York
13,500 square feet
Steel frame
Glazed and non-glazed block and aluminum panels

233 机械公司和 176 梯子公司的消防站

设计/竣工　1983/1985
布鲁克林，纽约
纽约市
13500 平方英尺
钢结构
玻璃和铝壁板
光滑的以及粗糙的墙体和铝板

The building's design responds to its urban site, where an elevated rail line marks a shift in grid patterns, by incorporating these two grids within the structure. This two-story firehouse contains fire-fighting equipment, battalion chief's offices, company offices, and sleeping accommodation.

The structural roof beam members of the superimposed grid contain red laser lights that symbolically illuminate the structure at night. In addition, a beacon of red light shines out when the fire engines are on-call.

1

该建筑周围有一条标志着网格模式转变的铁路线，建筑通过在结构中融汇两个网格系统，从而与其城市环境相呼应。这个两层高的消防站包括消防设备、部队领导办公室、公司办公室以及休息膳宿。

有条理的网格系统的屋顶结构梁上带有昼夜通明的灯，该灯在晚间的时候将结构照亮。而且，当有火警的时候，一盏红色的信号灯会照射出去。

1 View from the west
2 First level plan
3 Second level plan
4 Detail view from the north-west
5 Interior view

1 从正西方向看
2 一层平面
3 二层平面
4 从西南方向看细部
5 室内实景

4

5

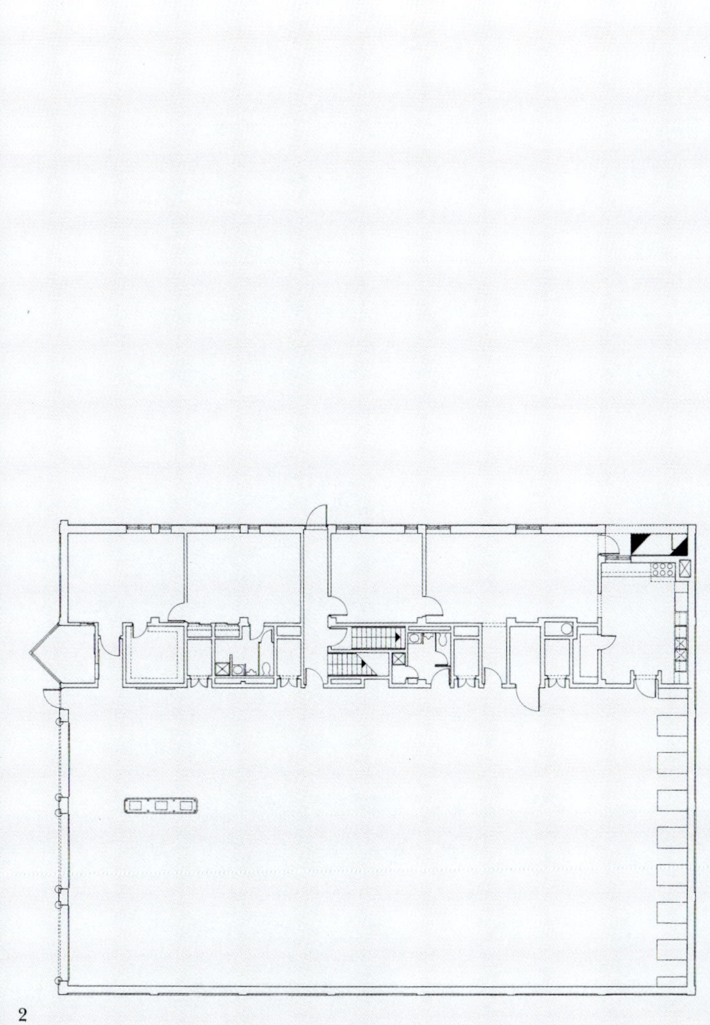

2

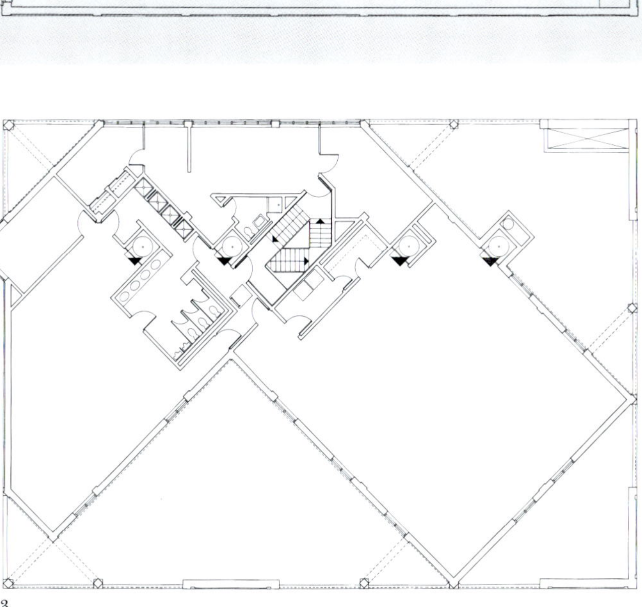

3

6

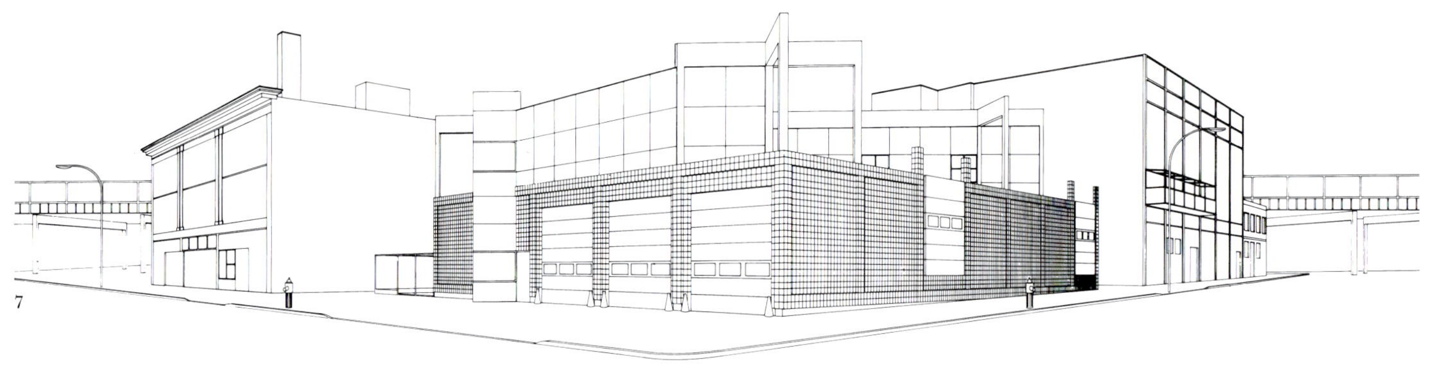
7

6 Detail view from the west
7 Perspective from the south-west
8 South elevation
9 North elevation
10 East elevation
11 West elevation
12 View from the south
13 Roof view from the east

6 从西向看细部
7 西南方向透视
8 南立面
9 北立面
10 东立面
11 西立面
12 从南向看
13 从东向看屋顶

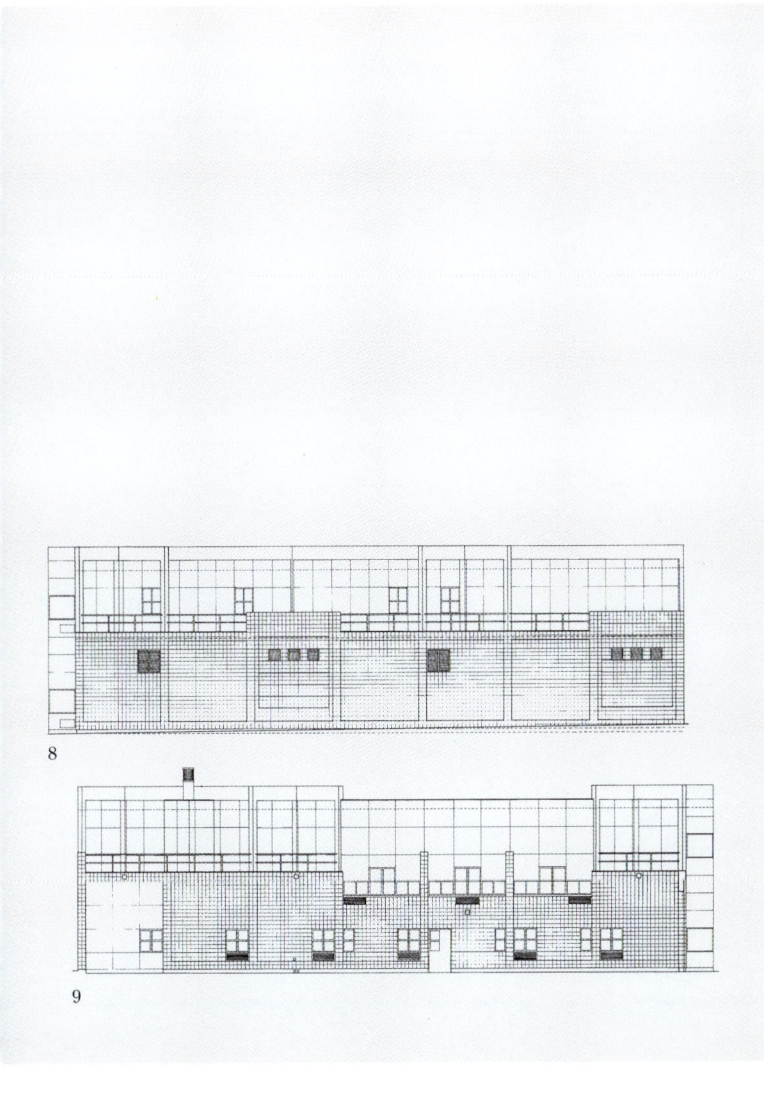

8

9

12

13

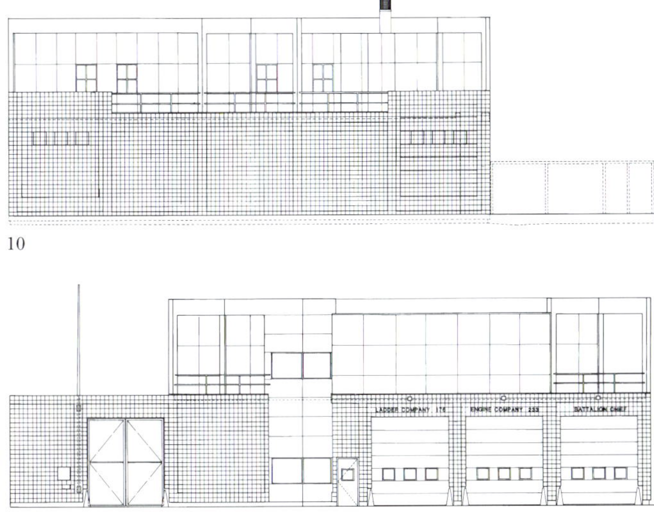

10

11

Fuller/Toms Loft

Design/Completion 1984/1987
New York, New York
Emily Fuller and Newby Toms
4,800 square feet
Wood floor, wallboard walls and ceiling, sliding wood doors

富勒/汤姆斯阁楼

设计/竣工 1983/1985
纽约，纽约州
艾米丽·富勒和纽柏·汤姆斯
4800平方英尺
木楼板，墙板隔墙和顶棚，推拉木门

The ideas for this project are twofold: first, to explore the question of scale; second, to explore the question of an internal insertion.

The site for this project is a loft in Lower Manhattan. The space is essentially a rectangular parallelepiped with a proportion of 40 to 100. The short side faces Broadway, a diagonal in the orthogonal street grid of New York. The idea is to insert a foreign body into the existing context in such a way as to produce a disorienting relationship between old and new. The intersection of the two geometries produces a condition in which neither geometry is dominant, thereby displacing and destabilizing conventional devices for orientation.

该项目的设计理念有双重含义：首先是探究比例的问题，其次是探究一个室内插入体的问题。

该项目位于曼哈顿岛下部的一个阁楼内。其空间是一个比例为40:100的平行六面体。短边面对宽阔的公路，与纽约市直交的街道公路网格斜交。设计的理念是将一个外部的物体插入到现存的机理中，从而以这种方式在新旧之间产生一种关系。两个几何体的交接就会产生这样一种情形，哪一个几何体都不是主体，于是取代并打乱了通常的方向的设定。

1

1　Living/dining room
2–3 Axonometric views

1　　起居室/餐厅
2－3　轴测图

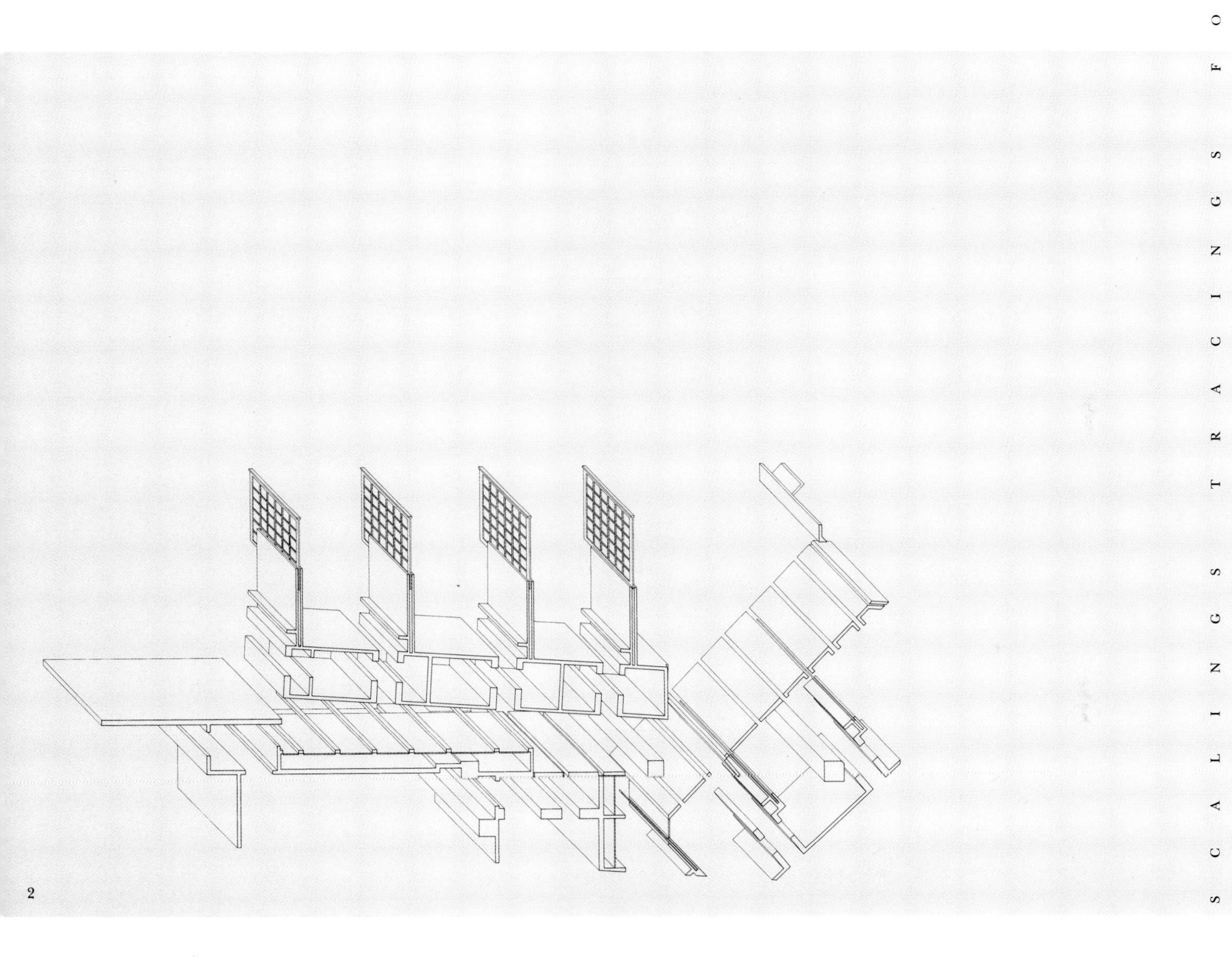

2

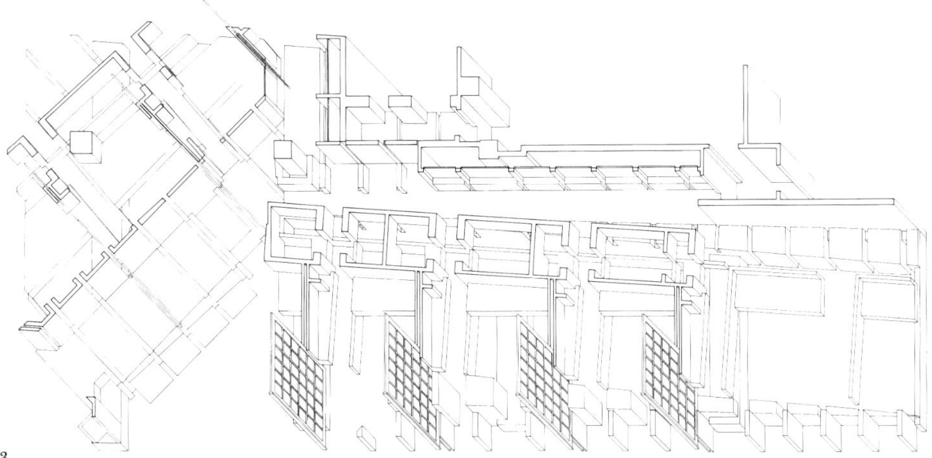

3

4

4 Detail
5 Floor plan
6 Axonometric
7 View along the window wall
8 Detail

4 细部
5 楼层平面
6 轴测图
7 沿窗墙看
8 细部

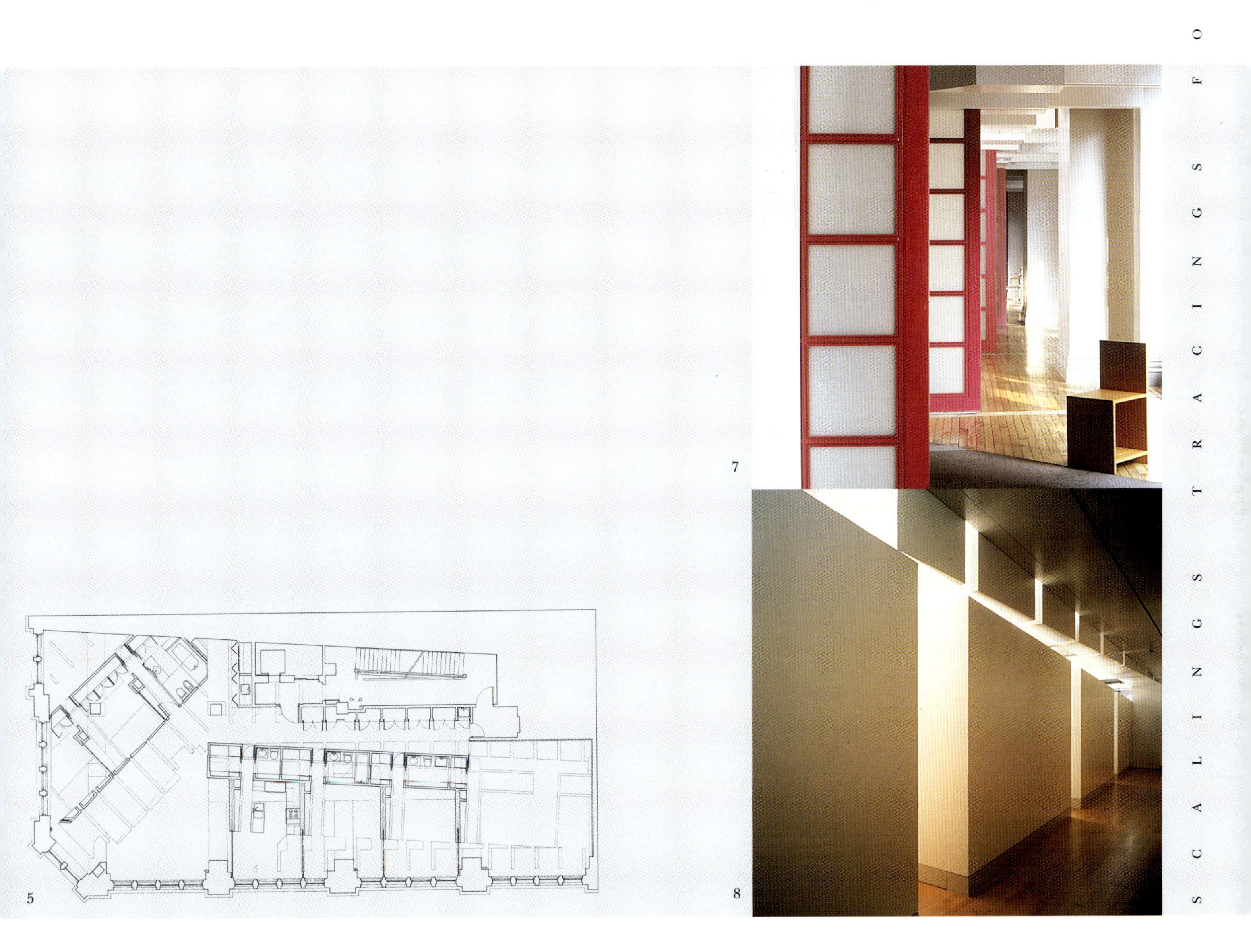

7

5

8

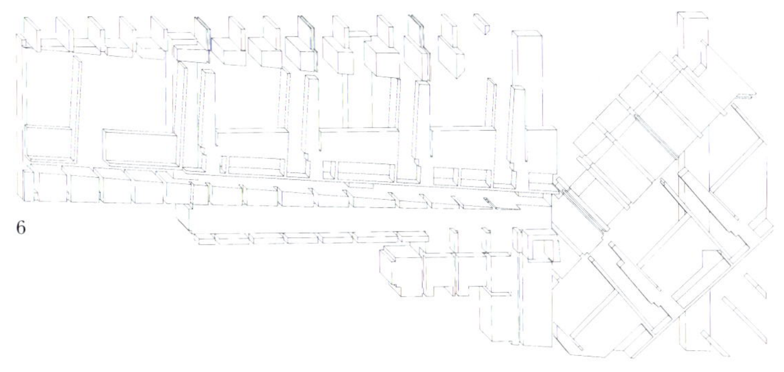

6

GRIDDINGS SCALINGS TRACINGS FOLDINGS

Romeo and Juliet Castles

Design 1985
Verona, Italy

罗密欧与朱丽叶城堡

设计　　1985
维罗纳，意大利

The program for this project was to present the dominant themes of the stories of Romeo and Juliet in architectural form at the site of the two castles. There are three important versions of the story which were taken as the basis of the architectural "program."
Each narrative is characterized by three structural relationships: division (the separation of the lovers/the balcony); union (the marriage of the lovers/the church); and their dialectical relationship (the togetherness and apartness of the lovers/Juliet's tomb). The project responds to fundamental cultural changes that have taken place in the last century, by using an architectural discourse that is founded in a process called scaling.

该项目的意图是在这两个城堡的位置上，在建筑形式中展现罗密欧与朱丽叶这个突出的主题。罗密欧与朱丽叶的故事有三个重要的看法，这些看法是建筑设计“意图”的基础。每一个故事情节被三个结构的关系赋予特征：分离（情人之间的分隔/阳台）；结合（恋人的婚姻/教堂）；他们之间辨证的关系（恋人的相距与分离/朱丽叶的坟墓）。项目通过建筑的缩放过程呼应了一个世纪前发生的基本的文化变化。

1

1 Presentation model, view from above
2 Axonometric
3–4 Plans

1 展示模型，从上部看
2 轴测图
3－4 平面

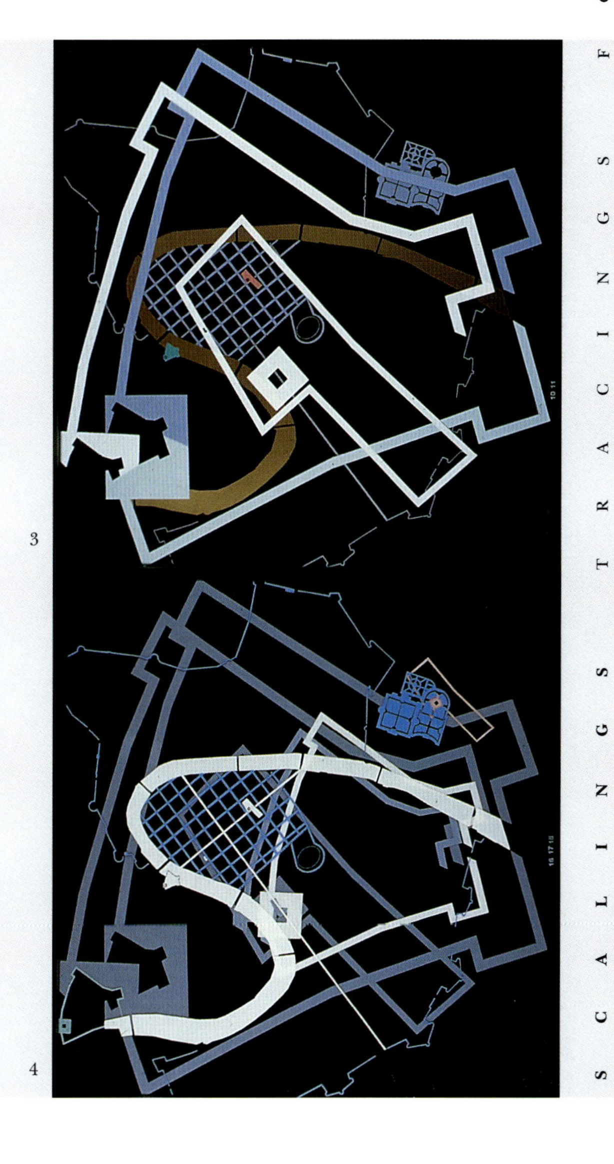
3
4

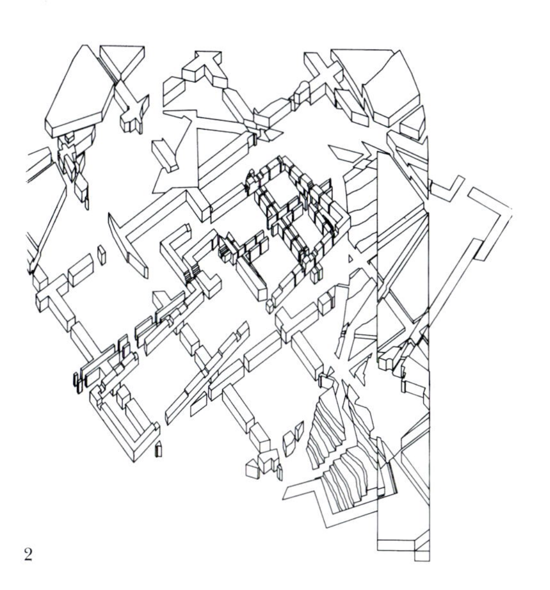
2

5

5 Plan
6 Axonometric
7–8 Plans

5 平面
6 轴测图
7-8 平面

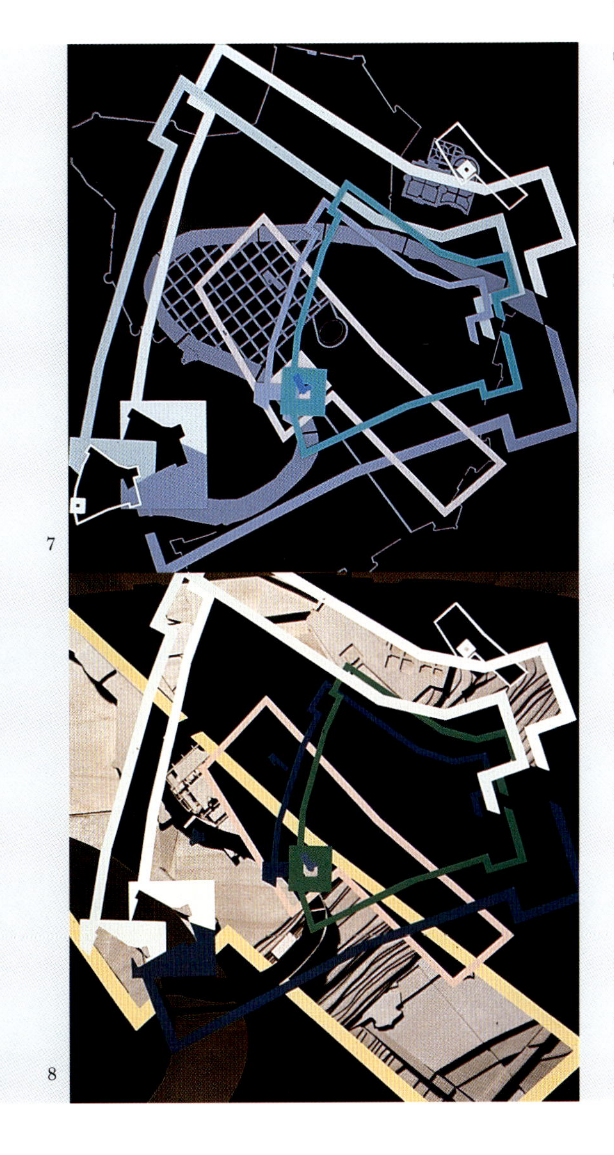
7
8

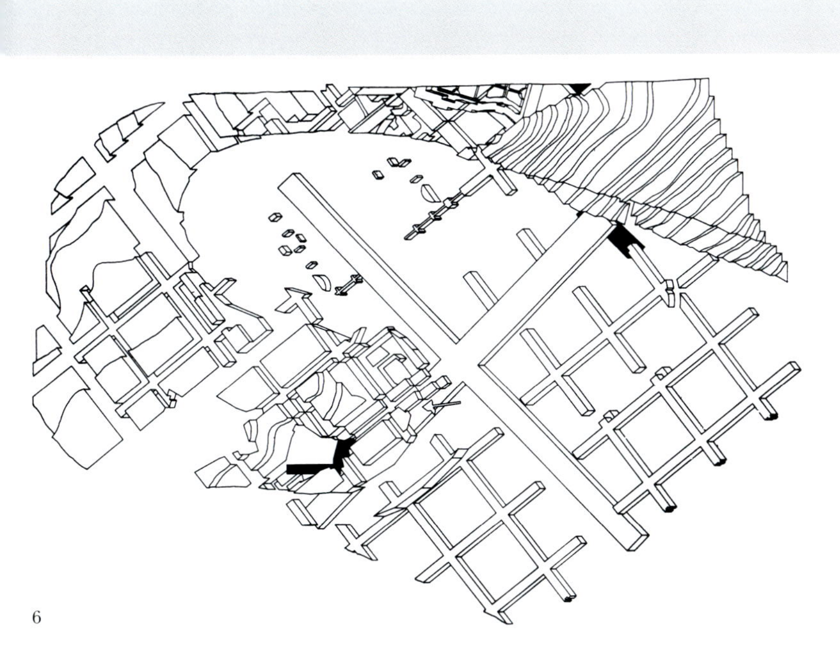
6

Tokyo Opera House

Design 1985
Tokyo, Japan
City of Tokyo

东京歌剧院

设计　　1985
东京，日本
东京市

This competition entry for the design of the new National Theater of Japan includes three theaters (a black-box theater, a 1,000-seat performance space, and an 1,800-seat opera house), a rehearsal space, office space, and underground parking.

This project attempts to establish an analogical relationship between the proposed site of a new center for culture and the old center of culture in Tokyo, which was traditionally the Noh Theater, located in the courtyard of the Emperor's realm. To symbolize this, a series of analogous relationships between old and new were established that ultimately refer back to the Emperor's realm in Kyoto.

这个为日本新的国家剧院所进行的竞赛设计包括3个剧场（多功能音乐厅，一个1000座位的演出大厅，还有一间1800座位的剧场），一间排练场、办公空间和地下停车场。

该项目试图展现场地内的新旧文化中心之间的相似关系，旧的文化中心是传统的能乐剧场，坐落于皇室的内院。为了表明这一关系，一系列的新旧文化中心的相似关系被展现，并最终归属于京都皇室。

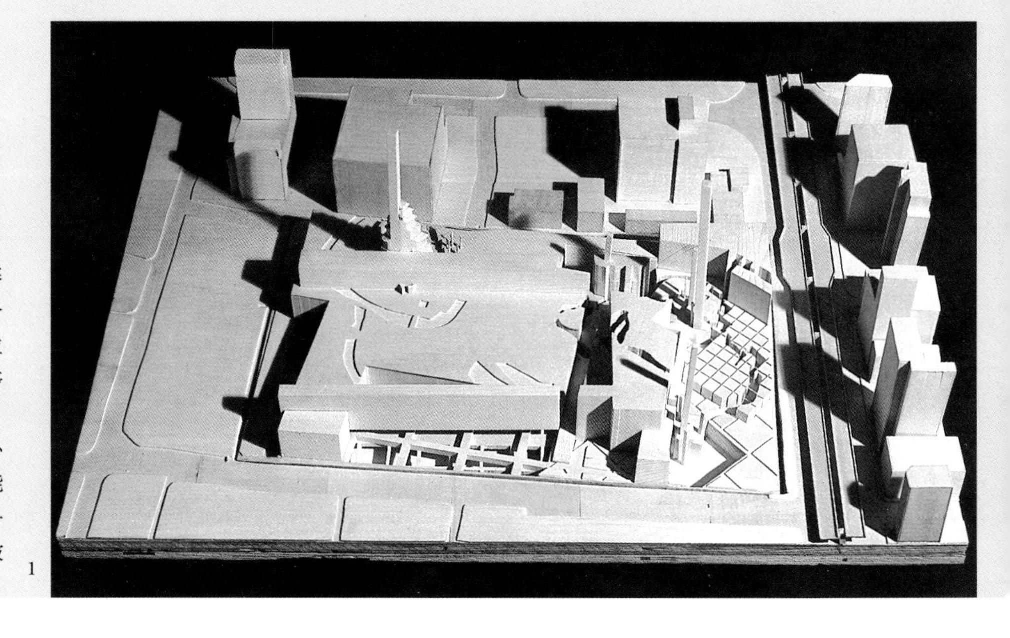

1

1 Presentation model, view from the west
2 Ground level plan
3 Small theater level plan
4 Medium theater level plan

1 展示模型，从正西方向看
2 首层平面
3 小剧场平面
4 中剧场平面

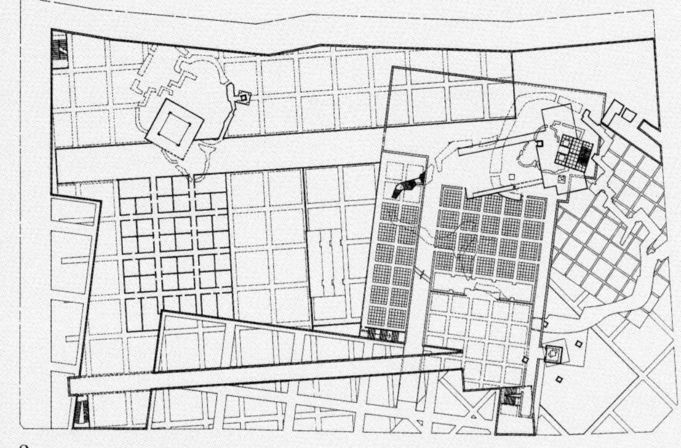

2

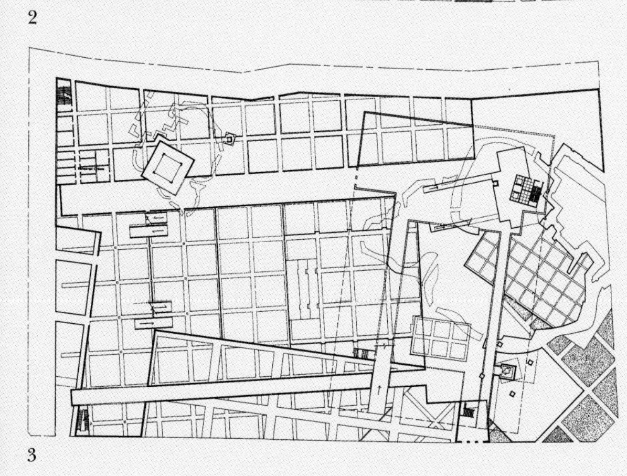

3

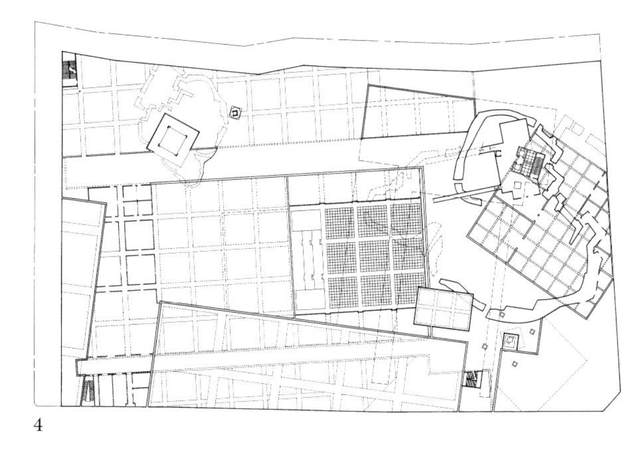

4

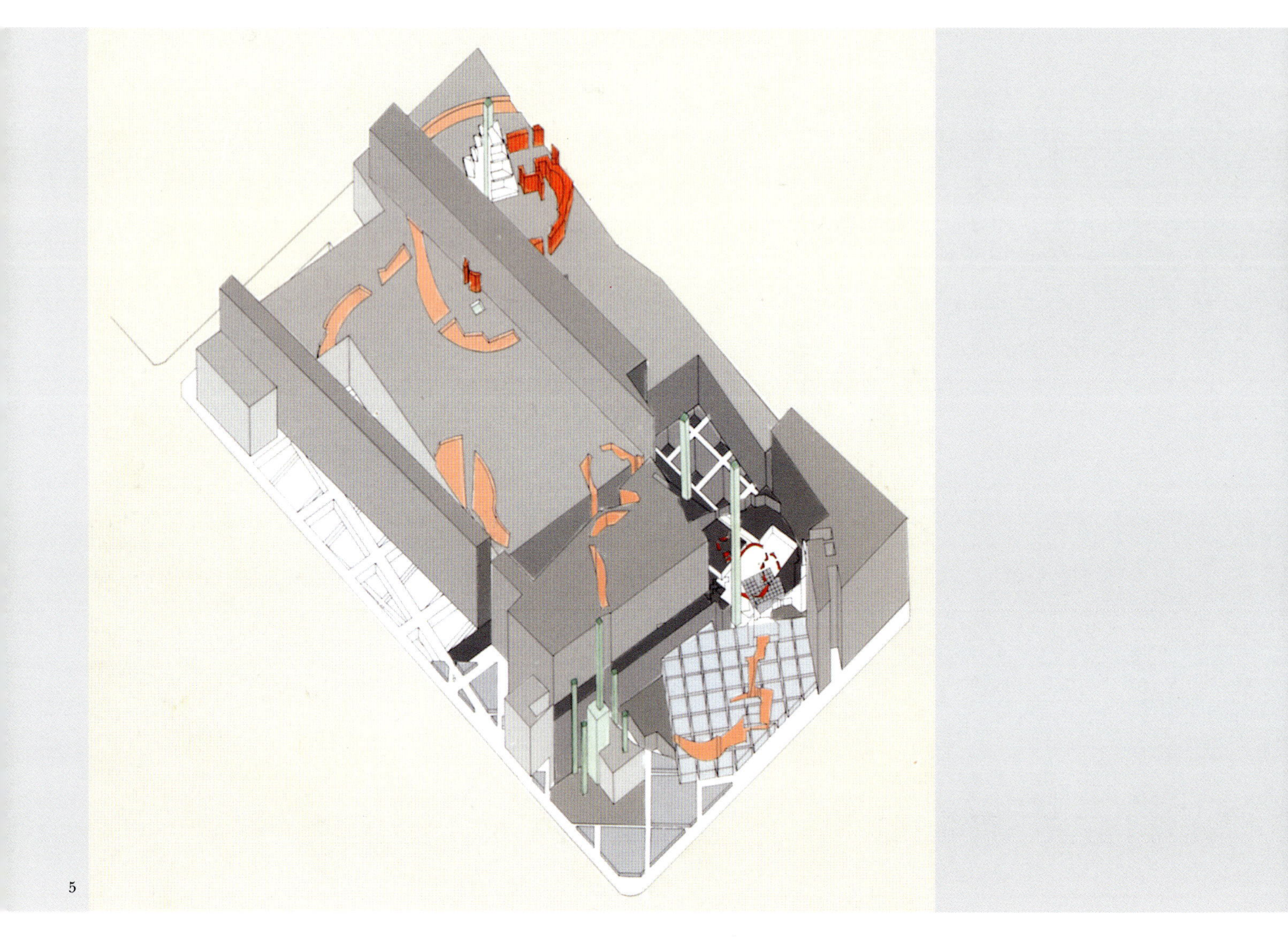
5

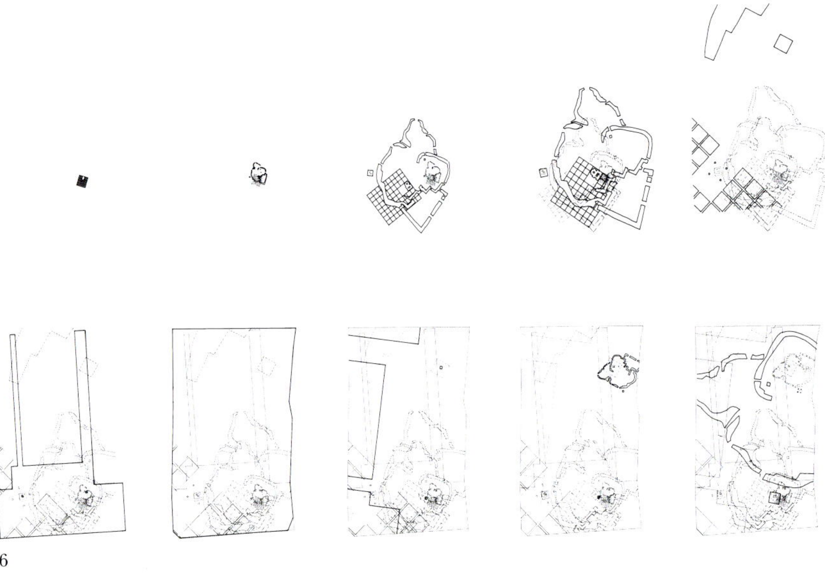
6

5 Axonometric, view from the south-west
6 Concept diagrams
7 South elevation
8 West elevation
9 Transverse section
10 Longitudinal section
11 Presentation model, view from the north-east
12 Site plan

5 轴测图，从西南方向看
6 设计概念示意图
7 南立面
8 西立面
9 横向剖面
10 纵向剖面
11 展示模型，从东北方向看
12 总平面

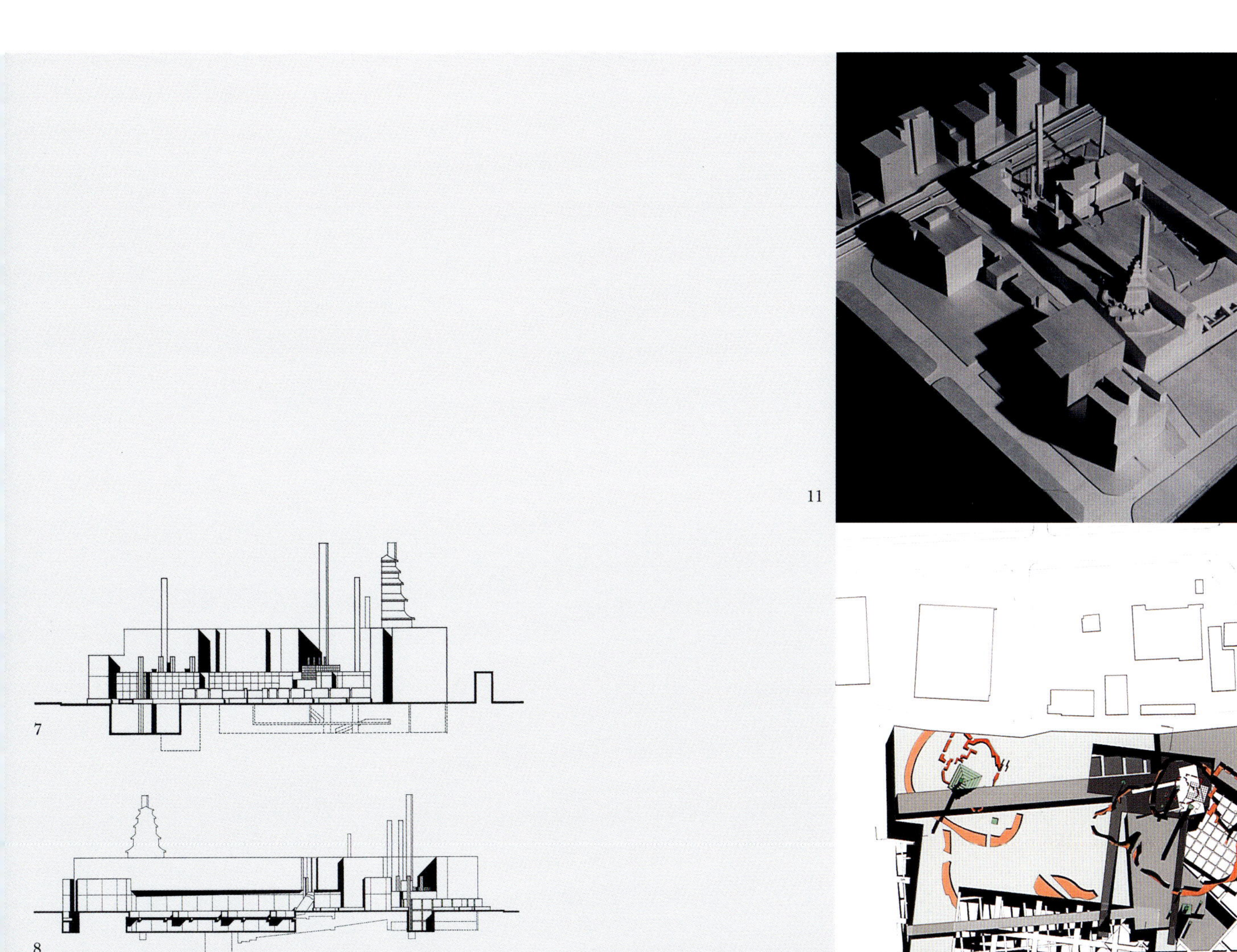

11

7

8

12

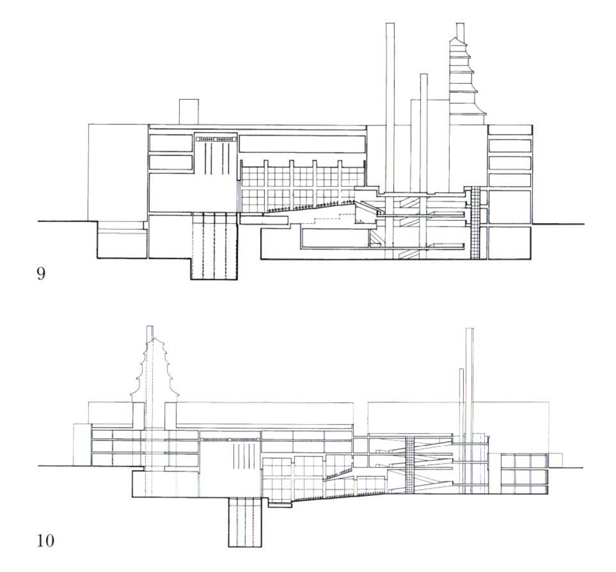

9

10

Biocentrum

Design 1987
Frankfurt am Main, Germany
J.W. Goethe University
350,000 square feet

生物研究中心

设计　　1987
美因河畔，法兰克福，德国
J·W·歌德大学
350000 平方英尺

This expansion of existing biotechnology research laboratories and support spaces was approached by considering the foundations of biology as an analogy for development of the scheme. DNA is used as a model of a logical sequence with infinite possibilities for expansion, change, and flexibility. Within this model, the design of the laboratory incorporated certain key technical design goals: providing a safe environment which protects the researchers and other building occupants from the various hazards encountered; heating, ventilating and air conditioning design which reduces the hazards of cross-contamination of experiments, and the spread of odors, toxic materials and other foreign agents.

对生物工艺研究实验室和配套用房的扩建，将生物学的基础看作是设计意图的相似部分，从而展开设计的。DNA 被当作一个模型来使用，它带有逻辑的顺序并有无限的扩展可能性、可变性和适应性。在这一模型中，实验室的设计融合了某些重要技术设计的目标：提供一个安全的环境，这一环境可以保护研究人员和其他建筑中居住者免受各种意外危险伤害；热能、通风、空气调节的设计，这些降低了试验、气体、有毒物质以及其他外部的化学试剂之间的交叉污染之危害。

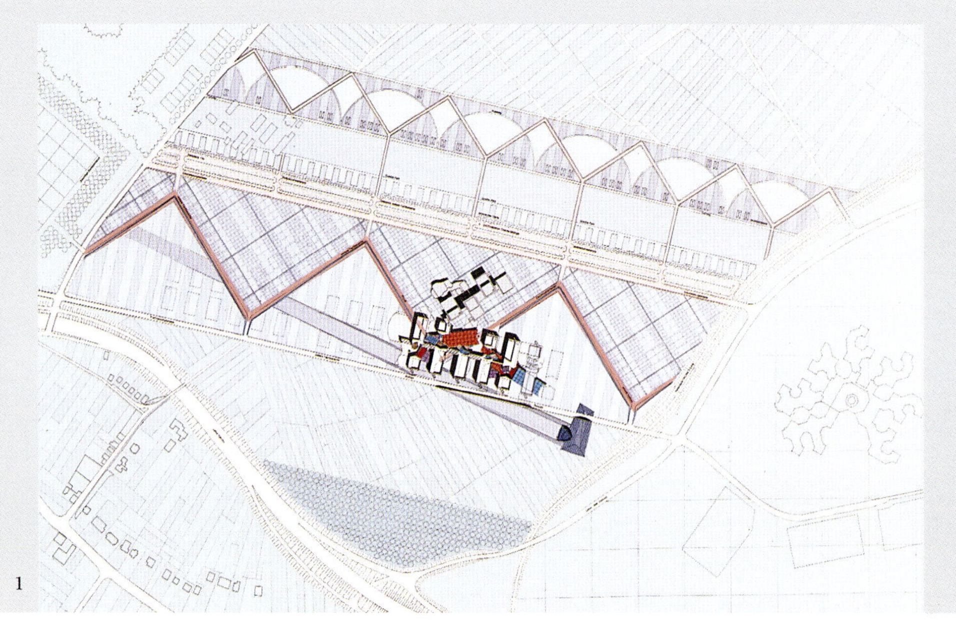
1

1 Site plan
2 Section AA
3 Section BB

1 总平面
2 AA 剖面
3 BB 剖面

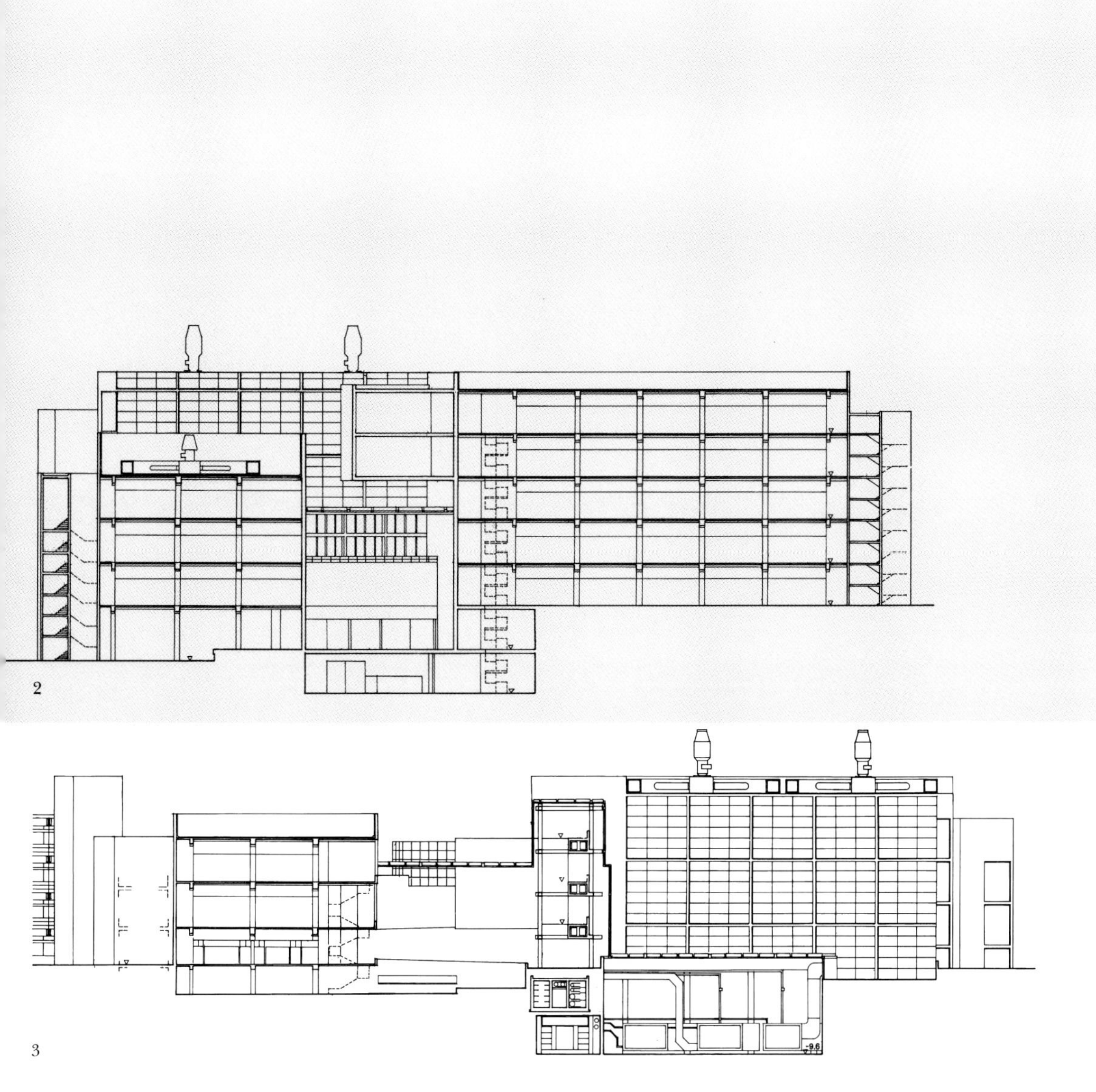

4

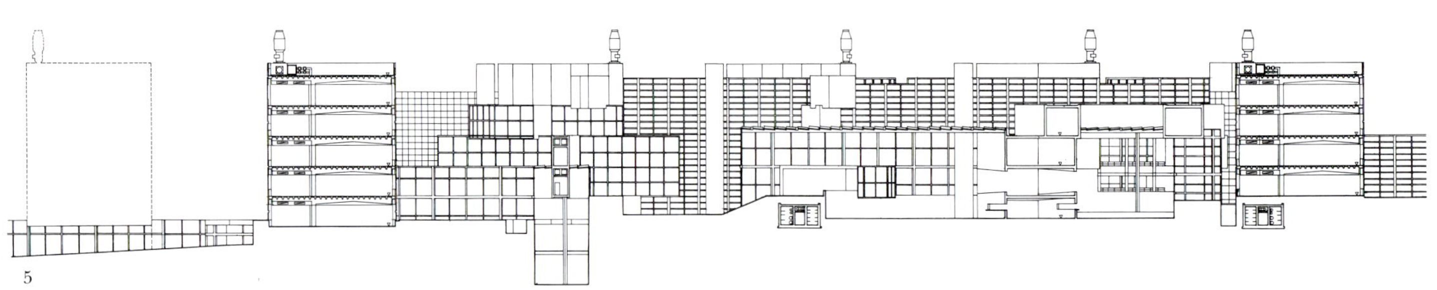

5

4 Presentation model, view from the east
5 Section CC
6 Second basement level plan
7 First basement level plan
8 Study model

4 从东部看展示模型
5 CC 剖面
6 地下二层平面
7 地下一层平面
8 研究模型

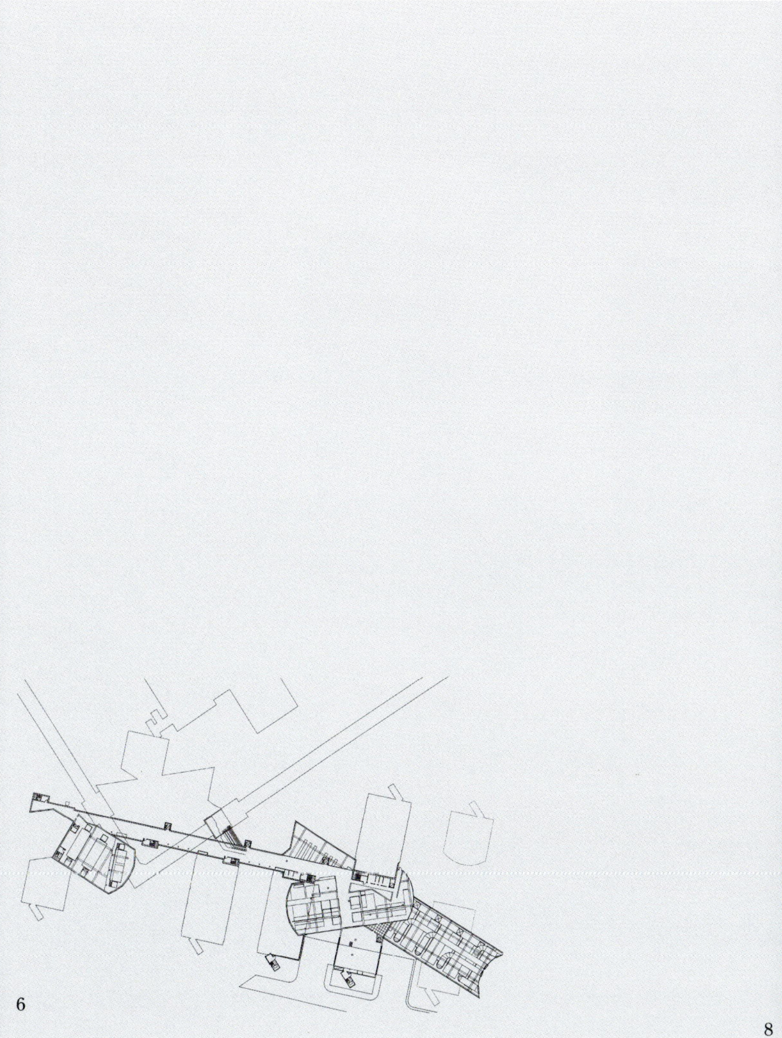

6

8

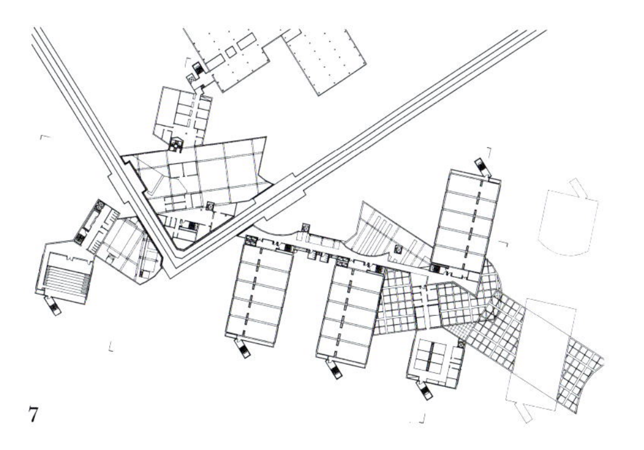

7

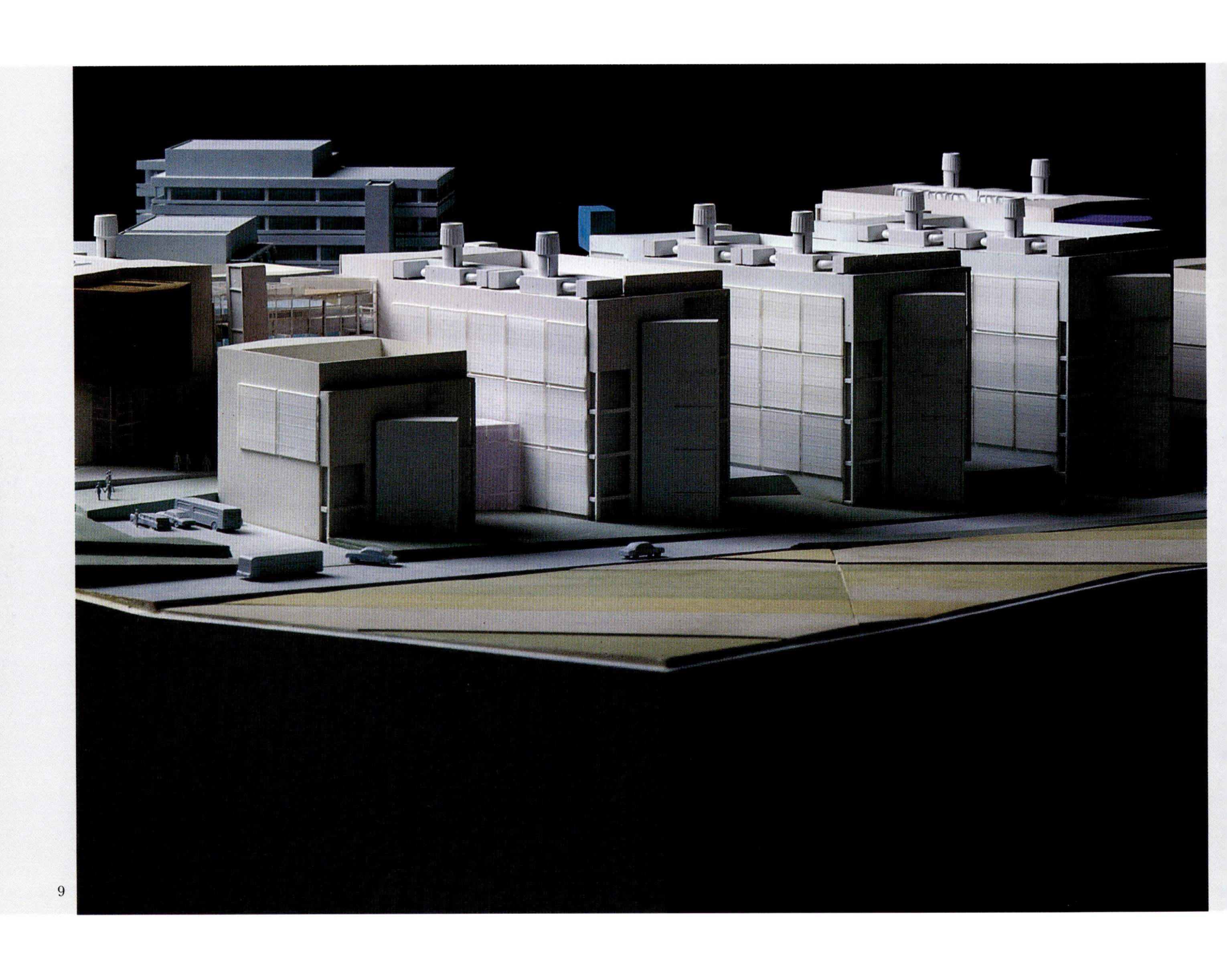

9

9 Presentation model, view from the south-east
10 Ground level plan
11 First level plan
12 Axonometric, view from the north-east
13 Concept diagrams

9 从东南方向看展示模型
10 首层平面
11 二层平面
12 从东北方向看轴测图
13 设计概念示意图

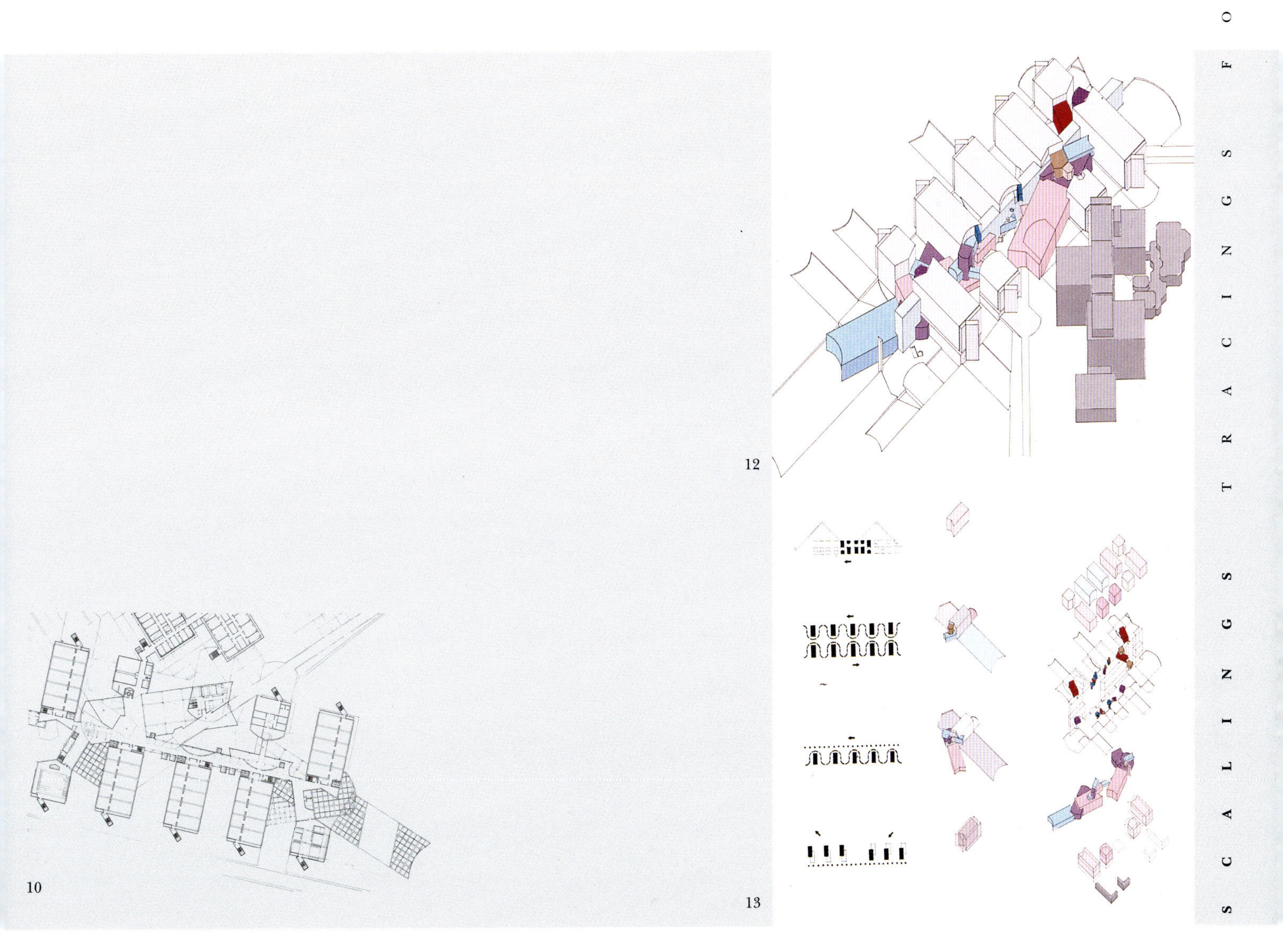

12

10

13

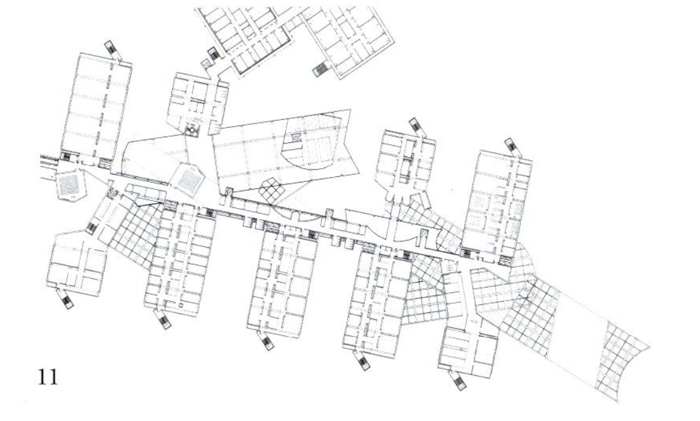

11

14

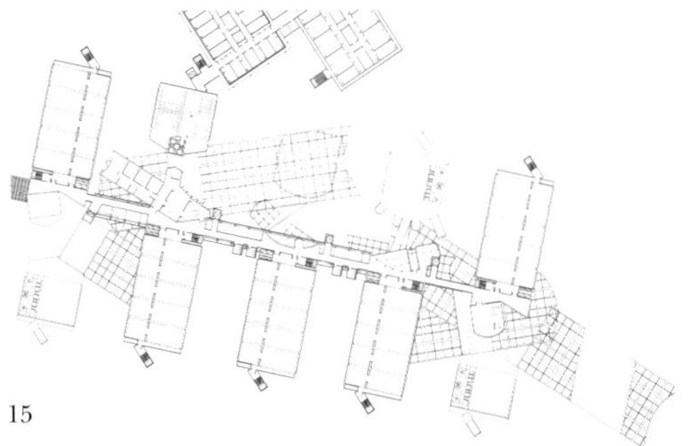

15

14 Presentation model, view from the south
15 Second level plan
16 South and north elevations
17 Third level plan

14 展示模型，从南向看
15 三层平面
16 南北立面
17 四层平面

16

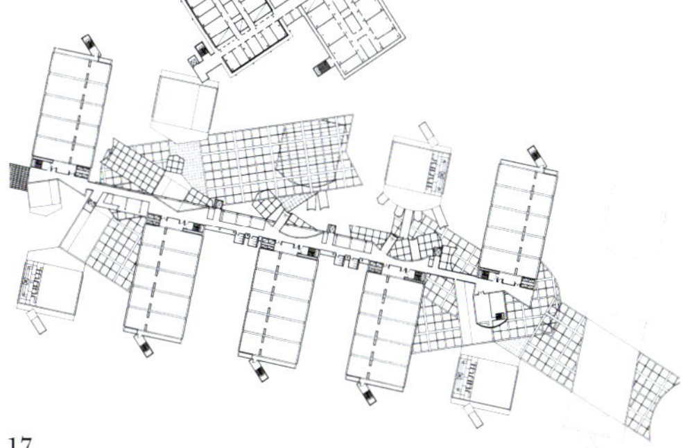

17

La Villette

Design 1986
Parc de La Villette, Paris, France
Establissement Public du Parc de la Vilette
4,300 square feet

拉维莱特

设计　1986
拉维莱特公园，巴黎，法国
拉维莱特公园机构
4300平方英尺

Parc de la Villette is a study of time—past, present, and future—and a questioning of representation in architecture. It replaces the actual conditions of time, place, and scale with analogies of these conditions. While the site exists in the present, it is also made to contain allusions to the present, the past, and the future. Analogies are made between the conditions that existed at the site in 1867, when an abattoir occupied the site; in 1848, when the site was covered by the city walls; and at the present, the time of Bernard Tschumi's La Villette project. The resultant ambiguous nature of time and place suggests an architecture that does not exist only in the present, but reverberates, suggesting an ever-increasing set of references.

拉维莱特公园是对时空的一种研究——过去、现在、未来——以及建筑表现问题上的研究。它用与之相似的情形代替了真实的时间、场地和比例。尽管场地是现实存在的，但是它也包含对过去、现在、将来的暗示。相似性产生于1867年和1848年场地的情形，1867年场地上是一个角斗场，1848年则是城墙；如今，则是贝尔纳时代的拉维莱特项目。时空和场地的合成的模糊属性暗示着一栋建筑，该建筑不应只存在于今天，而是应该反射、暗示一个不断增长的状态。

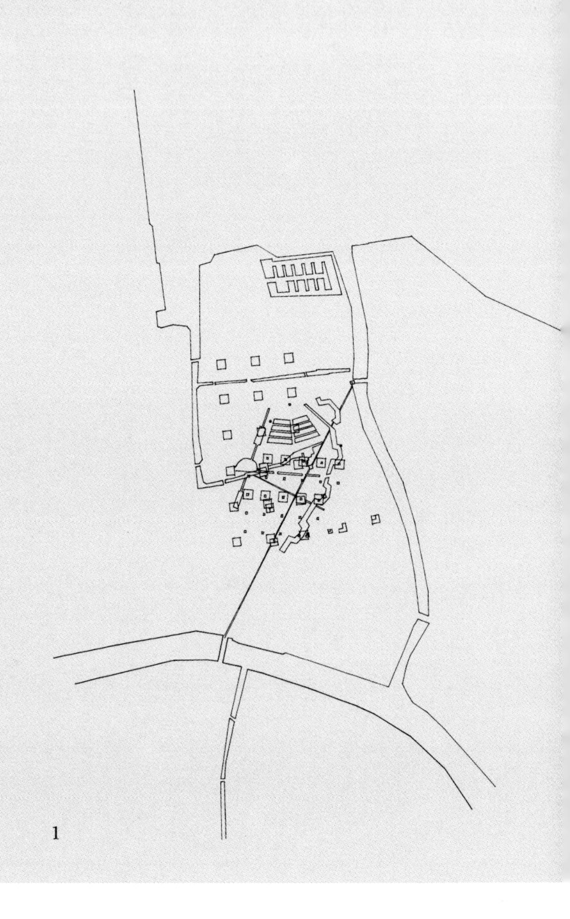

1

1 Presentation drawing, plan of scheme for site 3
2 Presentation drawing, axonometric of scheme for site 1
3 Plan, second scheme
4 Presentation model, second scheme

1 展示图，场地 3 的方案平面
2 展示图，场地 1 的方案轴测图
3 平面，方案 2
4 展示模型，方案 2

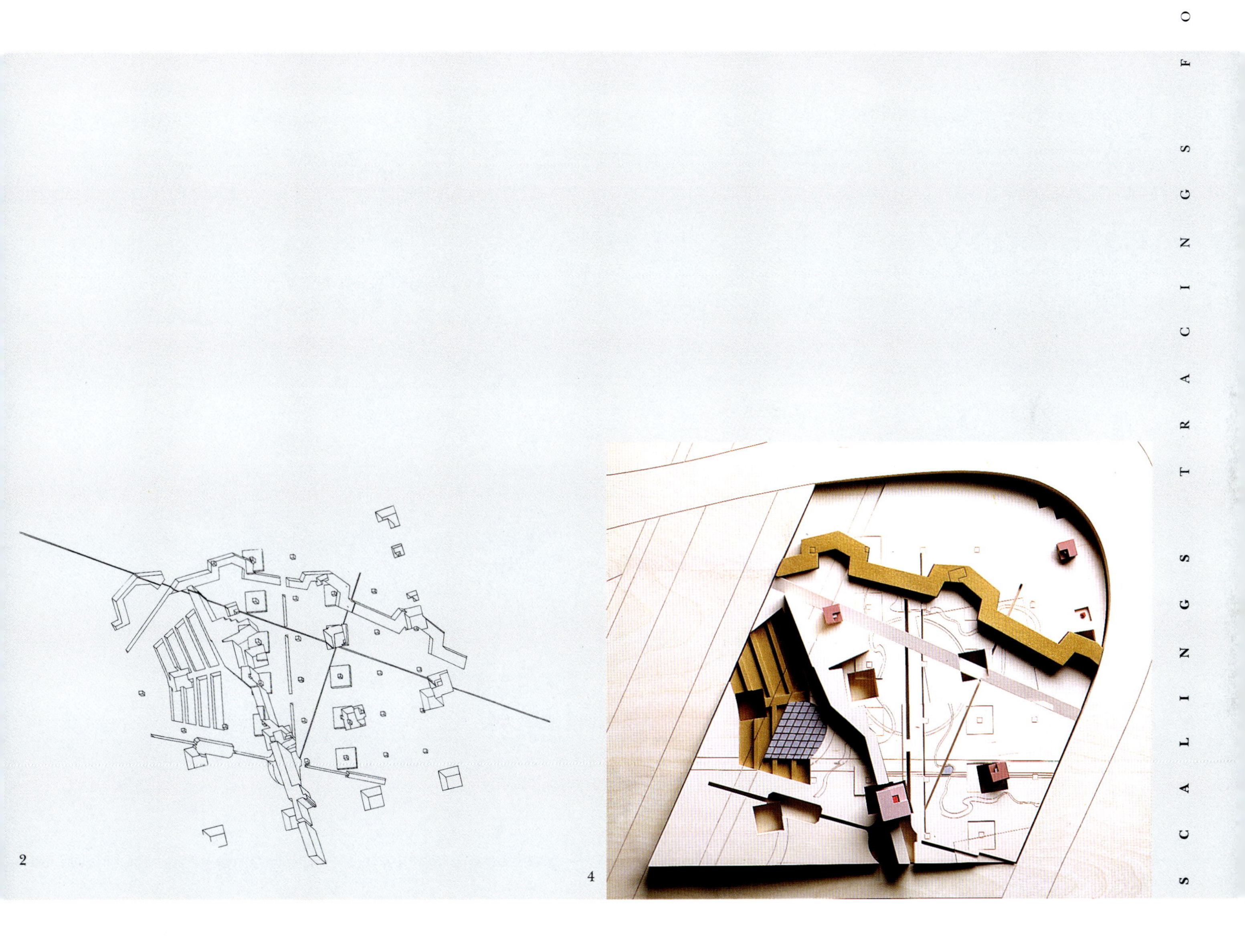

2

4

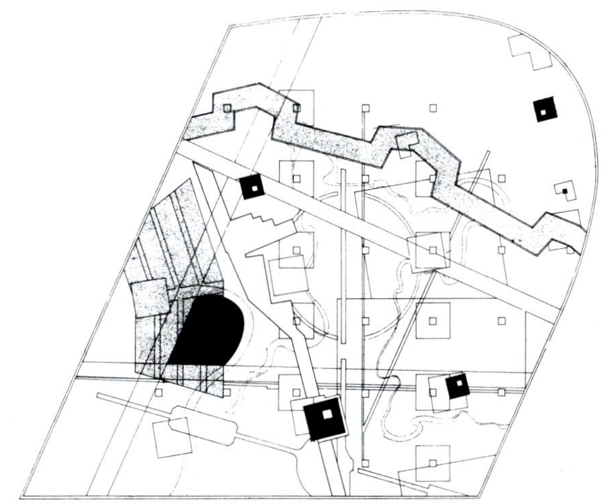

3

5

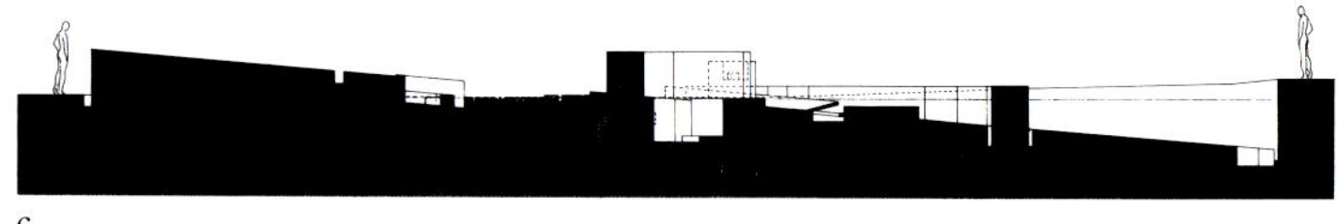
6

5 Presentation model, second scheme
6 Section 1, second scheme
7 Section 2, second scheme
8 Section 3, second scheme
9–10 Presentation models, first scheme

5 展示模型，方案 2
6 剖面 1，方案 2
7 剖面 2，方案 2
8 剖面 3，方案 2
9–10 展示模型，方案 1

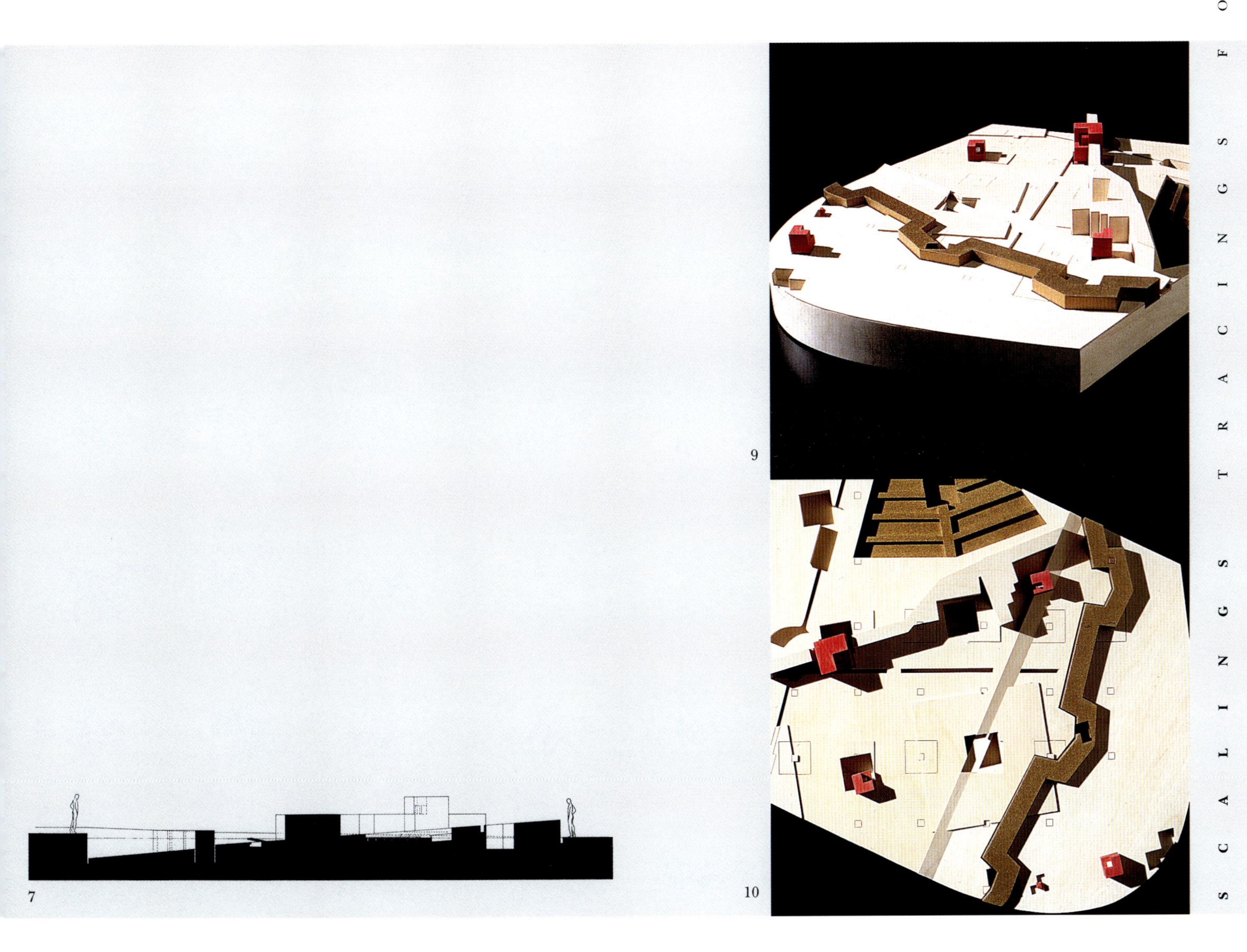

9

7

10

8

University Art Museum

Design 1986
Long Beach, California
University of California, Long Beach
67,500 square feet

大学艺术博物馆

设计　　1986
长滩，加利福尼亚州
加利福尼亚大学，长滩
67500 平方英尺

The master plan and museum design is the result of a history given to the project, compiled from a series of significant dates: the settlement of California in 1849; the creation of the campus in 1949; and the rediscovery of the museum in 2049.

The building consists of four major exhibition spaces: an audiovisual installation gallery, a black-box theater/auditorium, a cafe, a conference space, a library, administrative offices and storage areas, and a series of exterior terraced sculpture courtyards. The arboretum will contain a 2-acre artificial pond, botanical gardens, terraces, and seating areas. An elevated walkway provides a link between the northern and southern portions of the arboretum.

总体规划和博物馆的设计是这个项目的历史之结果，这是由一系列的重要日期汇编而成：1849 年加利福尼亚州的成立；1949 年校园的创建；2049 年博物馆的重新发现。

建筑包含了四个主要的展示空间：一个有视听装置的展廊，一个多功能音乐厅/一个礼堂、咖啡厅、问询处、阅览室、管理用房和储藏间，以及一系列的室外庭院雕塑平台。植物园将包含一个占地 2 英亩的人工池塘、植物园、露台以及休息区域。架空的人行道连接着植物园的南北两部分。

1

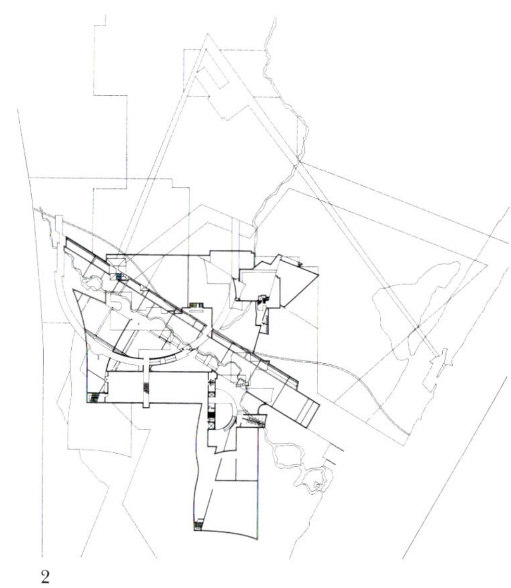

2

1 Presentation model, view from the west
2 Ground level plan
3 Site plan
4 Roof plan
5–6 Presentation models, view from above

1 从西向看展示模型
2 首层平面
3 总平面
4 屋顶平面
5-6 从上往下看展示模型

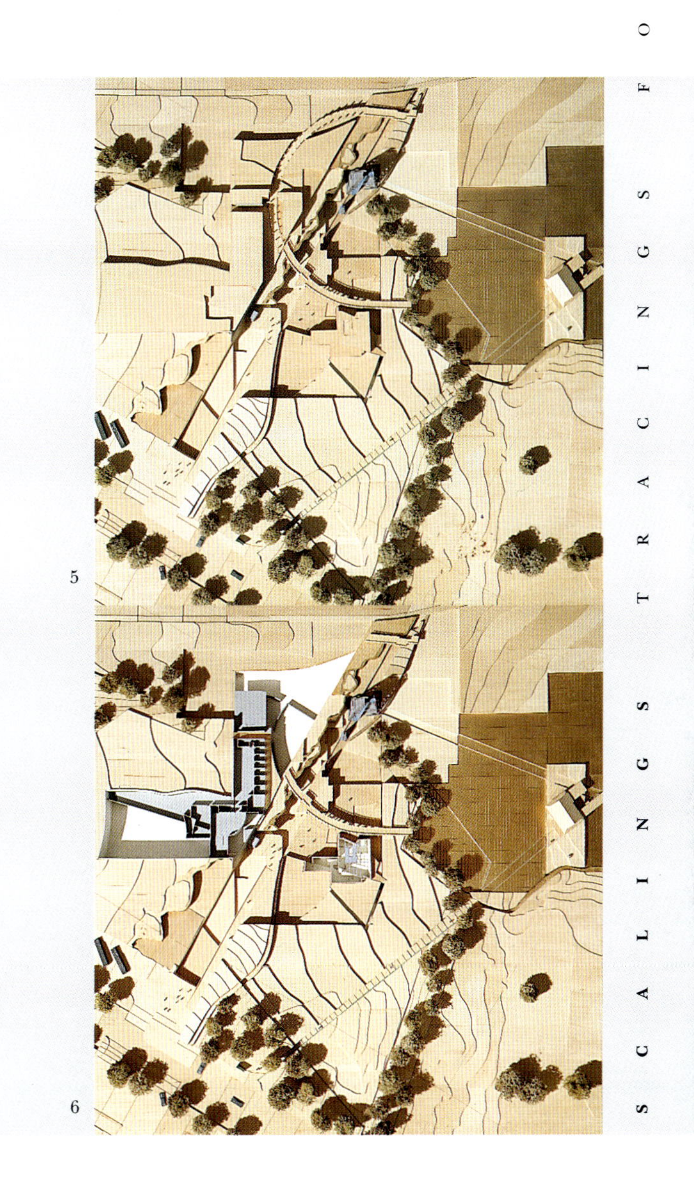

5

6

3

4

7

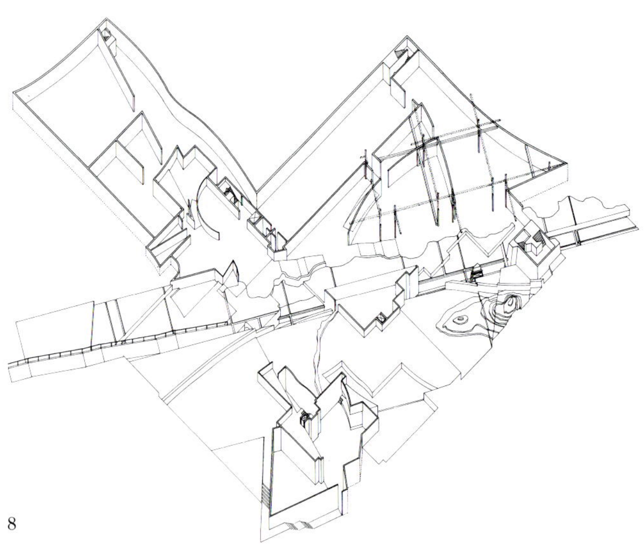
8

7 Site plan
8 Axonometric, view from the north
9 Lower level plan
10 Axonometric, view from the north
11 Presentation model, view from the north-west
12 Presentation model, view from the south-west

7 总平面
8 从北向看轴测图
9 地下层平面
10 从北向看轴测图
11 从西北方向看展示模型
12 从西南方向看展示模型

9

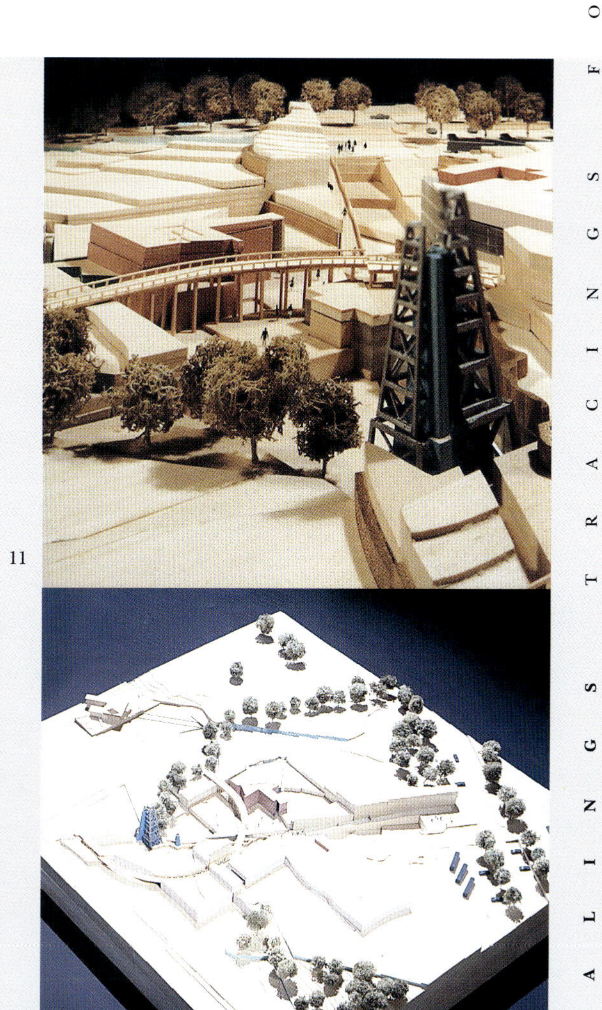

11

12

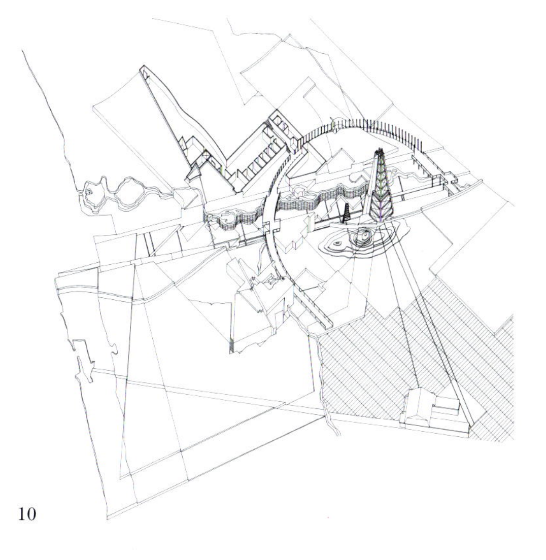

10

Progressive Corporation Office Building

Design 1986
Cleveland, Ohio
Progressive Corporation

进步党总部办公楼

设计　1986
克利夫兰市，俄亥俄州
进步党总部

The site was developed from the superposition of aspects of the geographical history of the state of Ohio and the city of Cleveland: the 18th century boundary of the Connecticut Western Reserve; the 1903 Daniel Burnham plan; and the Greenville Trace—surveys of the state carried out simultaneously from the north and south. These elements were altered in size and superposed on one another. All of the conditions, fictitious and real, artificial and natural, exist simultaneously in this reinvented site. The buildings sit on the site like huge chisels, breaking the pieces open to reveal the many-faceted layers of their history.

建筑所处的位置是从俄亥俄州和克利夫兰市的地理历史发展的重合方面发展而来的：18世纪的康涅狄格州西部保护边界；1903年丹尼尔·伯纳姆的规划；以及格林维尔市的轮廓——对该州从南到北的俯瞰。这些因素是在不断变化并且是相互重叠的。所有的情形，虚幻的亦或真实的，人造的亦或自然的，都同时存在于这个再次投资利用的场地中。建筑立在场地中就像一个大的凿子，将各个部分劈开，从而揭示了他们历史的多面性。

1

1 Presentation model, view from the north-west
2 Site plan
3 Axonometric, view from the west

1 展示模型，从西北方向看
2 总平面
3 从西向看轴测图

2

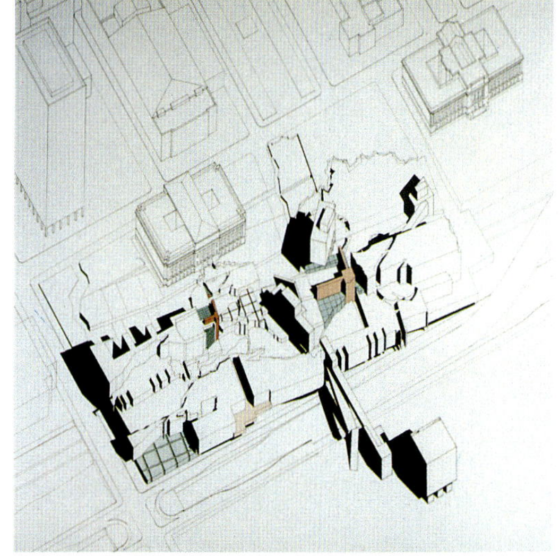

3

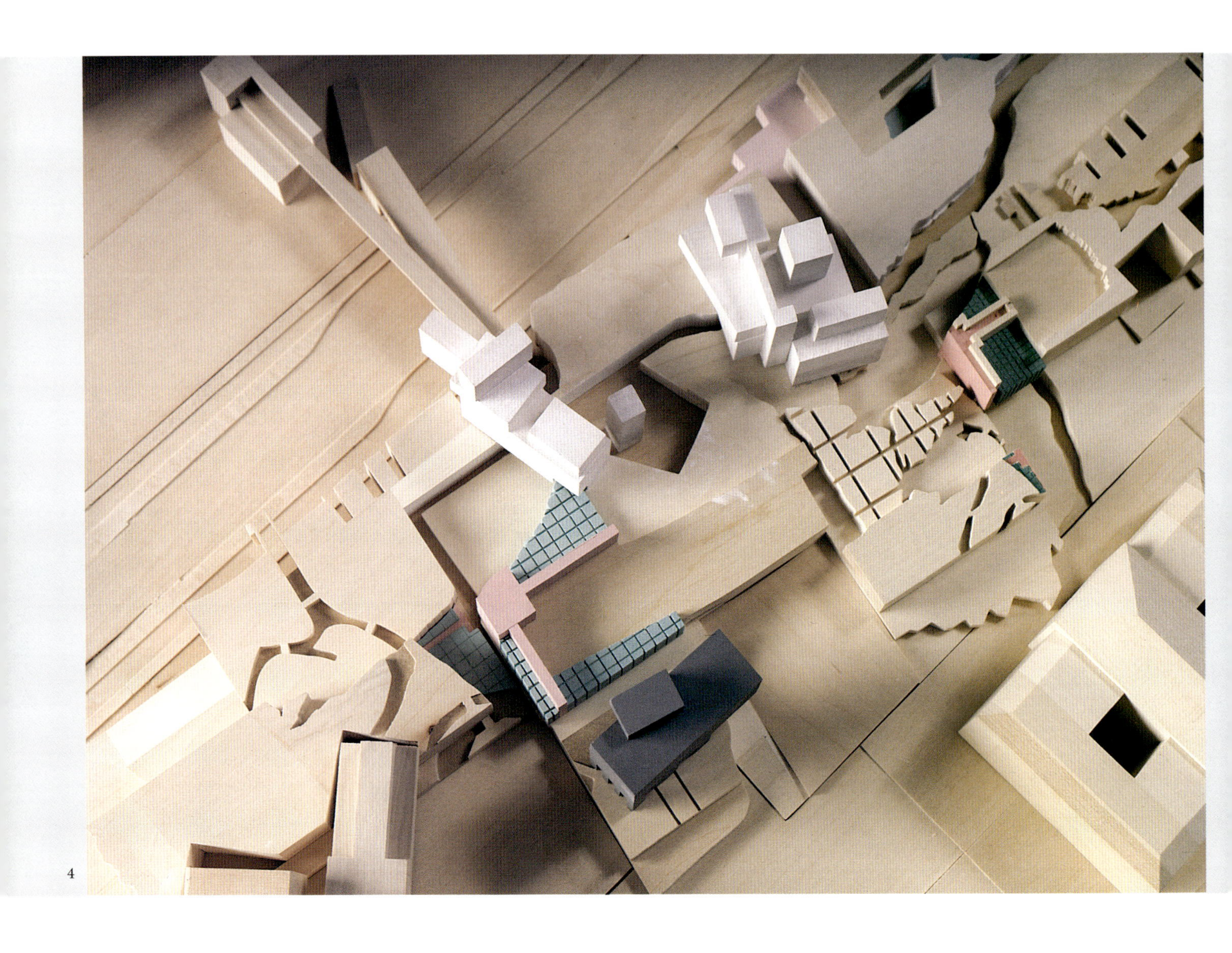
4

4 Presentation model, view from the south
5 Site plan
6 Presentation model, view from the west
7 Presentation model, view from the south-west

4 展示模型，从正南方向看
5 总平面
6 从西向看展示模型
7 从西南方向看展示模型

6

7

5

Wexner Center for the Visual Arts and Fine Arts Library

Design/Completion 1983/1989
Columbus, Ohio
The Ohio State University, State of Ohio
140,000 square feet

韦克斯纳视觉艺术中心

设计/竣工　1983/1989
哥伦布，俄亥俄州
俄亥俄州立大学，俄亥俄州
140000平方英尺

Instead of selecting any of the obvious building sites on the campus, a site was created by locating the Center between several proposed sites and existing buildings. This can be described as a non-building, an archaeological earthwork whose essential elements are scaffolding and landscaping.

The scaffolding consists of two intersecting three-dimensional gridded corridors which link existing buildings with the new galleries and arts facilities. One part of scaffolding is aligned with the Columbus street grid, the other with the campus grid, so the project both physically and symbolically links the campus with the city beyond. The Center acts as a symbol of art as process and idea.

在一些被建议的场地与现有的建筑之间的场地放置了该中心，取代了校园的明显的建筑场地。这可以被形容为非建筑、一个考古工程，其基本元素是脚手架和景观美化。

脚手架由两个有趣的三维网格走廊组成，该走廊将新的展廊以及艺术用房与现有的建筑物相连。脚手架的一部分与哥伦布市的街道平行，另一部分与校园的网格相平行。所以该项目具体性地并象征性地将校园与城市相连。该中心就像过程和设想一样起着艺术象征的作用。

1

1 View from the south
2 Site plan
3 Ground level plan

1 从正南方向看
2 总平面
3 首层平面

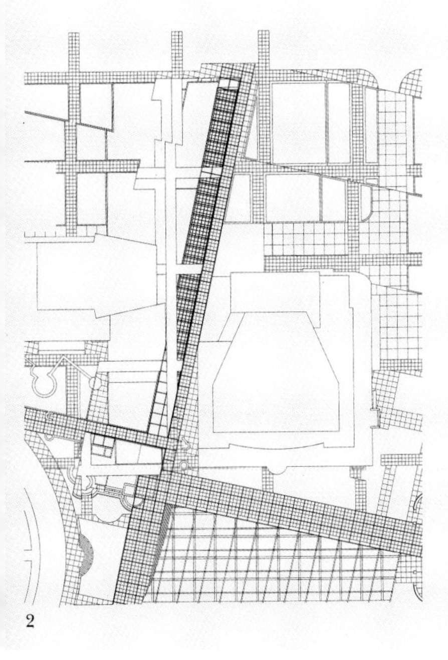

2

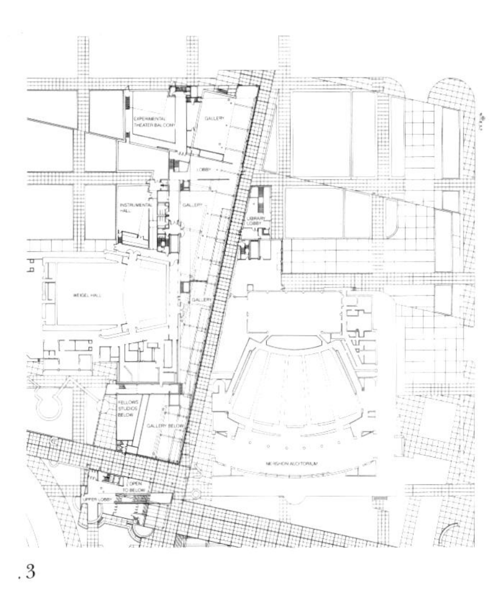

3

4

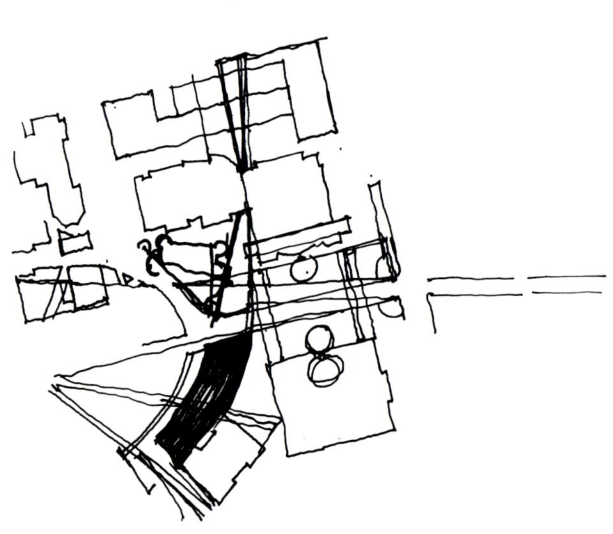
5

4 View from the south-east
5 Conceptual sketch
6 Section through lobby and offices, view from the south
7 Section through upper lobby and moat, view from the south
8 Scaffolding, detail view from the south-east
9 Scaffolding, detail view from the south

4 从东南方向看
5 概念草图
6 从南向看大厅和办公室剖面
7 从南向看大厅上空和城壕剖面
8 脚手架，从东南方向看细部
9 脚手架，从南向看细部

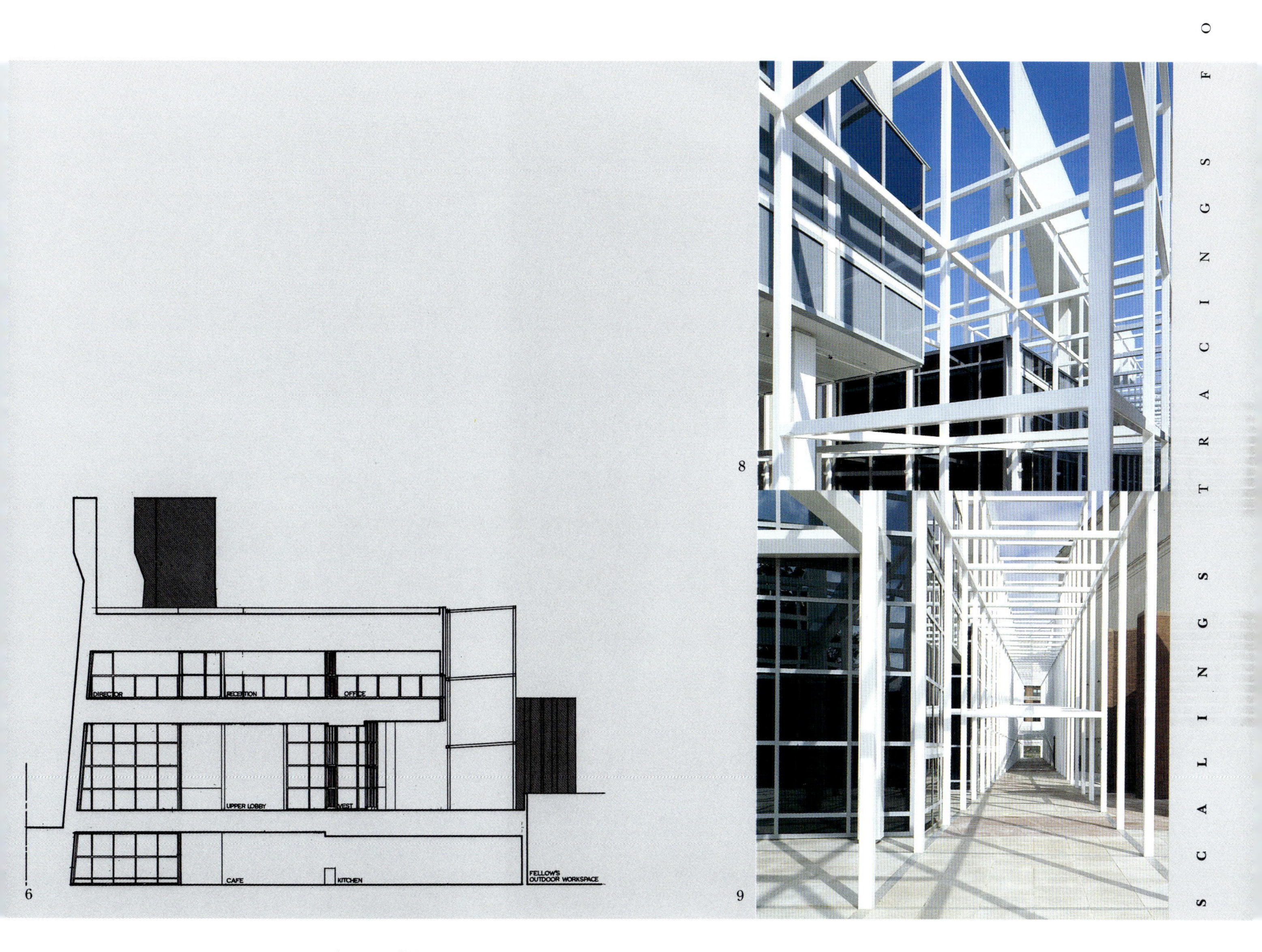

6

8

9

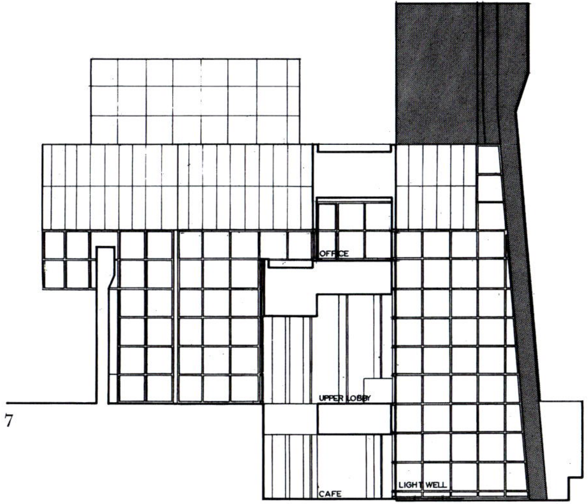

7

10

10 Lobby
11 Section through gallery ramp, view from the east
12 Section through gallery ramp, view from the west
13 West elevation
14 South elevation

10 大厅
11 从东向看展廊坡道剖面
12 从西向看展廊坡道剖面
13 西立面
14 南立面

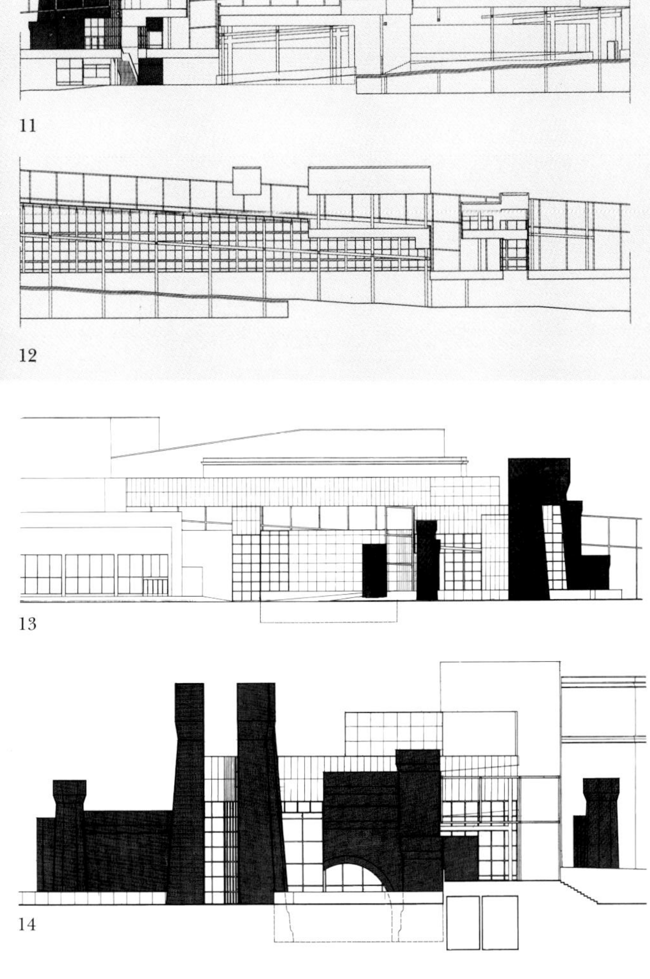

11

12

13

14

15

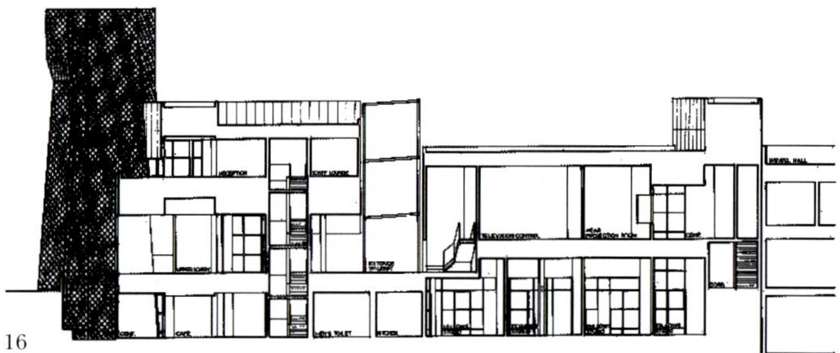

16

15 Detail view
16 Section through lobby, view from the east
17 Section through gallery ramp, view from the west
18 Section through gallery ramp, view from the east
19 Gallery, view from the north
20 Black-box theater

15 细部
16 从东向看大厅剖面
17 从西向看展廊坡道剖面
18 从东向看展廊坡道剖面
19 从北向看展廊
20 多功能音乐厅

17

19

20

18

GRIDDINGS SCALINGS TRACINGS FOLDINGS

Carnegie Mellon Research Institute

Design 1988
Pittsburgh, Pennsylvania
Carnegie Mellon University
85,000 square feet

卡内基·梅隆研究学会

设计　1988
匹兹堡，宾夕法尼亚州
卡内基·梅隆大学
85000平方英尺

Eisenman Architects was selected to develop a master plan for the Pittsburgh Technology Center and design a new facility for the Carnegie Mellon Research Institute. The design had to address the "knowledge revolution," and represent Pittsburgh's revitalization as the first post-industrial city.

The fundamental structure for this development is the "Boolean cube," a geometric model for computer processing. Each building is made up of pairs of cubes. Each pair contains two solid cubes and two frame cubes corresponding to office and laboratory modules. Each pair can be seen as containing the inverse of the other as solid and void.

埃森曼建筑事务所被委托为匹兹堡科技中心开发一项总体规划并为卡内基·梅隆研究学会设计一栋新的办公用房。设计必须强调"知识革命"，并体现匹兹堡作为后工业城市的复兴。

这一发展的基本结构是"布利安立方体"，它是计算机程序的一个几何模型。每一栋建筑是由若干对立方体组成。每一组包括两个立方实体和两个立方体框架，其与办公室和图书馆模块相呼应。每一组可以被看成是实与虚的相互包含。

1

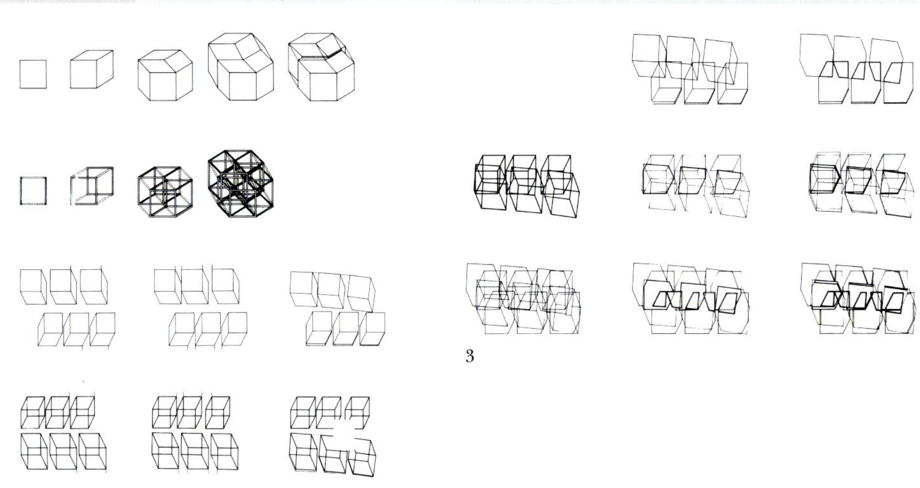

2

3

1 Study model, Boolean cube
2–3 Concept diagrams
4 Second level plan
5 Third level plan
6 Fourth level plan
7 Sixth level plan
8 Seventh level plan
9 Roof level plan

1 研究模型，布利安立方体
2–3 概念示意图
4 二层平面
5 三层平面
6 四层平面
7 七层平面
8 八层平面
9 屋顶平面

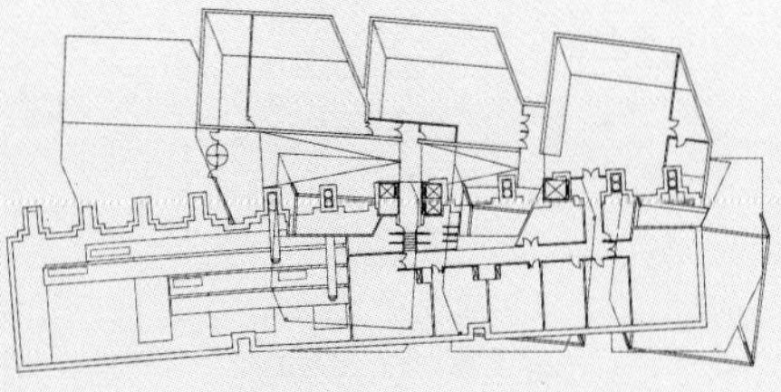

4

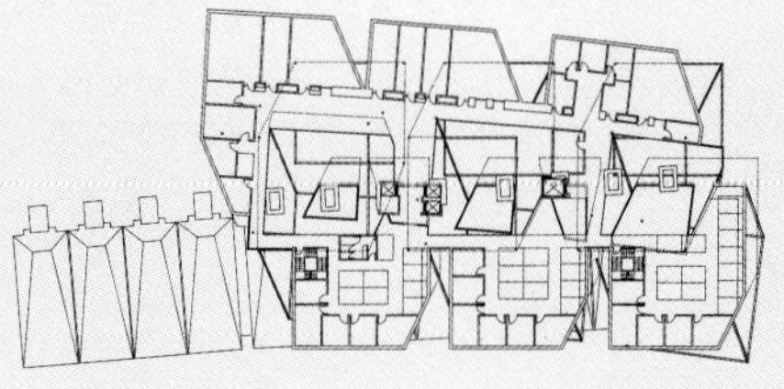

5

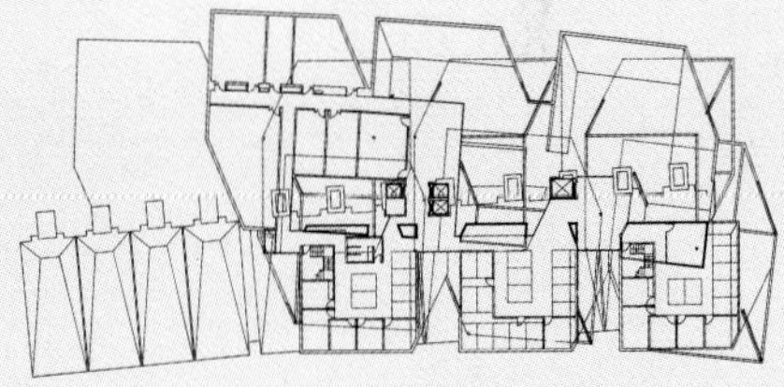

6

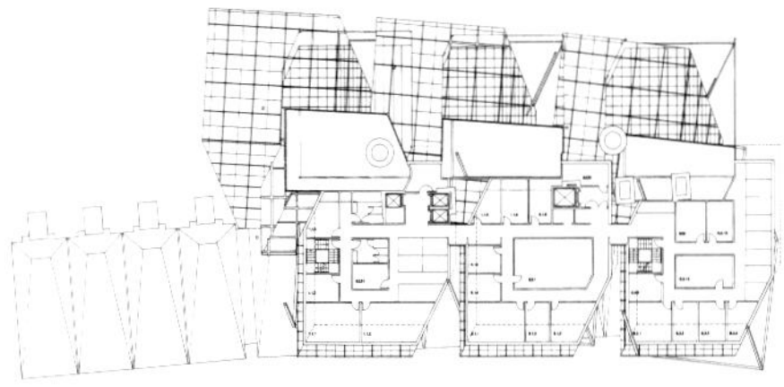

7

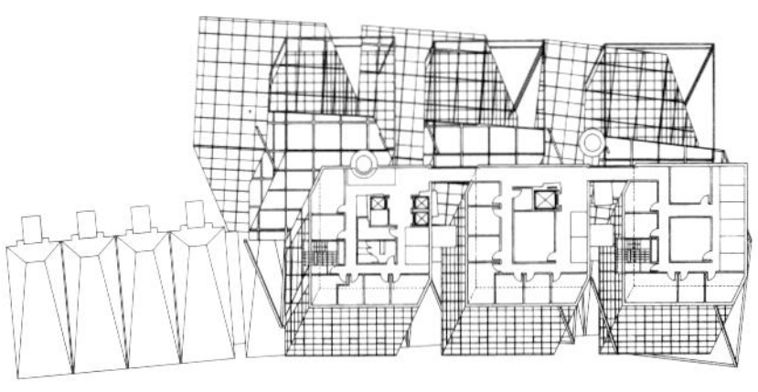

8

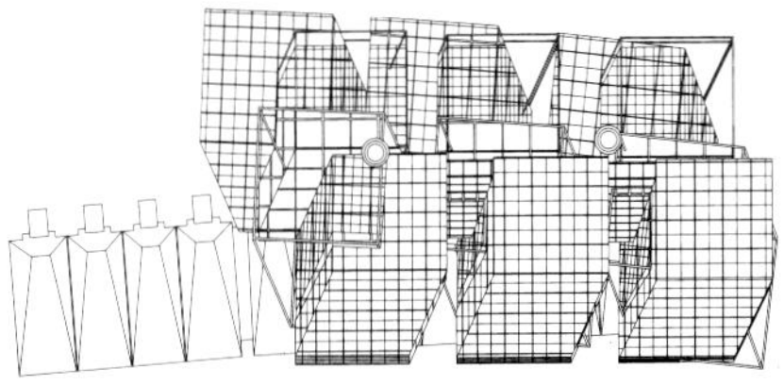

9

10

11

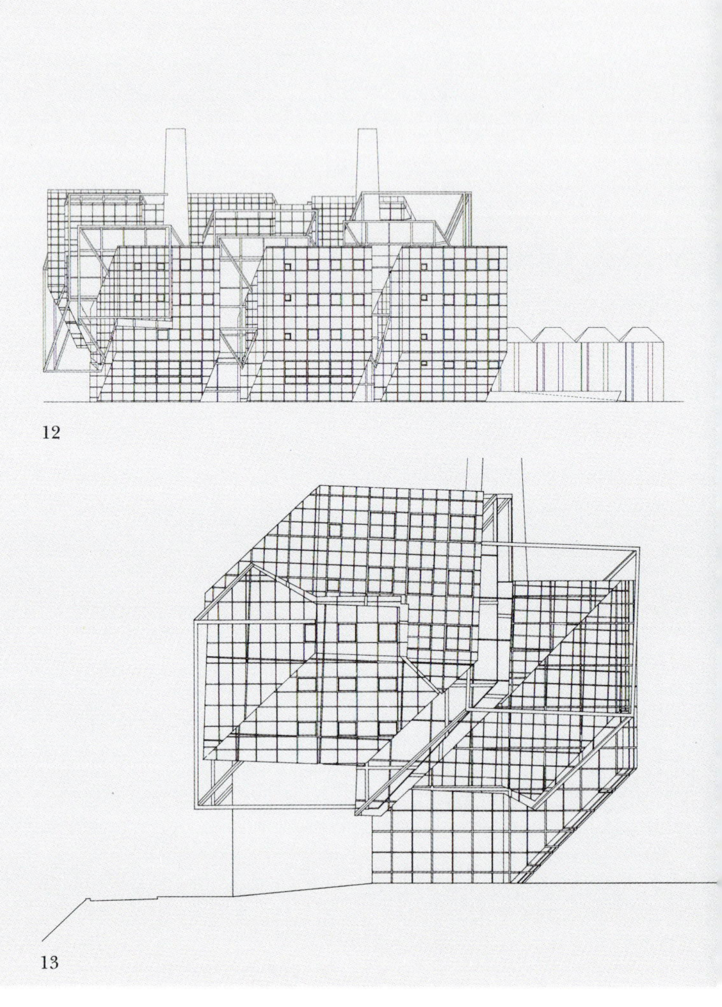

12

13

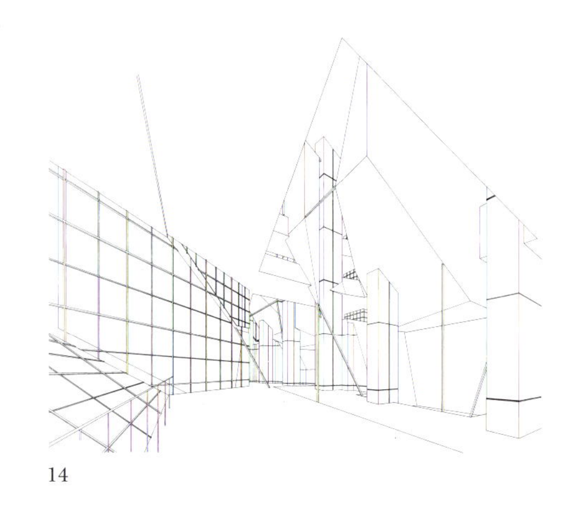

14

10–11 Study model
12 North elevation
13 East elevation
14 Interior perspective
15 Site model, view from the east
16 Transverse section, view from the east

10－11 研究模型
12 北立面
13 东立面
14 室内透视
15 从东向看场地模型
16 从东看横向剖面

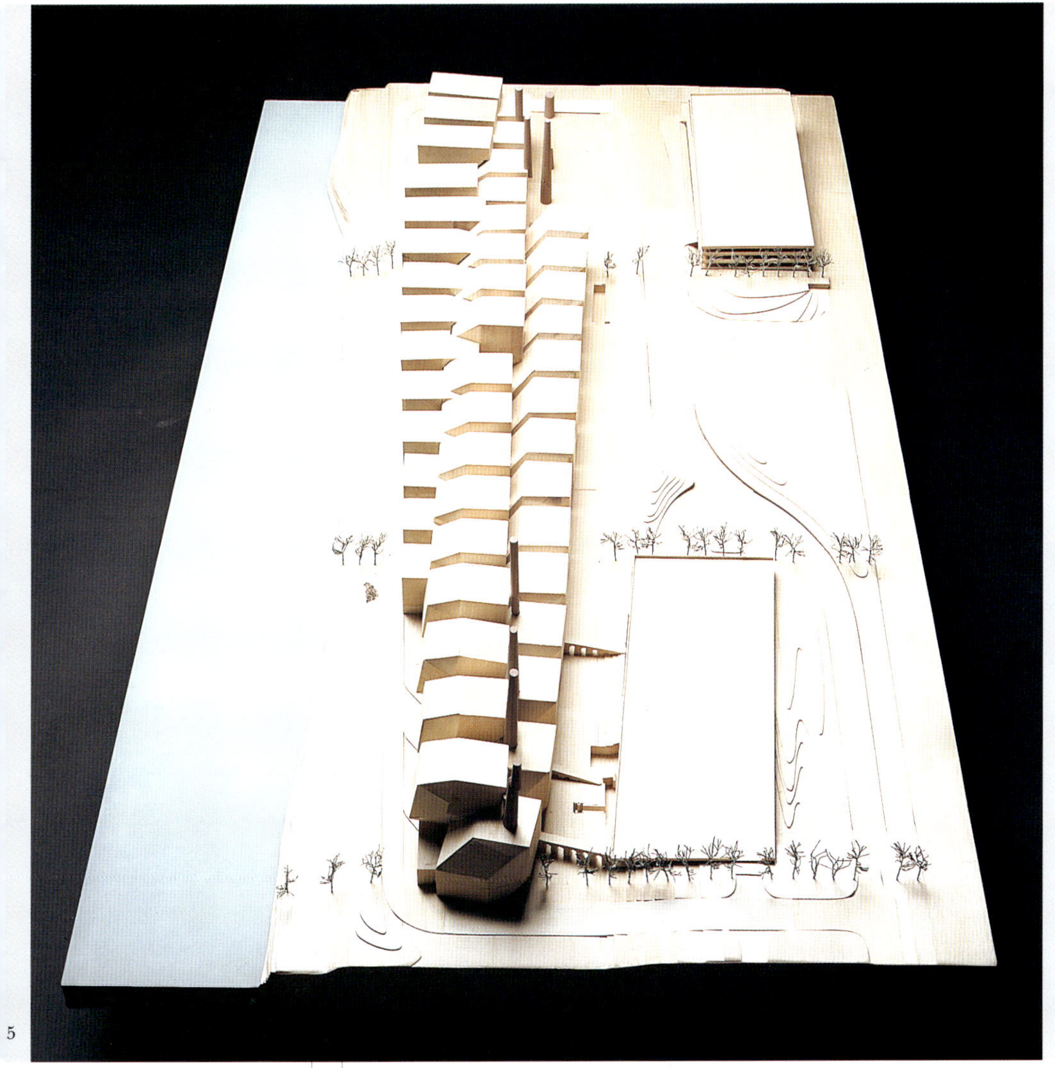

5

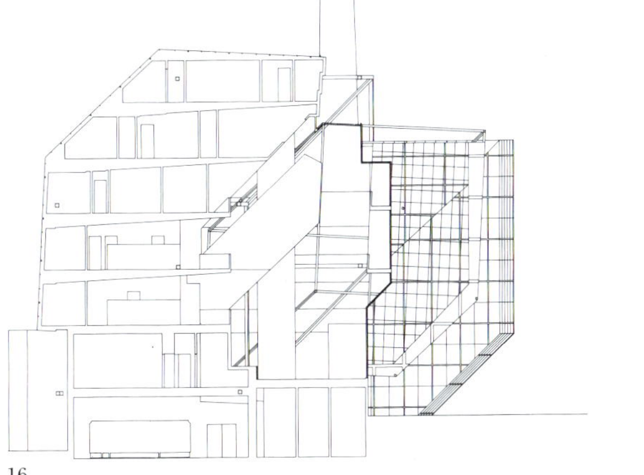

16

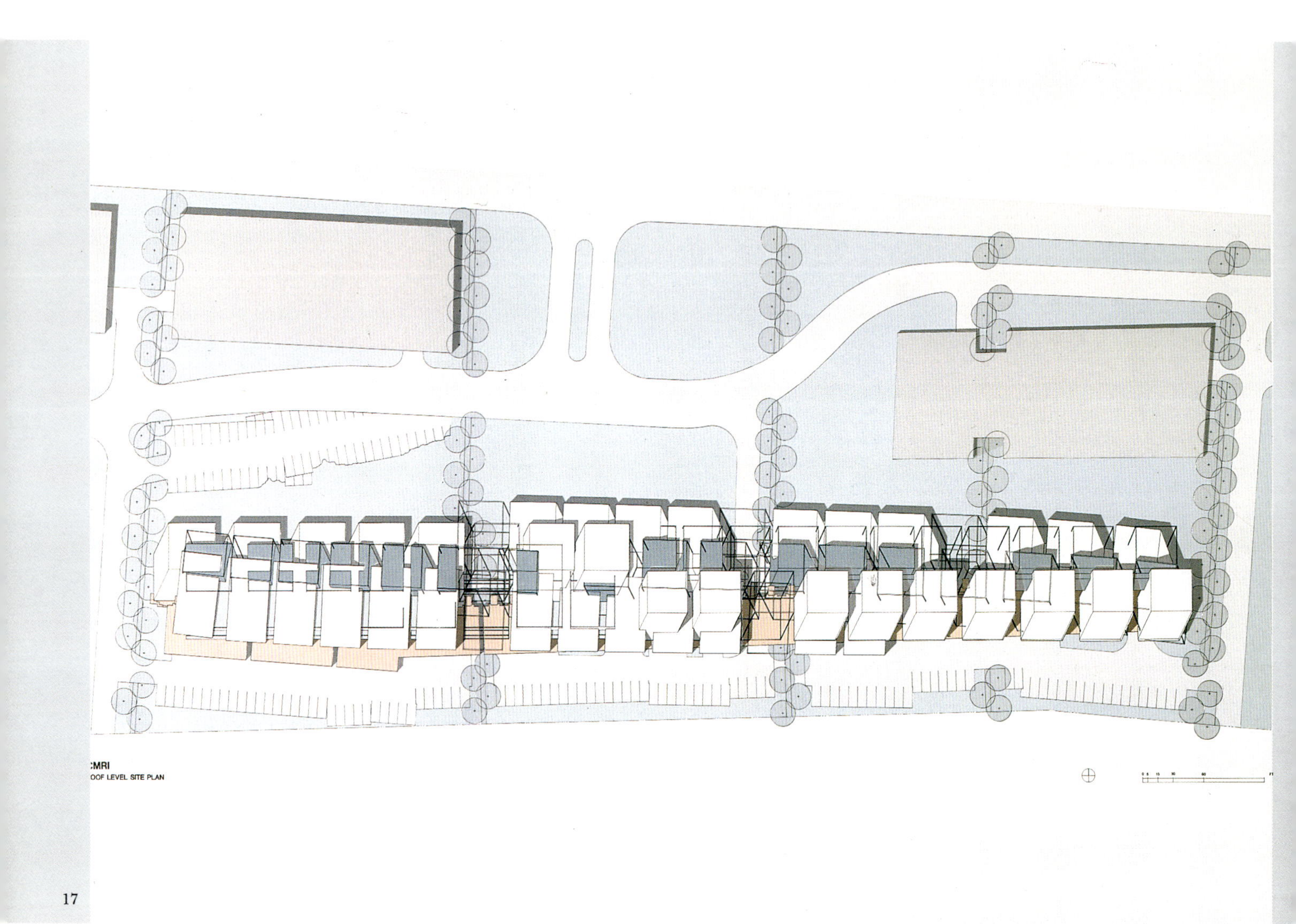

17

18

17 Site plan
18 Longitudinal section, view from the north
19 Roof plan
20 Longitudinal section, view from the south

17 总平面
18 从北看纵向剖面
19 屋顶平面
20 从南看纵向剖面

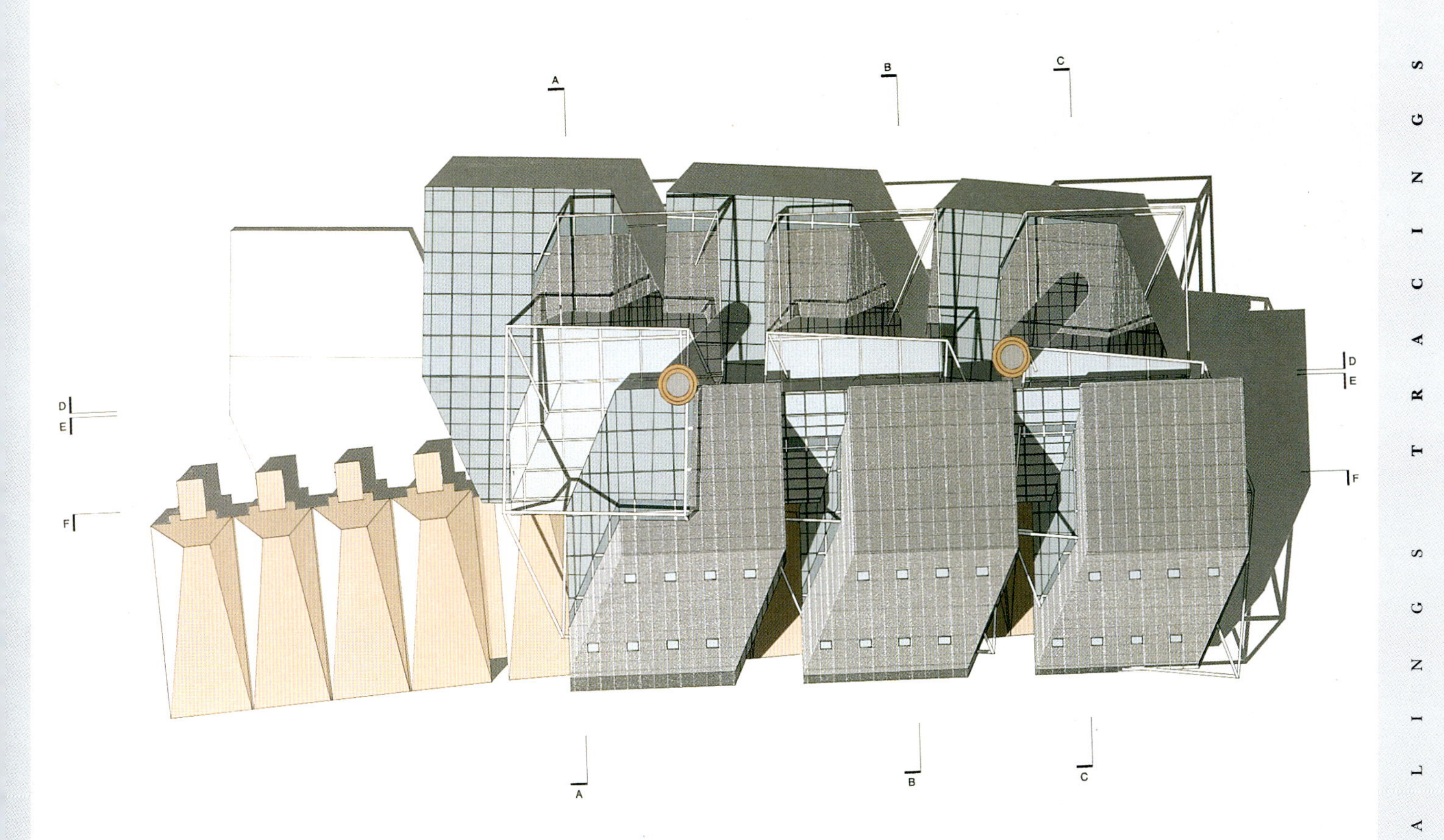

19

20

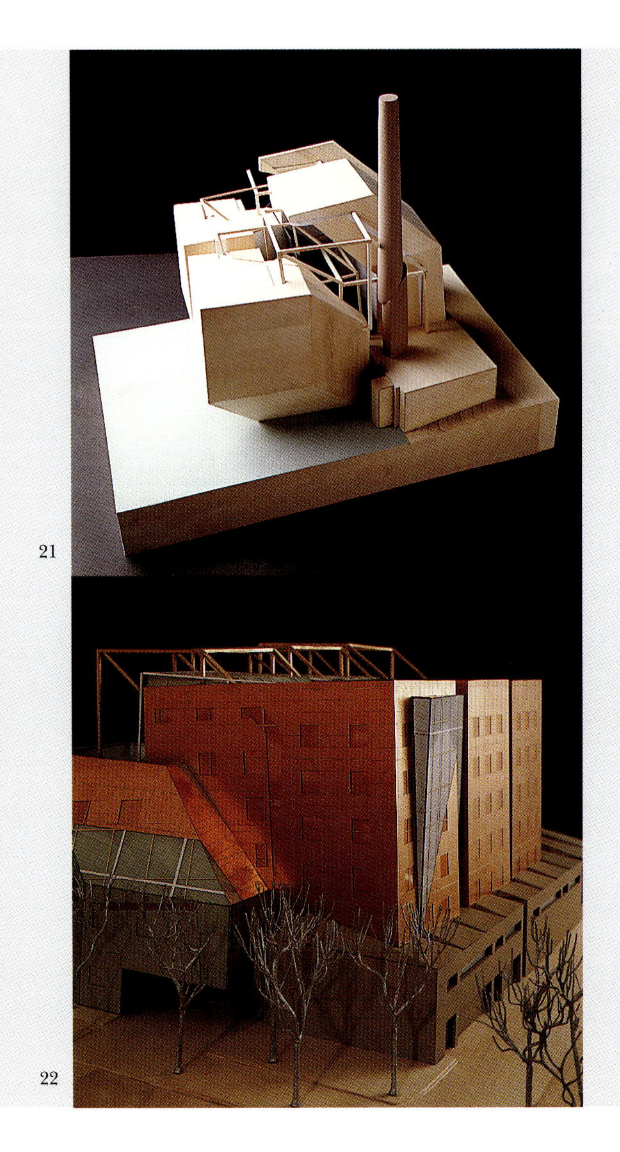

21

22

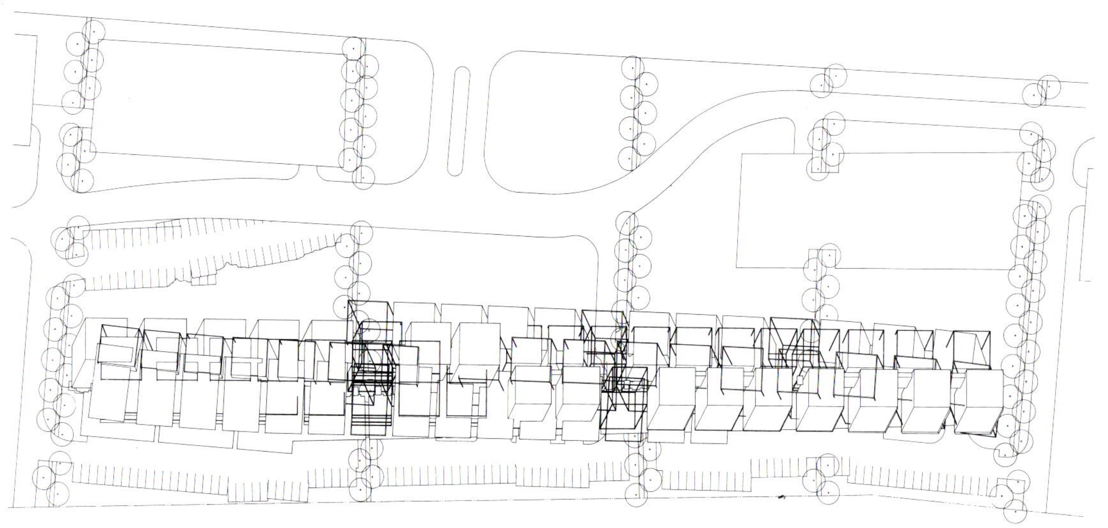

23

21 Study model
22 Presentation model (office building), view from the south-west
23 Site plan
24 Study model
25 Presentation model (office building), view from the north-west
26 Presentation model (office building), view from the north-east
27 Study model

21 研究模型
22 从西南看展示模型（办公楼）
23 总平面
24 研究模型
25 从西北看展示模型（办公楼）
26 从东北看展示模型（办公楼）
27 研究模型

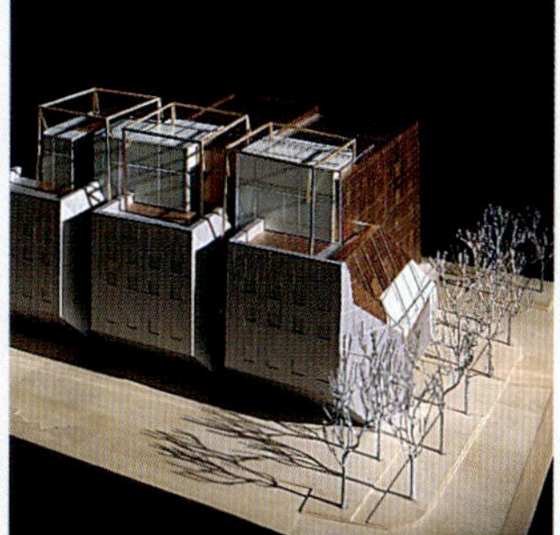
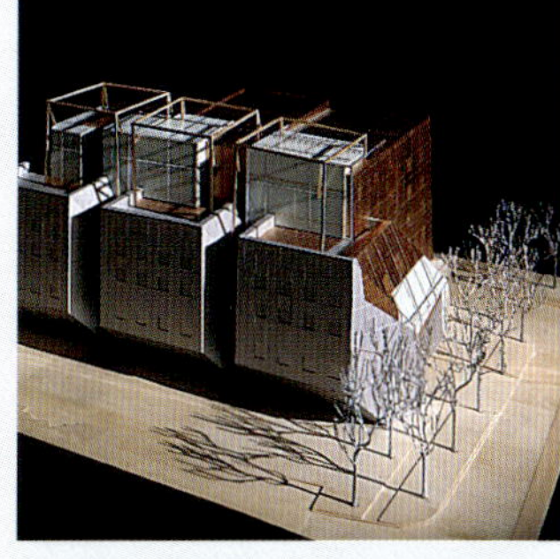

25

4

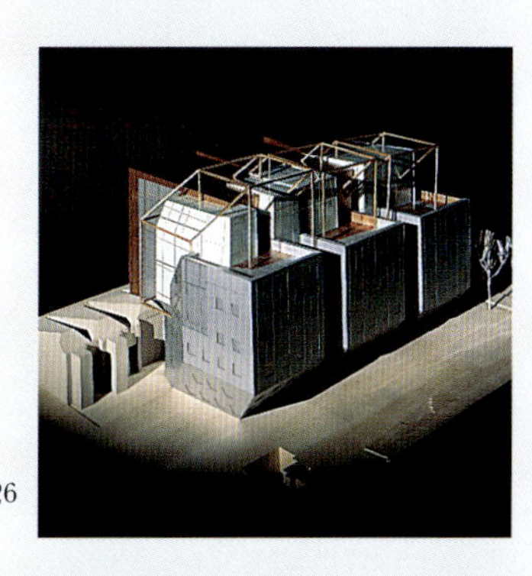

26

27

Guardiola House

Design 1988
Cadiz, Spain
Mr Javier Guardiola Sr
1,200 square feet

瓜迪奥拉别墅

设计　　1988
加的斯，西班牙
哈维尔·瓜迪奥拉先生
1200 平方英尺

This house can be seen as the manifestation of a receptacle in which traces of logic and irrationality are intrinsic components of the object/place. It exists between the natural and the rational, between logic and chaos; the arabesque. It breaks the notion of figure/frame, because it is figure and frame simultaneously. Its tangential el-shapes penetrate three planes, always interweaving. These fluctuating readings resonate in the material of this house which, unlike a traditional structure of outside and inside, neither contains nor is contained. It is as if it were constructed of a substance which constantly changes shape.

该住宅可以被看成是容器的展示，在这一容器中轨迹的逻辑性和非理性是物体/场所的固有成分。它存在于自然与推理之间，存在于逻辑和混乱之间，存在于蔓藤花纹中。它打破了数字/框架的观念，因为它同时是数字和框架。它的方形的直角形状穿透三个平面，并总是互相交织在一起。这些浮动的事物与建筑材料相呼应，它们不像传统的外部和内部的结构，既不包含也不被包含。看起来像是由不断变化形状的物质组成的。

1

1 Perspective view from the south-east
2 Concept diagrams, plan
3 Concept diagrams, elevation
4 Concept diagrams, plan
5 Concept diagrams, elevation

1 东南方向透视
2 概念示意图，平面
3 概念示意图，立面
4 概念示意图，平面
5 概念示意图，立面

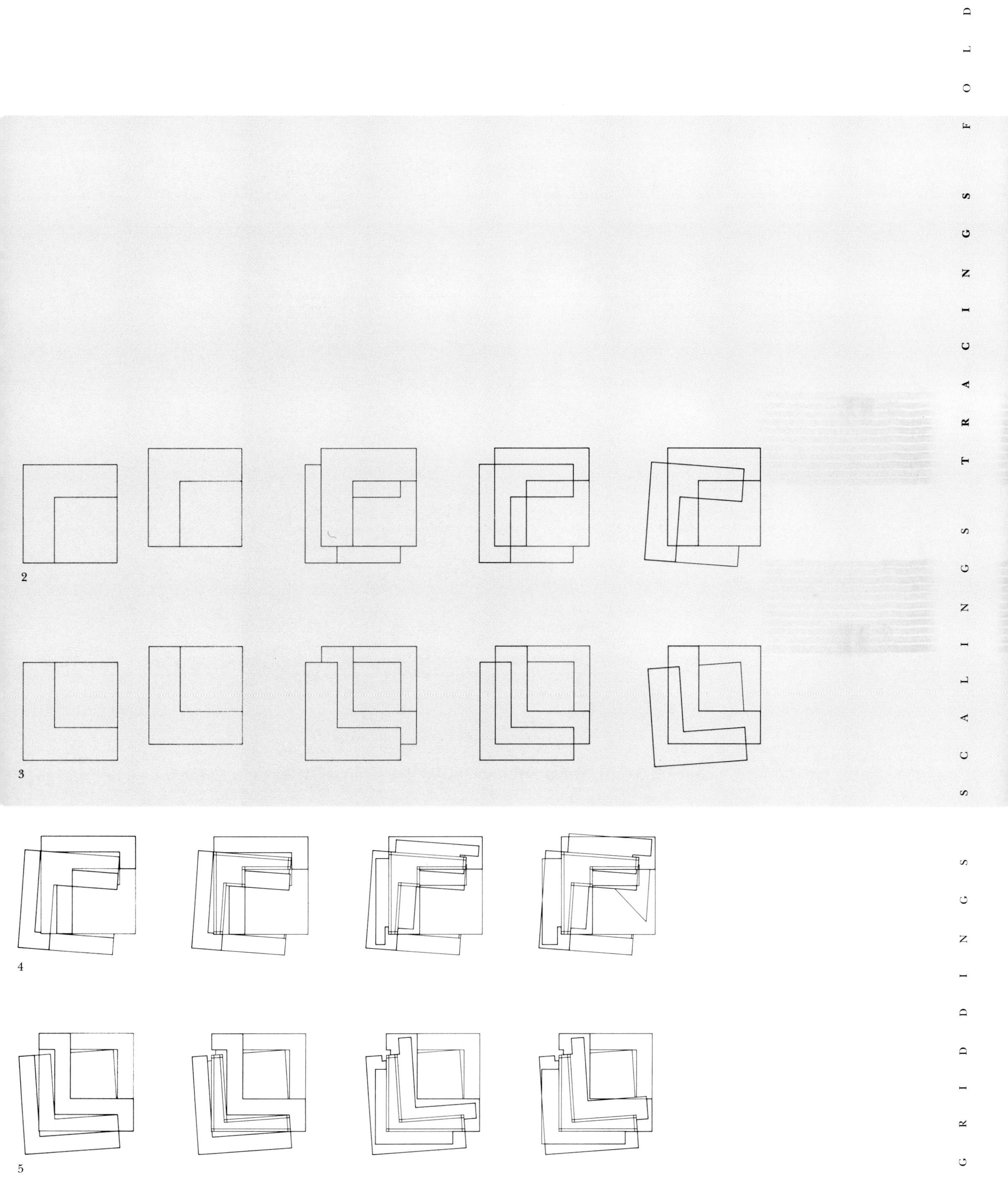

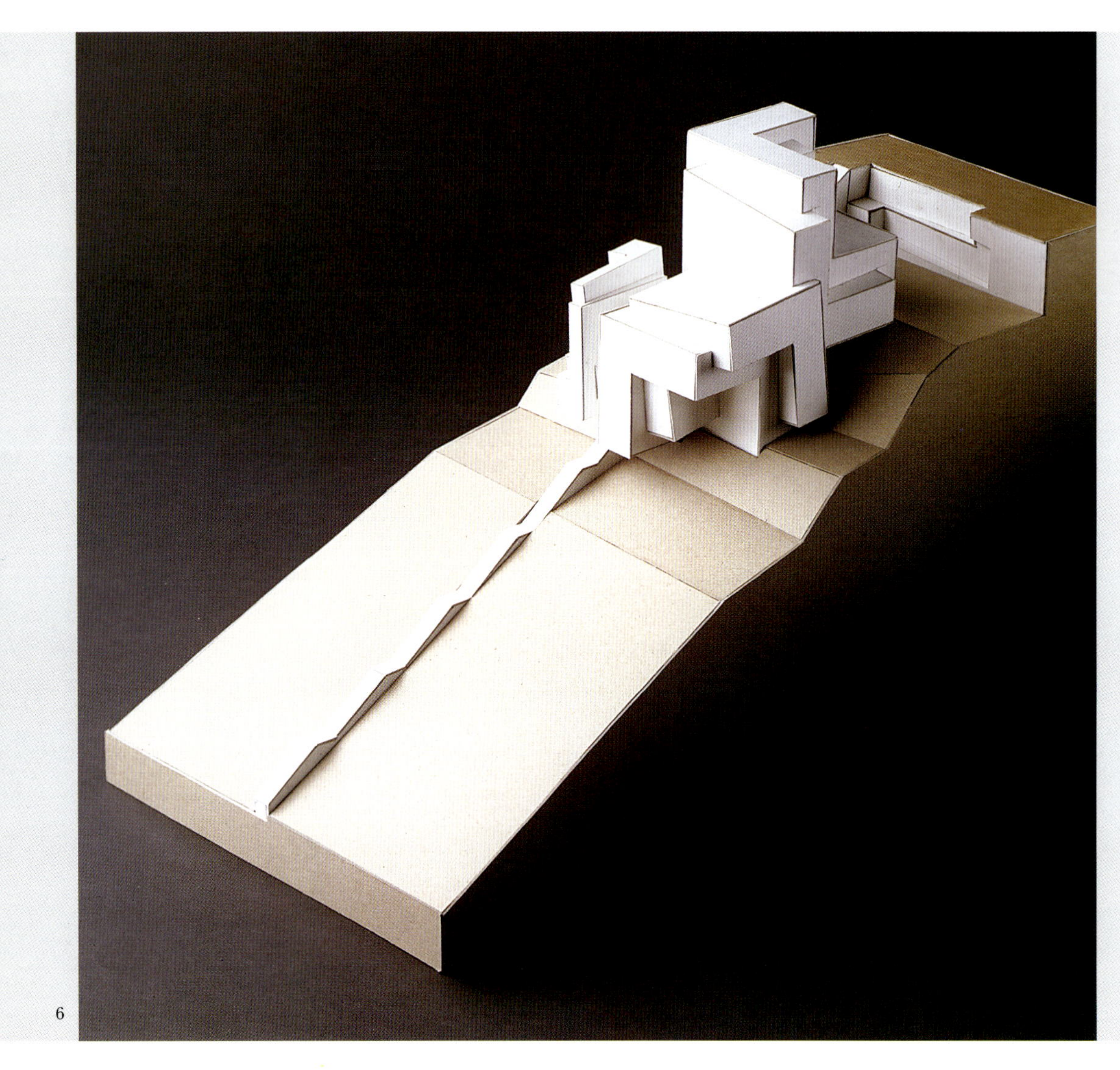

6

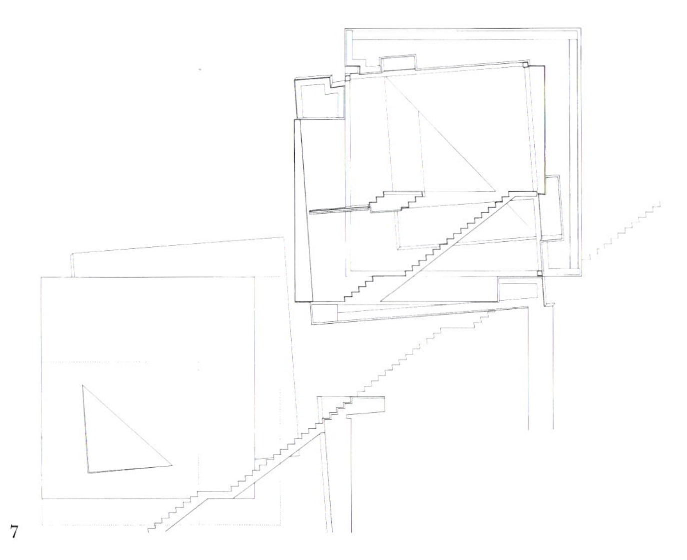

7

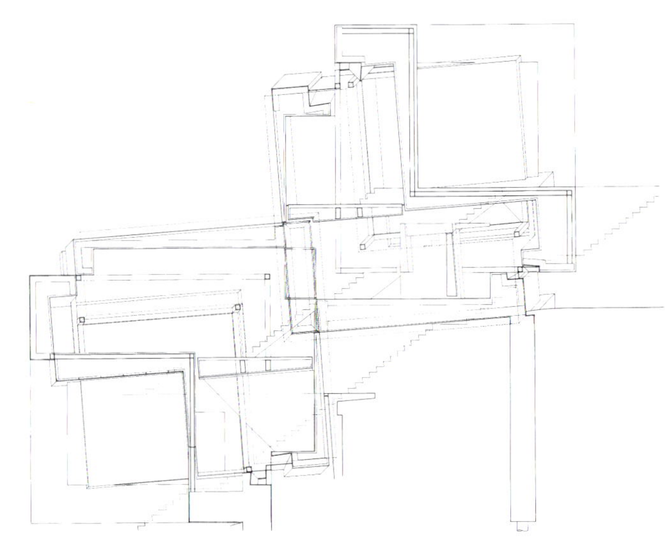

8

6 Study model, view from the south-east
7 Section BB
8 Section AA
9 West elevation
10 East elevation
11 Structural model, view from the south-east
12 Structural model, view from the east

6 从东南方向看研究模型
7 BB 剖面
8 AA 剖面
9 西立面
10 东立面
11 从东南看结构模型
12 从东看结构模型

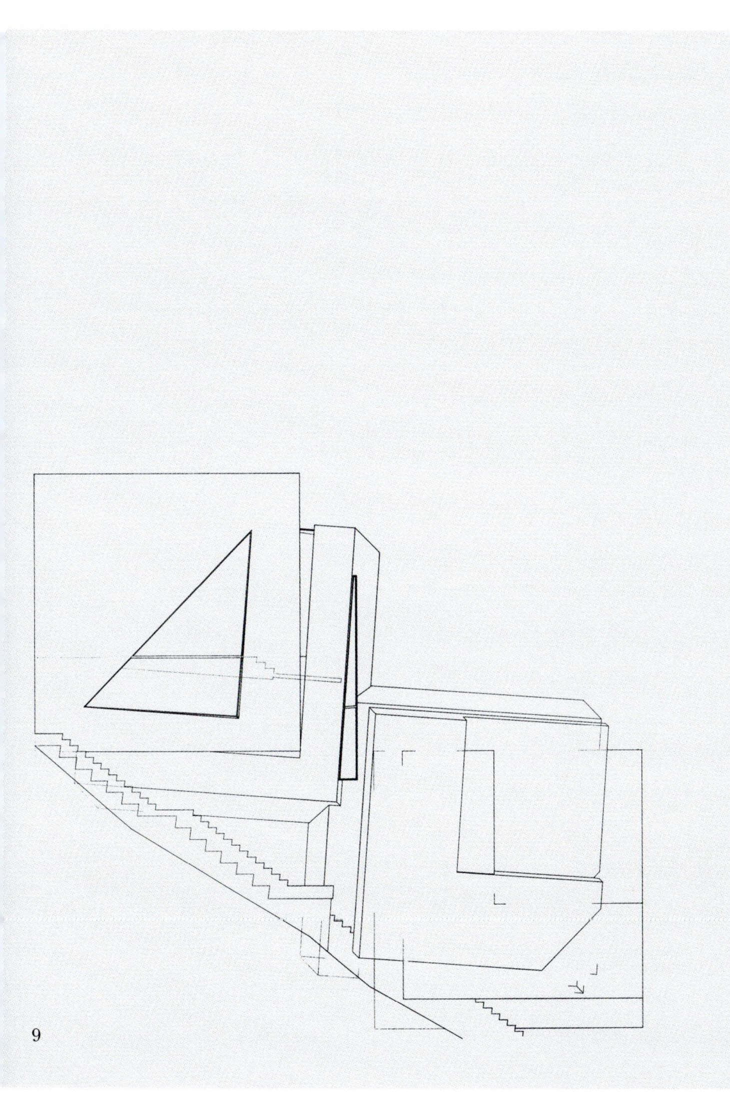

9

11

12

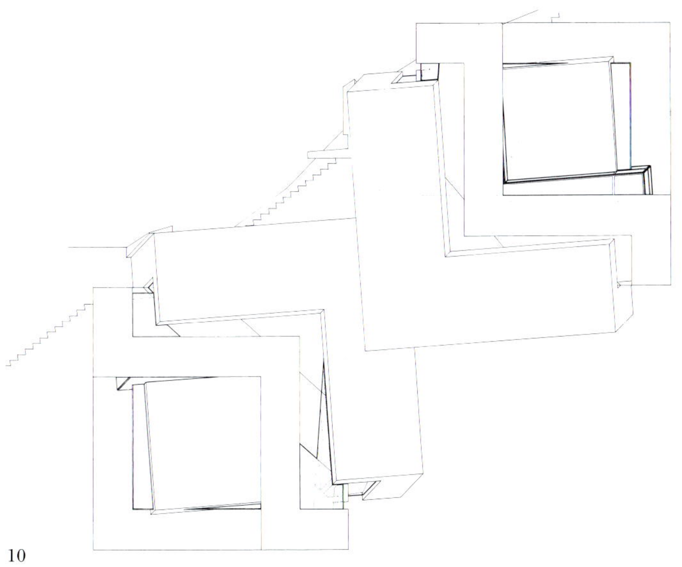

10

13

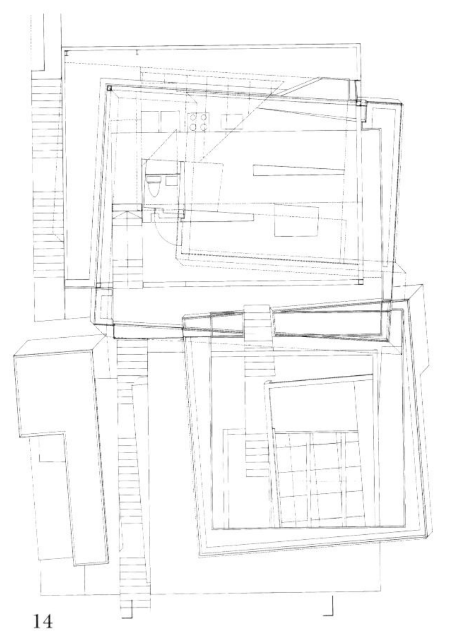

14

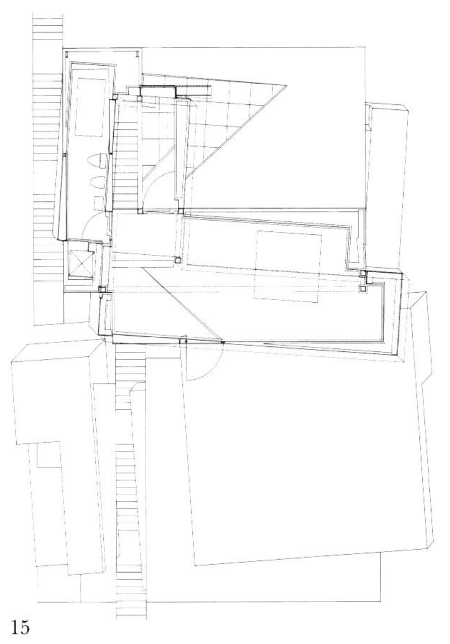

15

13 Study model, view from the south
14 Intermediate level plan
15 Upper level plan
16 Section EE
17 Section DD
18 Study model, view from the south-east
19 Study model, view from the north-west

13 从南看研究模型
14 夹层平面
15 顶层平面
16 EE 剖面
17 DD 剖面
18 从东南方向看研究模型
19 从西南方向看研究模型

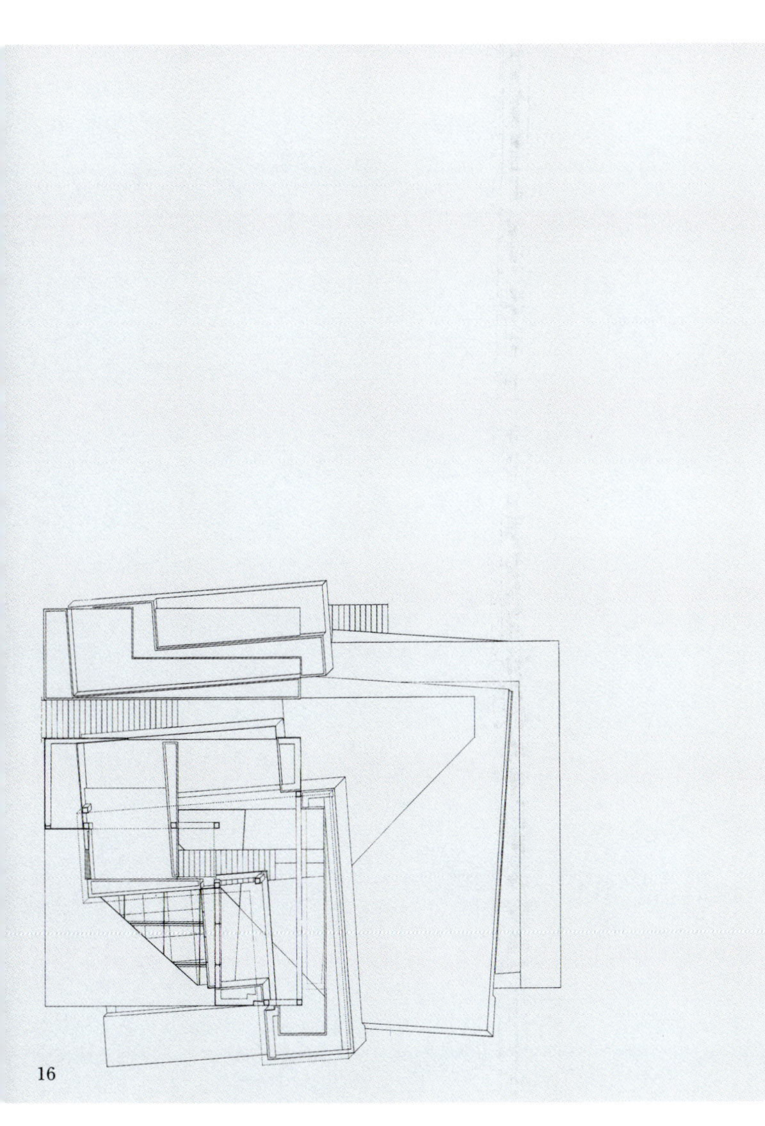

16

18

19

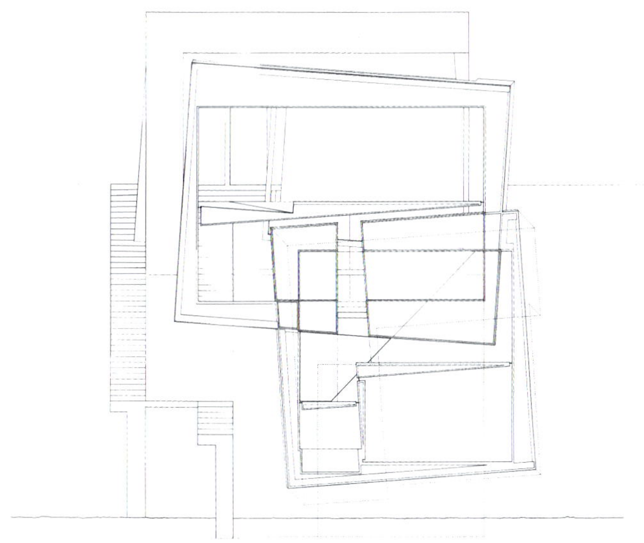

17

Aronoff Center for Design and Art

Design/Completion 1988/present
Cincinnati, Ohio
University of Cincinnati
273,000 square feet

阿诺夫艺术与设计中心

设计/竣工 1988/至今
辛辛那提，俄亥俄州
辛辛那提大学
273000 平方英尺

Design disciplines must assume a more important role in our media-dominated age of information than ever before. The Aronoff Center for Design and Art is programmed to be a model for this kind of leadership. For this project, we had to reconceptualize what a building has to be in order to house such inventive, contemporary activity.

The vocabulary of the building derives from the curves of the land forms and the chevron forms of the existing building; the dynamic relationship between these two forms organizes the space between them. We worked with the students, faculty, administrators, and friends of the College so that the building was an evolutionary process of work which everyone can say "was created by us."

设计规则在今天多媒体信息时代应该扮演着更重要的角色。阿诺夫设计和艺术中心堪称是这一领导阶层的典范。对该项目而言，我们必须给建筑重新定义，从而包含富于创造的当今的活动。

建筑的语汇来源于地形的曲线和现有建筑的臂章形式；两者之间生动的关系构成了它们之间的空间。我们和学生、教员、管理者以及大学的朋友们一起工作，所以建筑是一件演变过程的作品，每个人都可以说“这房子是我们创造的”。

1

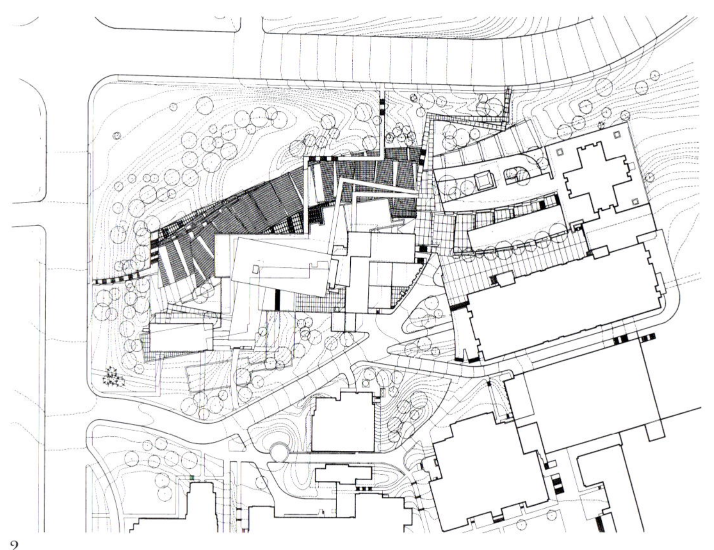

2

1 Presentation model, aerial view
2 Site plan
3–4 Concept diagrams, curved line
5 400-level floor plan
6 600-level floor plan

1 展示模型，鸟瞰
2 总平面
3–4 设计概念示意图，曲线
5 400 标高处平面
6 600 标高处平面

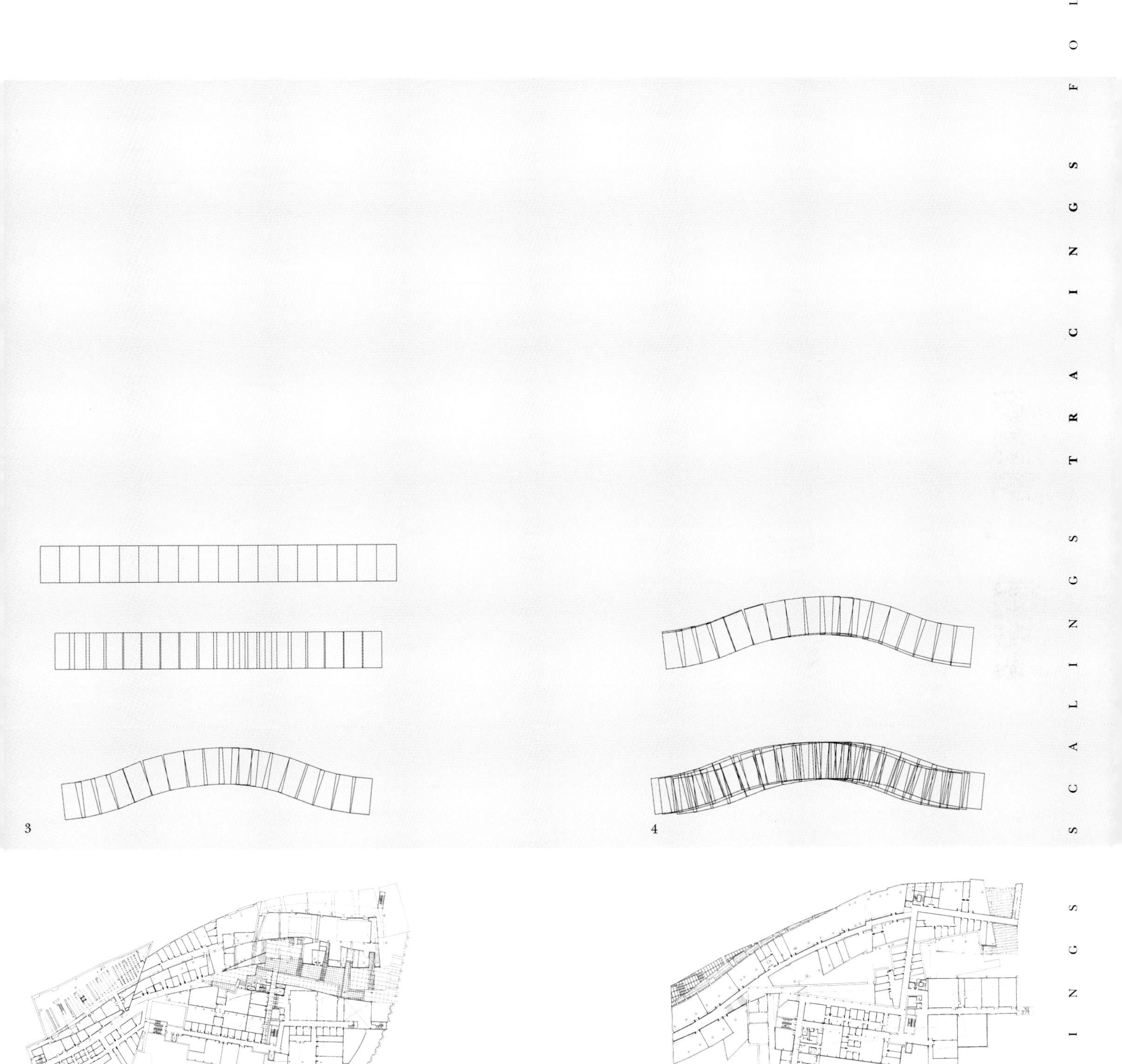

3

4

5

6

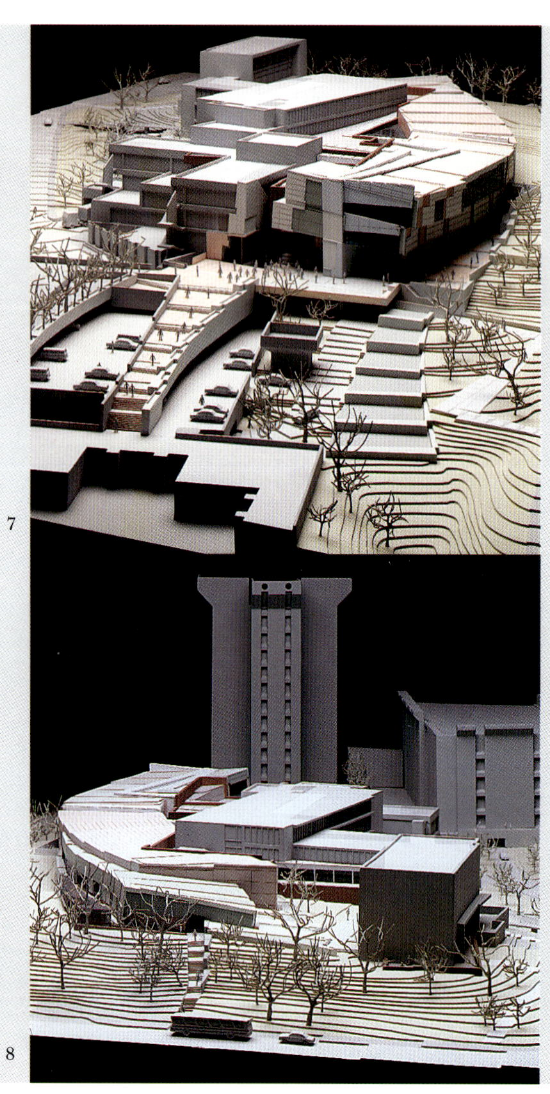

7

8

9

7 Presentation model, east view
8 Presentation model, west view
9 700-level floor plan
10 Concept diagram, tiled curve
11 Concept diagram, tiled curve trace
12 Nine-segment model, east view
13 Nine-segment model, east sectional view

7 展示模型，东向外观
8 展示模型，西向外观
9 700 标高处平面
10 设计概念示意图，平铺的曲线
11 设计概念示意图，平铺曲线轨迹
12 9 个部分的模型，东向外观
13 9 个部分的模型，东端剖视

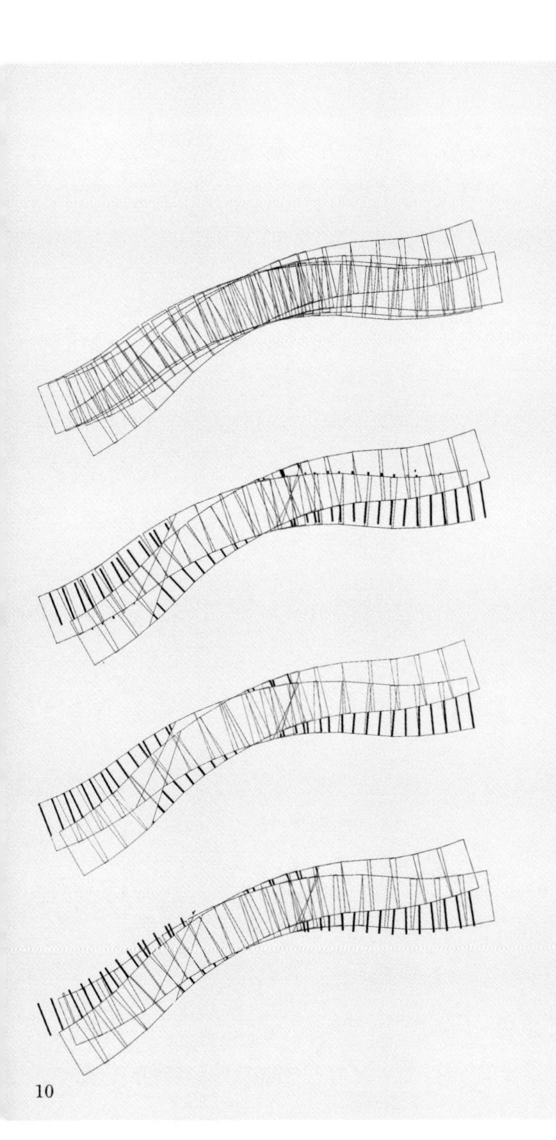

10

12

13

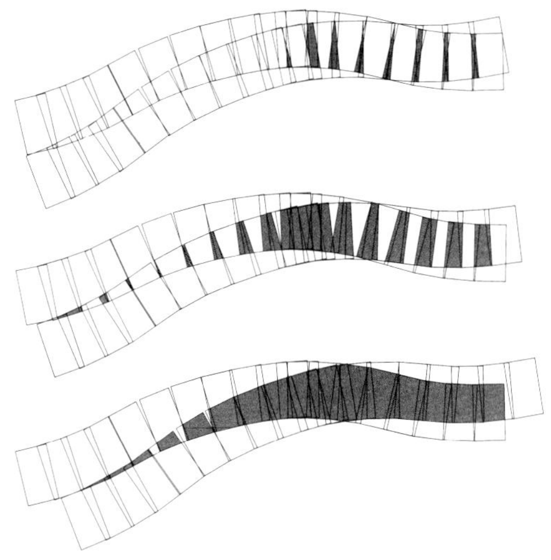

11

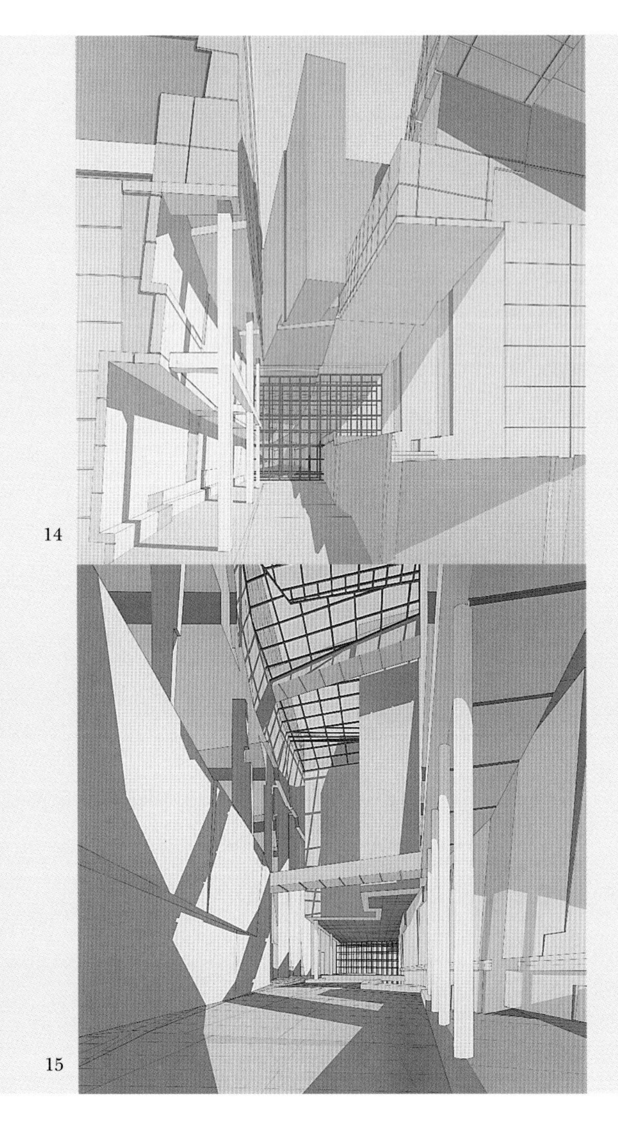
14

15

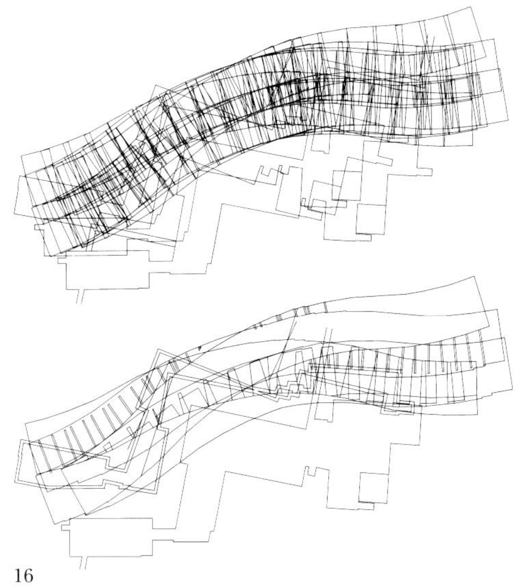
16

14 East entrance
15 College Hall, east view
16 Concept diagrams, composite curves and chevrons
17 Concept diagrams, chevron trace and imprint
18 Transverse section
19 Nine-segment model, east view
20 Nine-segment model, auditorium section

14 东入口
15 学院大厅，东向外观
16 设计概念示意图，相对的曲线和臂章形式
17 设计概念示意图，臂章轨迹及其烙印
18 横断面
19 9个部分的模型，东向外观
20 9个部分的模型，礼堂剖面

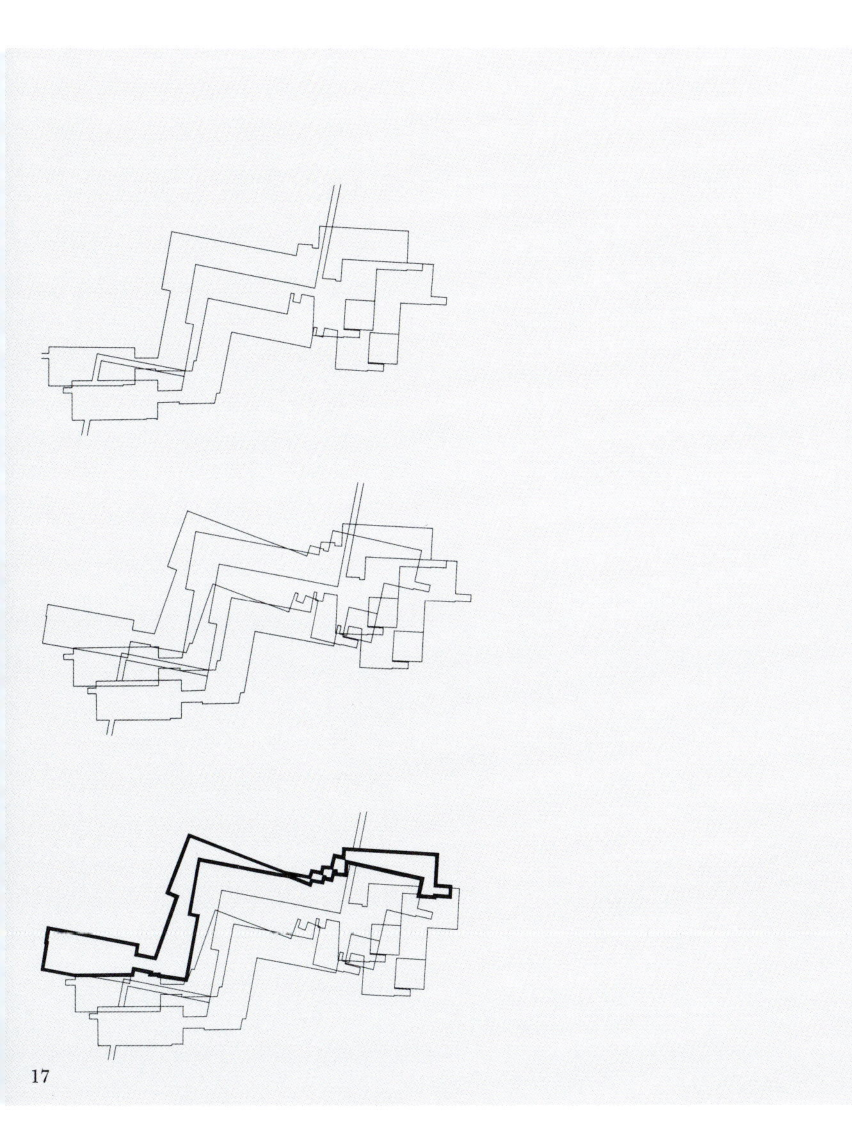

17

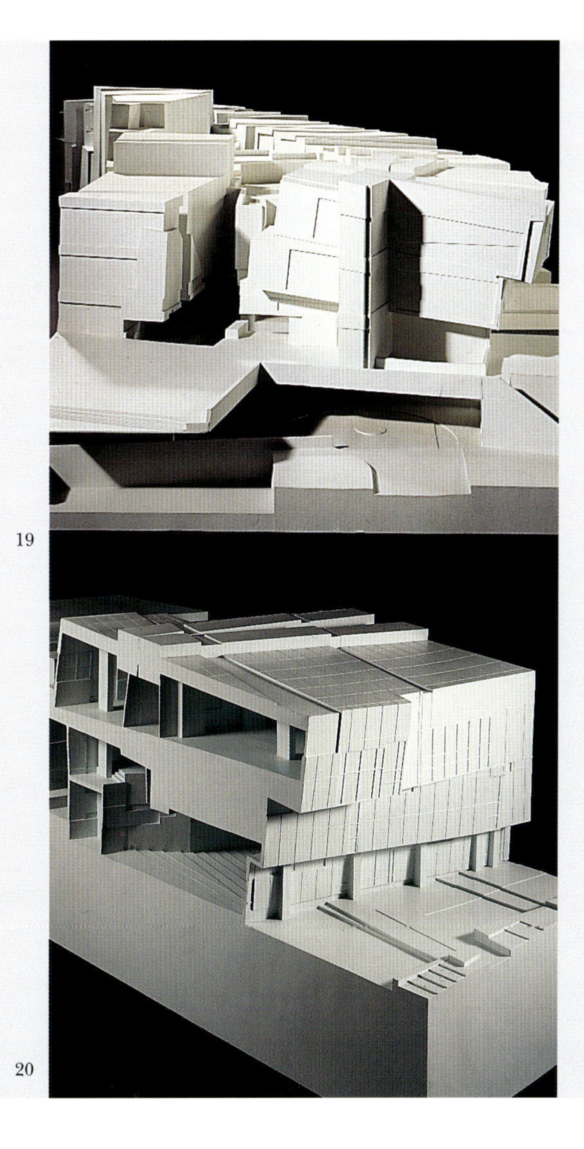

19

20

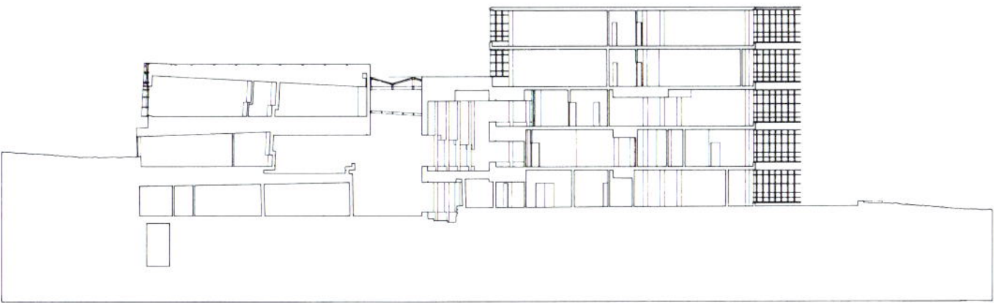

18

21

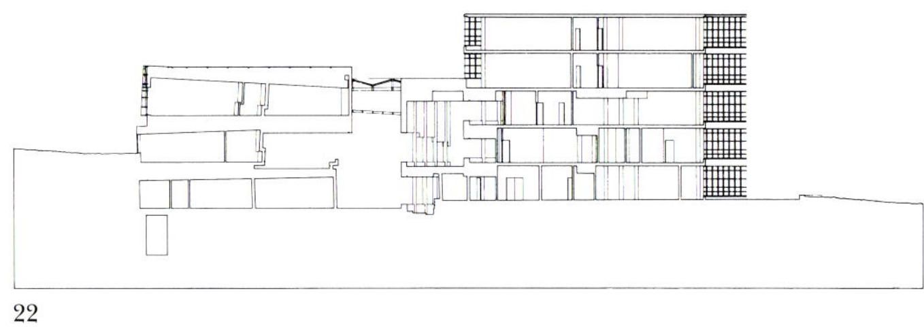

22

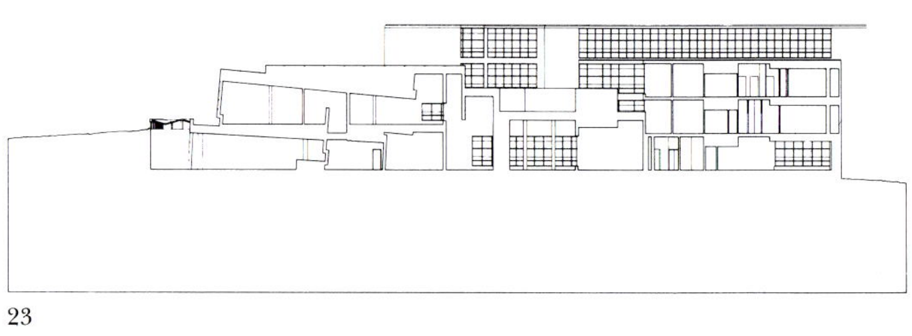

23

21 Roof plan
22–27 Transverse sections
28 Nine-segment model, auditorium section, east view
29 Nine-segment model, auditorium section

21 屋顶平面
22–27 横断面
28 9个部分的模型，礼堂剖面，东向外观
29 9个部分的模型，礼堂剖面

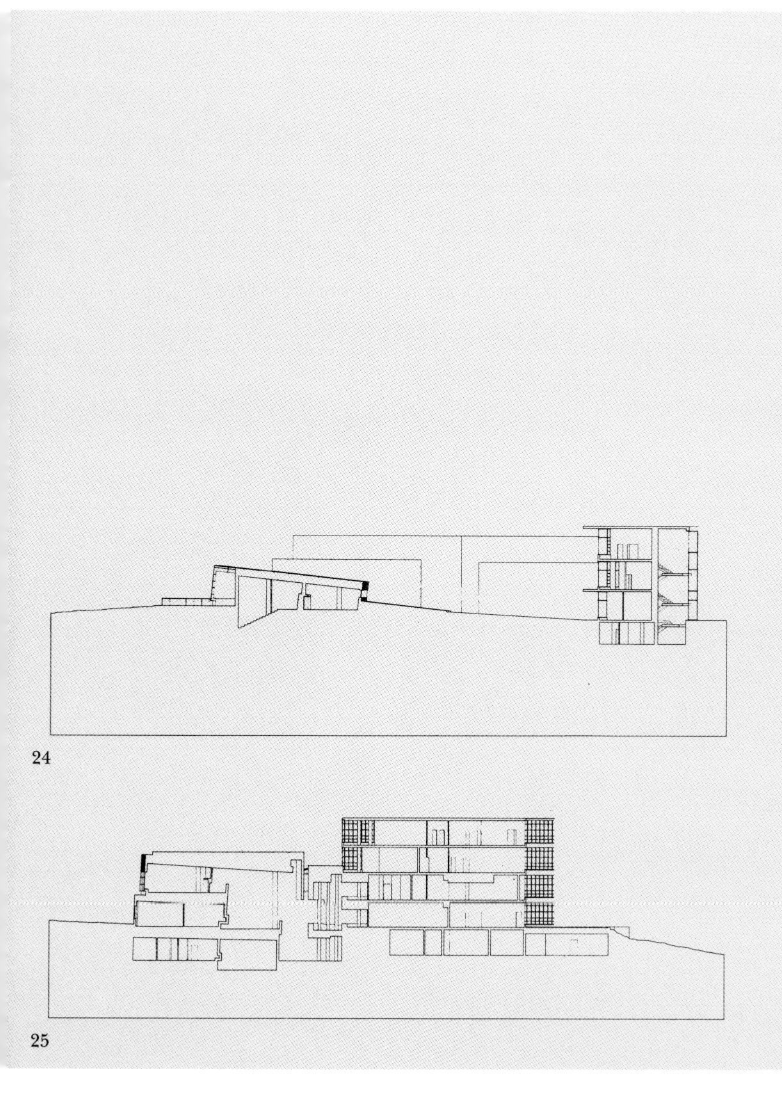

24

25

28

29

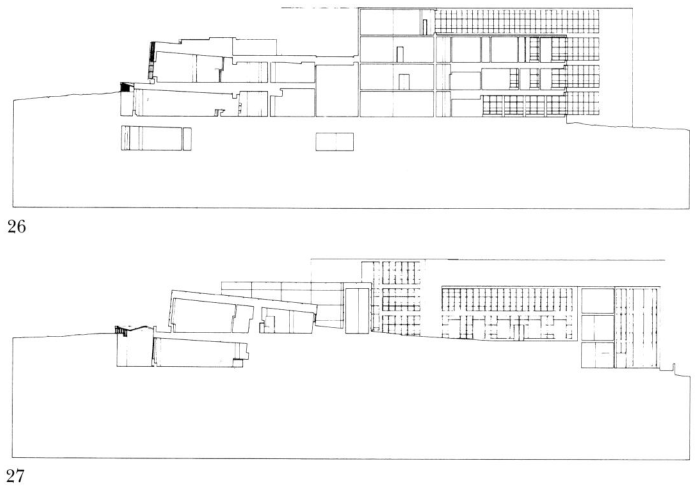

26

27

Koizumi Sangyo Office Building

Design/Completion 1988/1990
Tokyo, Japan
Koizumi Sangyo Corporation
43,000 square feet

小泉三洋办公楼

设计/竣工　1988/1990
东京，日本
小泉三洋公司
43000 平方英尺

In the West, the concept of place (*topos*) has always been pre-eminent. Less important, but latent or repressed in this topos, has been the concept of *atopia*, or no place. Tokyo can be seen as embodying a concept of atopia lying with topos. This project proposes that this "lying within" can be seen as another order, another potential structure. These ideas have always been a part of Japanese thought: the Japanese word *ma* stands for the notion of "the space between," and *ku* for "no place." In this project, the idea was not to build the place, but to build a place between. The project deals with the idea of imprint—the former presence of place—and trace—the absence of place—as the major components of any space.

在西方，场所（主题）的概念是由来已久的。变得不太重要但是潜伏或者被压制在这一主题中，就是 *atopia* 的概念，或者说是无场所。东京带着它的主题，体现着 *atopia* 的概念。该项目提出"蕴藏其中"可以被看成是另外一种顺序，另一种潜在的结构。这些理念始终是日本思想的一部分：日本语"*ma*"表示"之间（的空间）"，"*ku*"表示"无场所"。在该项目中，设计的理念不是创造场所而是创造场所之间。项目将烙印——以前场所的形式——和痕迹——场所的空缺作为任何空间的主要构成元素。

1

1 View from the east
2 Ground level plan
3 Showroom, ground level plan
4 View of exhibition gallery from the north-east
5 View of showroom from the west

1 从东向看
2 首层平面
3 展示房间，首层平面
4 从东北方向看展廊
5 从西向看展示房间

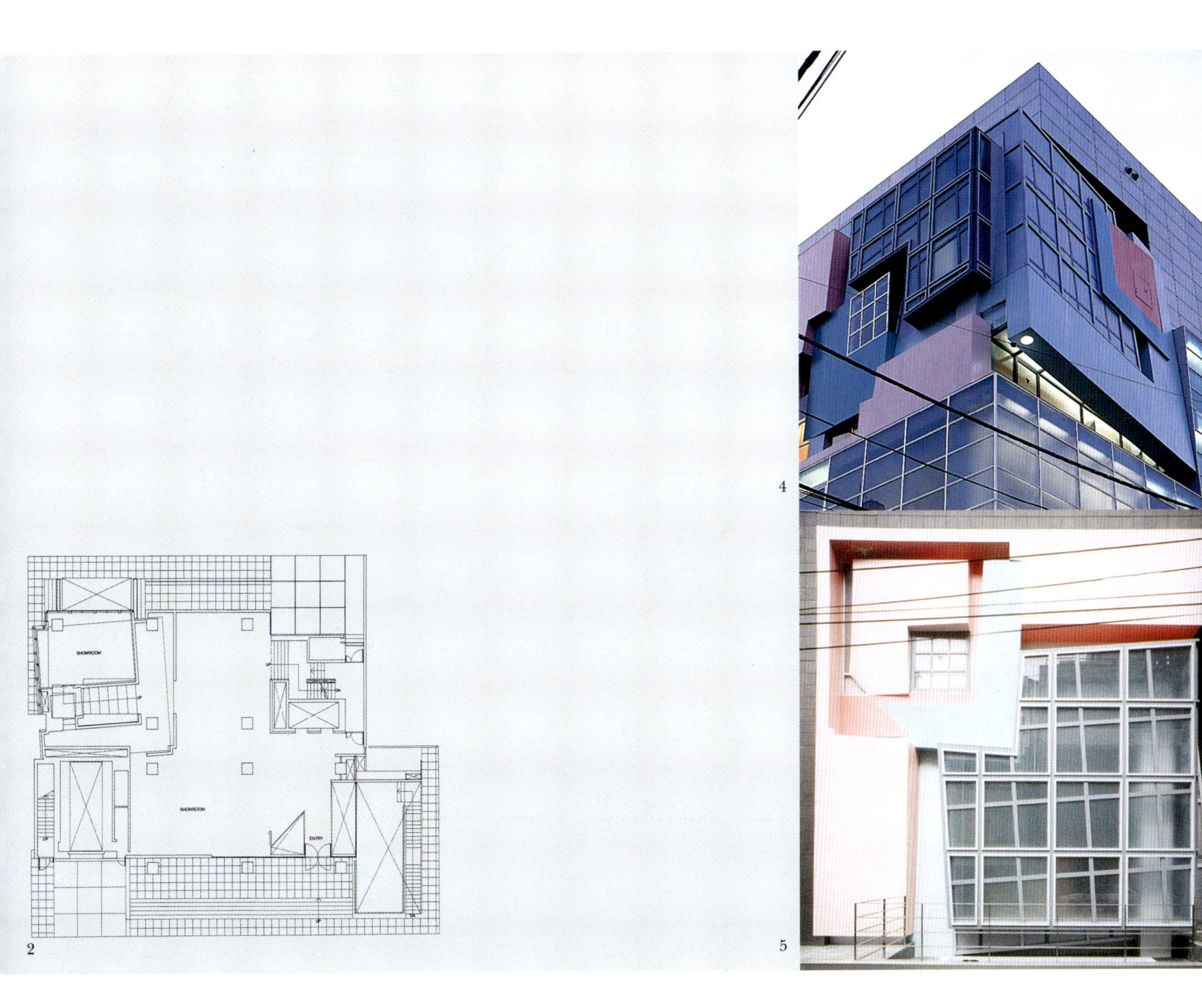

2 4 5

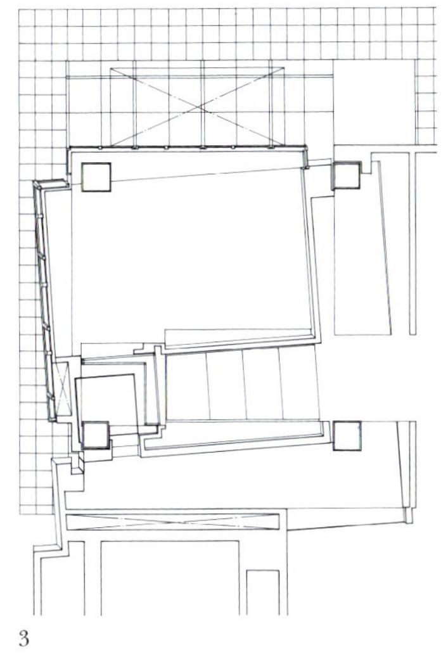

3

6

6 View from the north-east
7 Showroom, third level plan
8 Fifth level plan
9 Facade detail
10 View of exhibition gallery from the north-east

6 从东北方向看
7 展示房间，四层平面
8 六层平面
9 立面细部
10 从东北方向看展廊

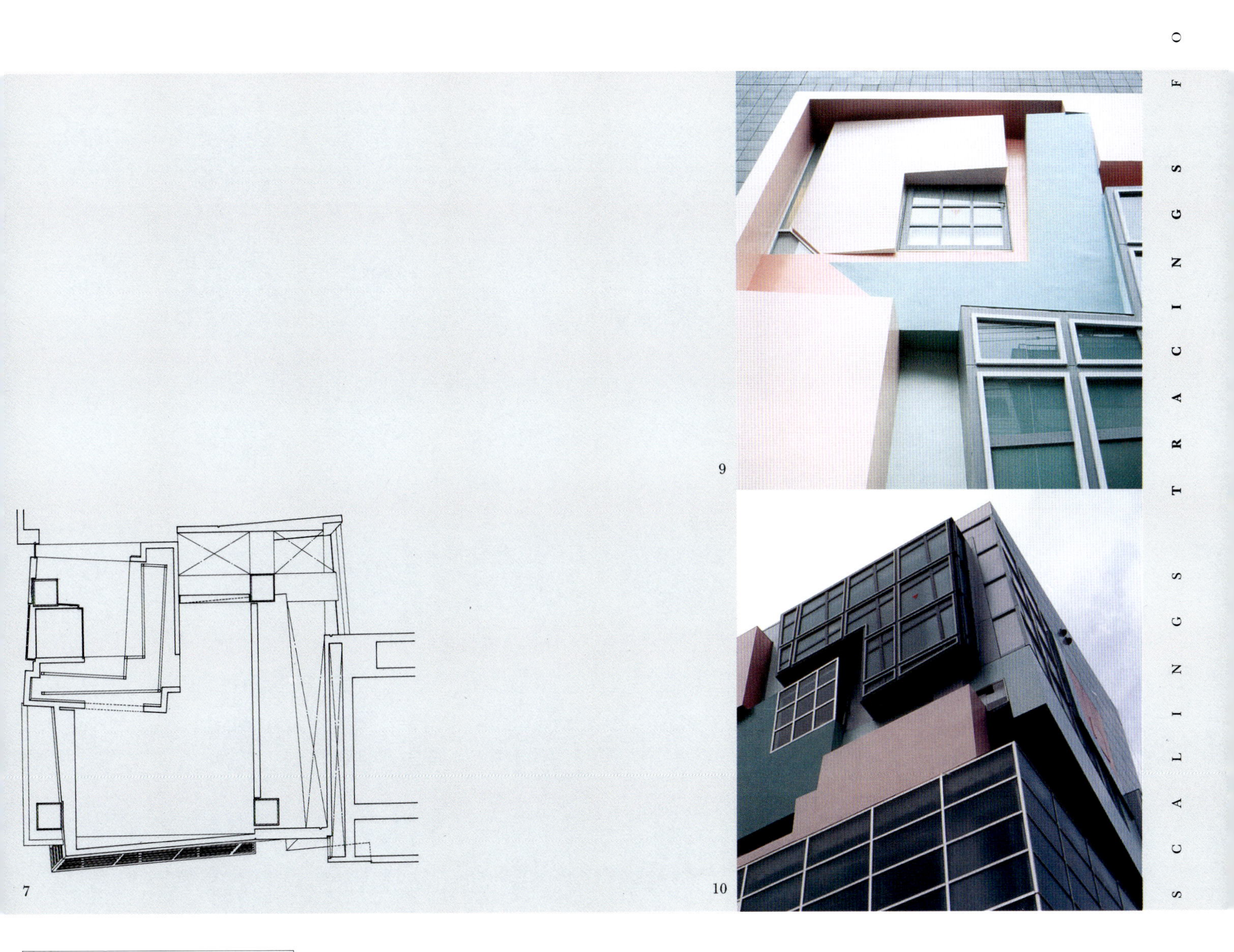

9

10

7

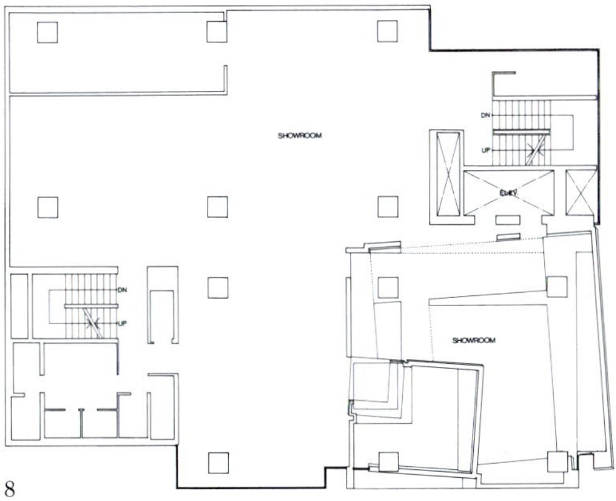

8

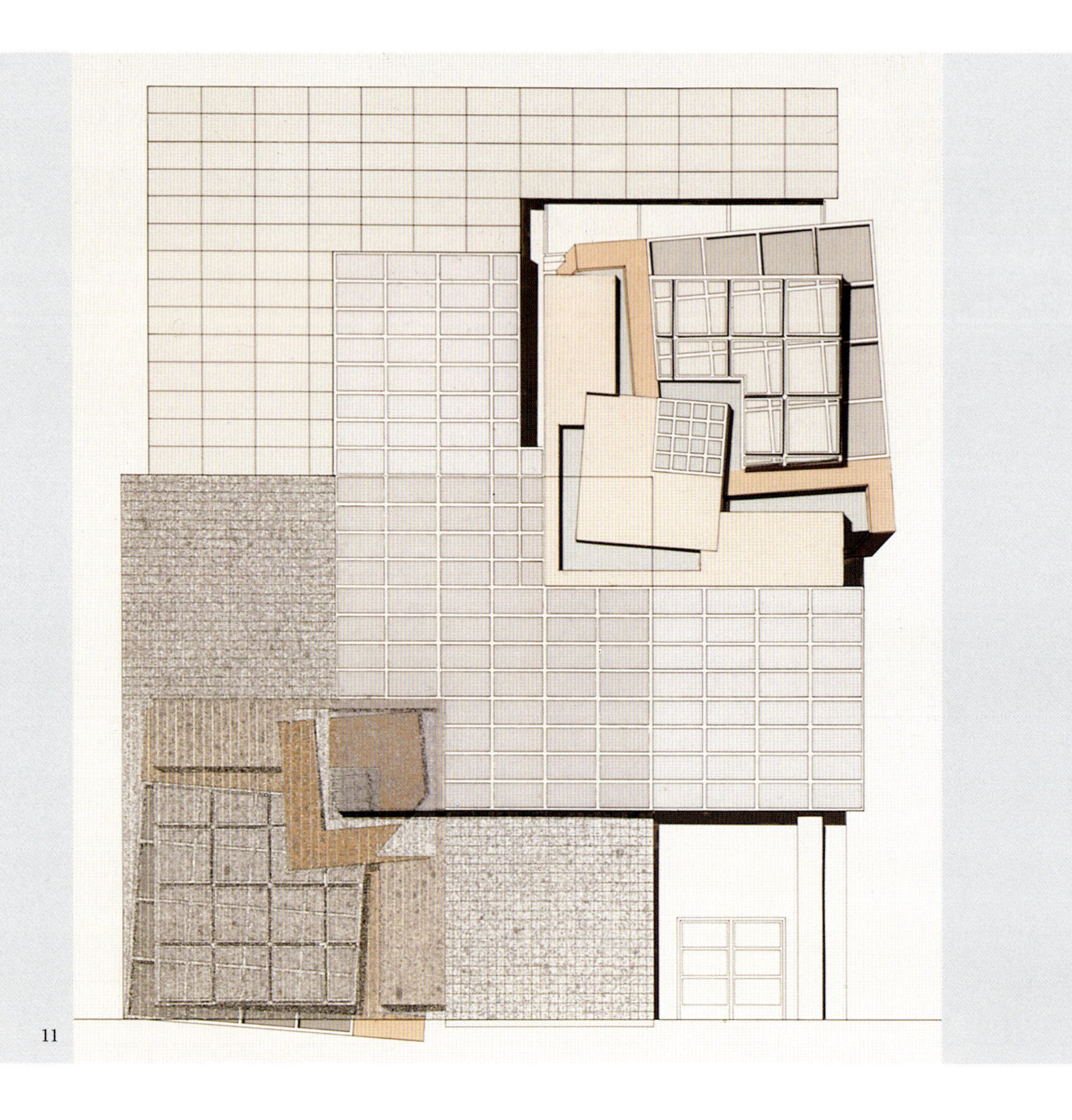

11

11 East elevation
12 Exhibition gallery, fifth level plan
13 Exhibition gallery, sixth level plan
14–16 Interior details

11 东立面
12 展廊，六层平面
13 展廊，七层平面
14–16 室内细部

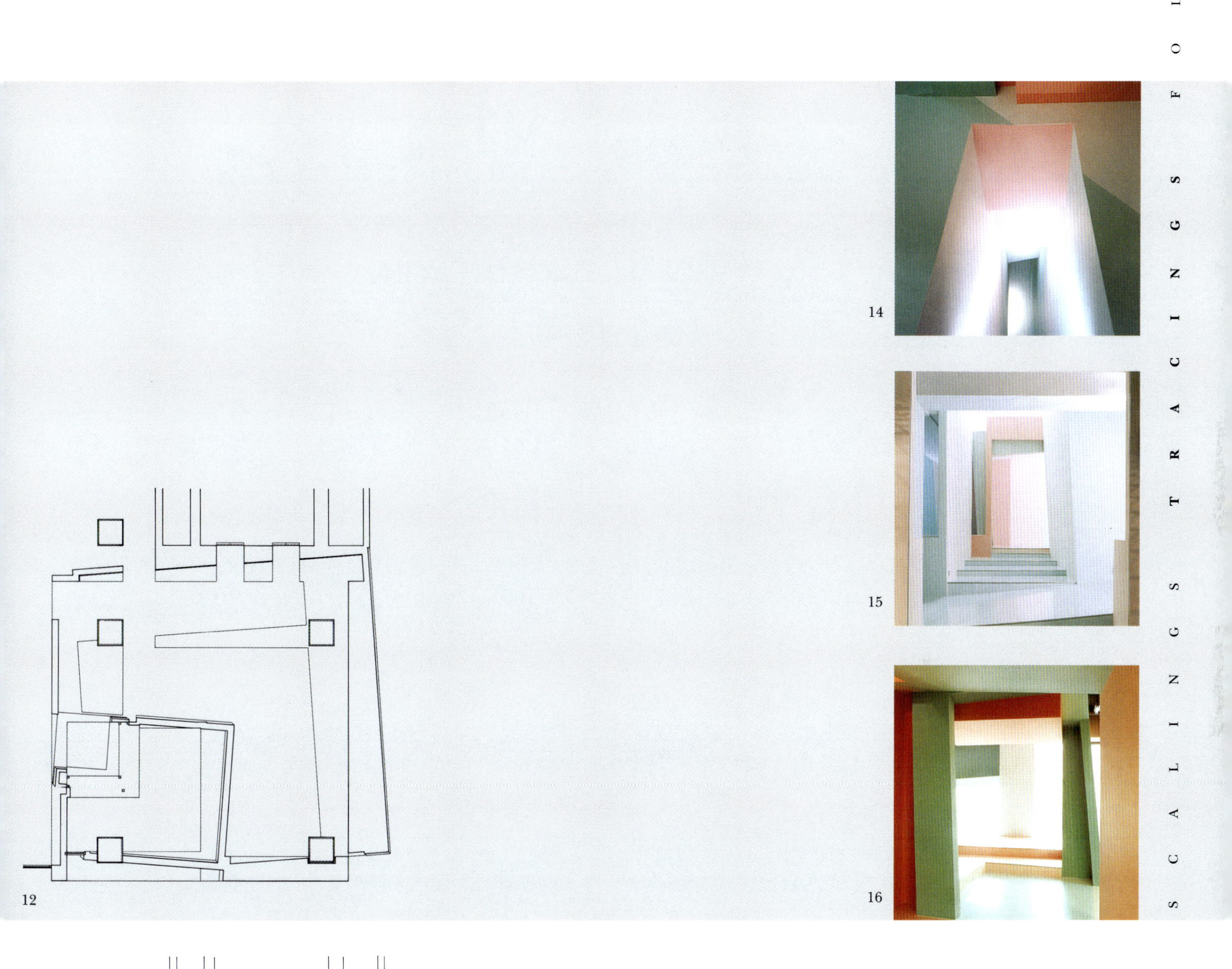

14

15

16

12

13

17

17 Night view from the east
18 Building section
19 East elevation
20–21 Interior details

17 从东向看夜景
18 建筑剖面
19 东立面
20－21 室内细部

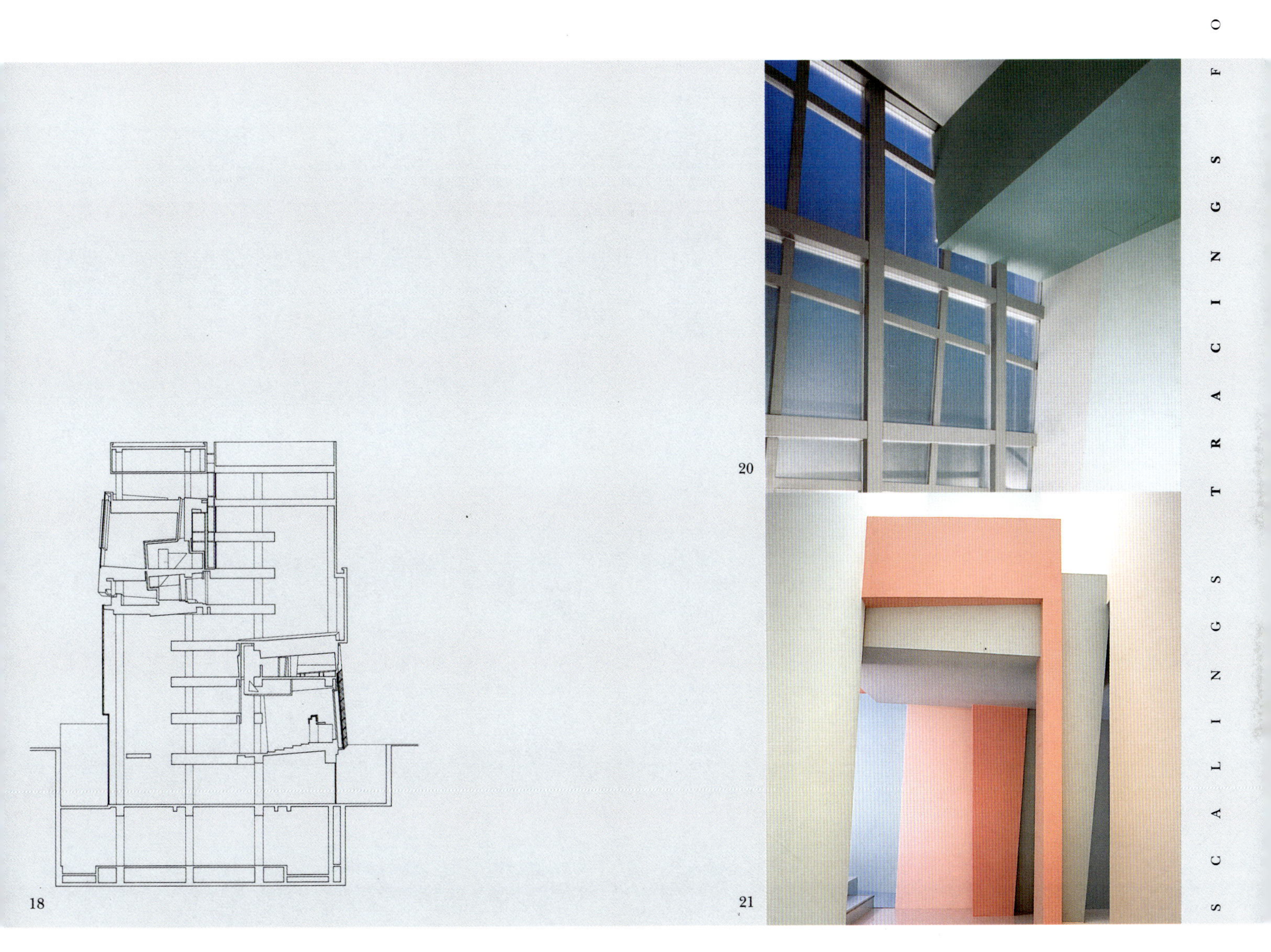
18
20
21

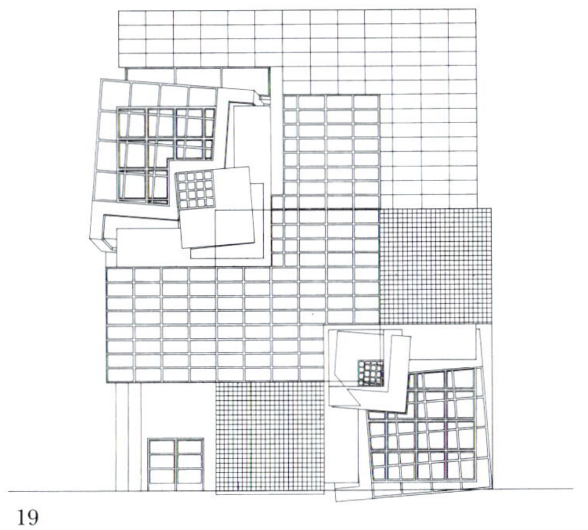
19

Siena Bank Master Plan

Design 1988
Siena, Italy
Monte Paschi Bank/Siena Chamber of Commerce
60,000 square feet

锡耶纳银行总体规划

设计　1988
锡耶纳，意大利
蒙特帕奇银行/锡耶纳商会
60000 平方英尺

The program for the Siena Bank Competition required the design of an office building, parking garage, and bus terminal, while unifying two adjacent piazzas on an elevated site in the center of this historic hill town. We examined the site for traces of political and geographical histories, looking for similarities in form which might lead to a different understanding and interpretation of the town and its past.

By moving the old city wall down and moving the line of the oval up to the level of the piazza, a link was created between the levels of the city, allowing the upper level to display the full range of its archaeological nature.

锡耶纳银行的设计竞赛需要的是一幢办公楼、停车场以及公交车终点站的设计，同时将在该历史山城中心的两个架空的比萨饼店统一进来。我们研究场地的政治和历史的痕迹，寻找形式中的相似性，这些相似性也许会引导出对该城镇及其过去的不同理解和阐述。

通过推翻旧的城墙、将椭圆地平线向上移动到比萨饼店所在的高度，城市各层面之间的一个连接就产生了，从而使得上部分可以展现它的天然风貌的完整轮廓。

1

1 Presentation massing model
2 Massing plan
3 Site section EE

1 规划模型
2 总平面
3 场地 EE 剖面

2

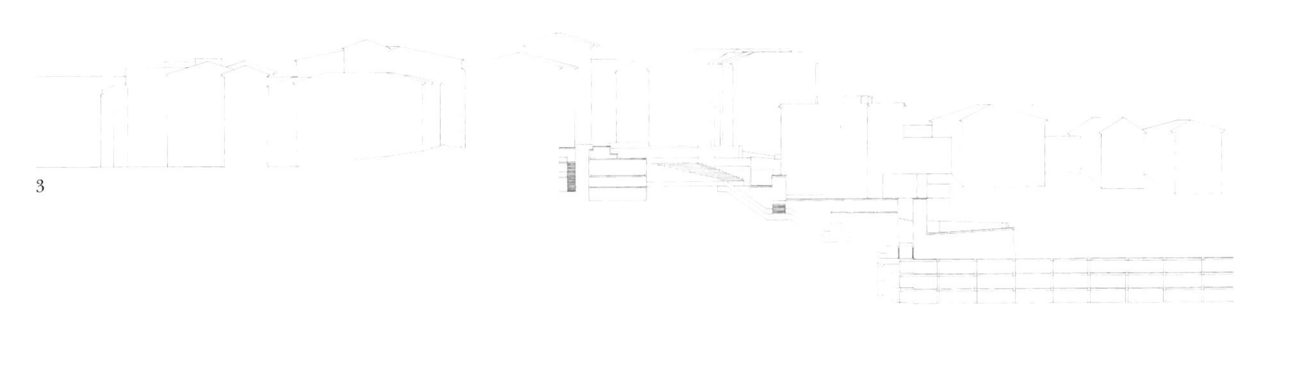

3

4

4 Presentation building model, view from the north-west
5 331 elevation level plan
6 Site section FF
7 331 elevation level plan
8 Roof level plan

4 从西北方向看建筑模型
5 331 标高处立面
6 场地 FF 剖面
7 331 标高处立面
8 屋顶平面

7
5
8

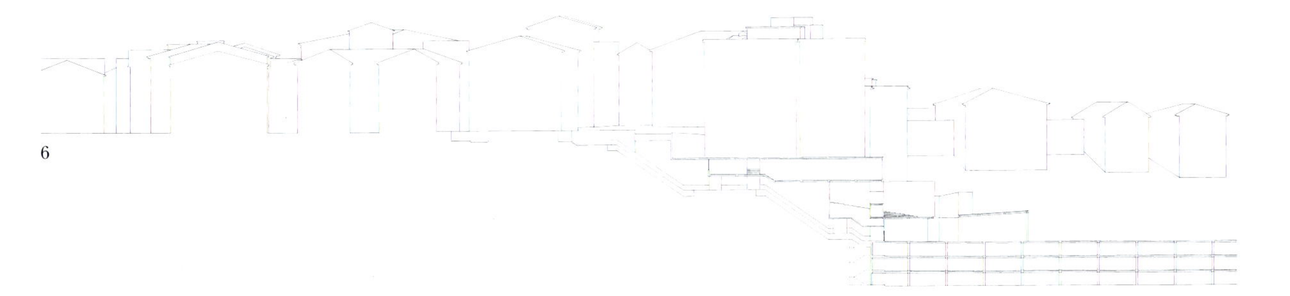
6

Greater Columbus Convention Center

Design/Completion 1989/1993
Columbus, Ohio
Greater Columbus Convention Center Authority
530,000 square feet

大哥伦布会议中心

设计/竣工　　1989/1993
哥伦布市，俄亥俄州
大哥伦布会议中心
530000平方英尺

The design for the Greater Columbus Convention Center is simultaneously suggestive of the railyards that once occupied the site, nearby highway ribbons, and overlays of delicate fiber optic cables that represent the information age. It reflects High Street's traditionally narrow structures with articulated facades that have been extruded away from the street.

The design also solves one of the most persistent problems in convention center design—diagrammatic clarity. Differences in forms clearly distinguish the various exhibition spaces and parts of the concourse. The strengths of the scheme are accomplished without relying on unsatisfying quotations from Columbus's past, or images typically found in "generic" convention halls.

大哥伦布会议中心的设计同时也是曾经坐落在该场地的铁路修车场的暗示，以及与公路带状物相邻，并覆盖了棘手的视觉光线电缆。它用相关的建筑立面映射了公路上传统的狭窄的结构，这些立面已经被挤出街道。

该设计也解决了长期以来一直存在于会议中心的一个问题——图表的清晰。形式上的差别清晰地将各种展示空间及部分大厅进行区分。方案的力量的展现不是依靠对哥伦布市历史的不悦的引用，也不是依赖于“一般的”会议厅中存在的典型景象。

1

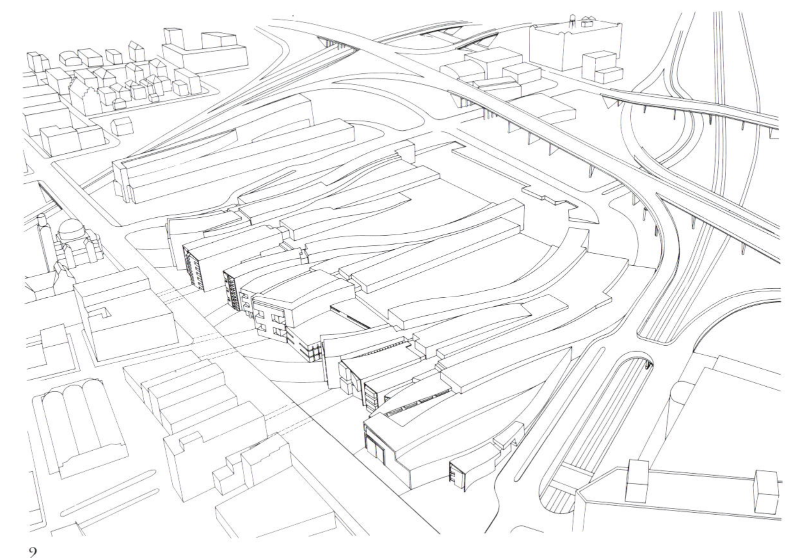

2

1 Aerial view from the south-west
2 Competition perspective, view from the south-west
3 West facade, view from the north
4 Site plan
5 Functional diagram, circulation concourse
6 Functional diagram, meeting rooms and ballroom
7 Functional diagram, concourse and prefunction
8 Functional diagram, administration and service

1 从西南方向鸟瞰
2 西南方向透视实景
3 从北向看西立面
4 总平面
5 功能示意图，流线设计
6 功能示意图，会议室和舞厅
7 功能示意图，大厅和前厅
8 功能示意图，管理和服务用房

3

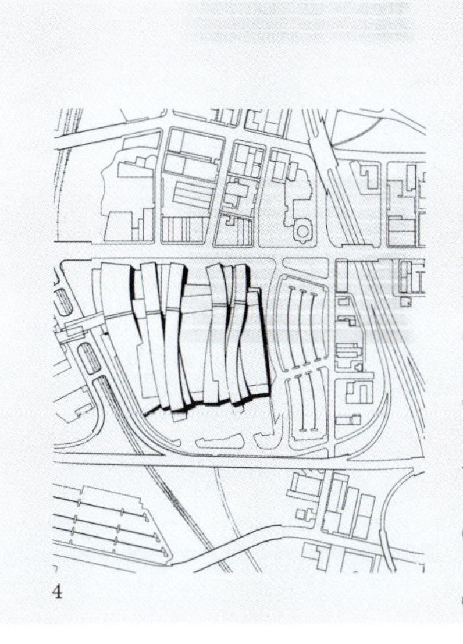
4

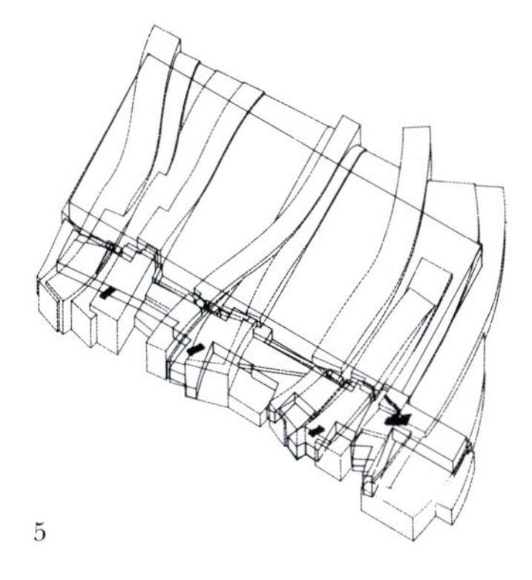
5

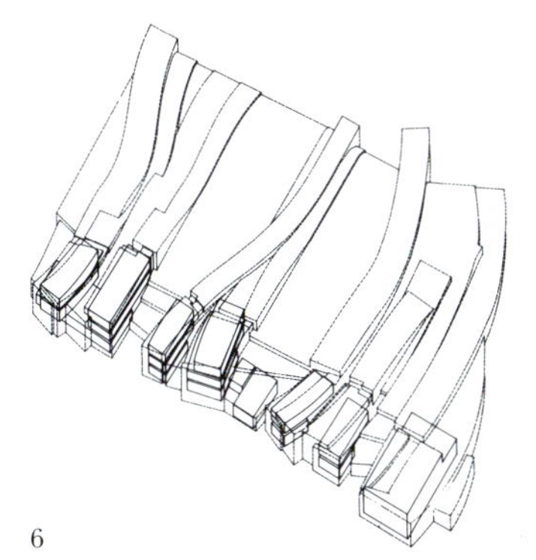
6

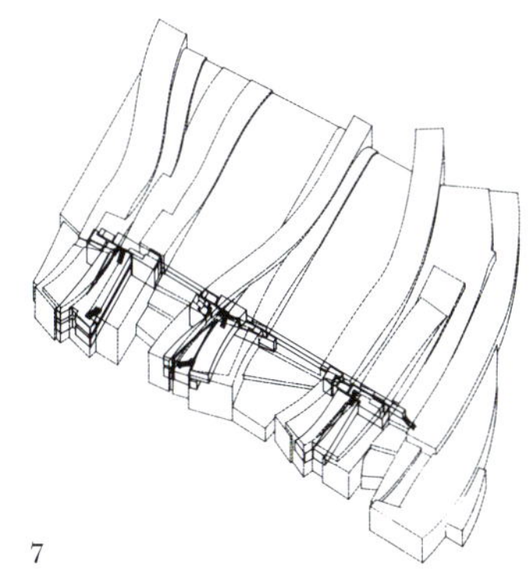
7

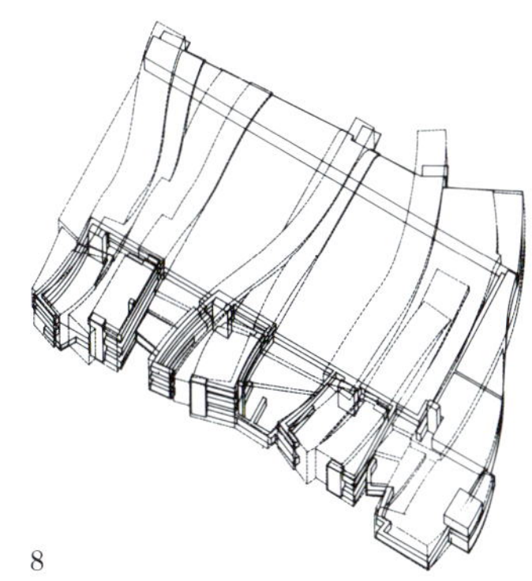
8

9

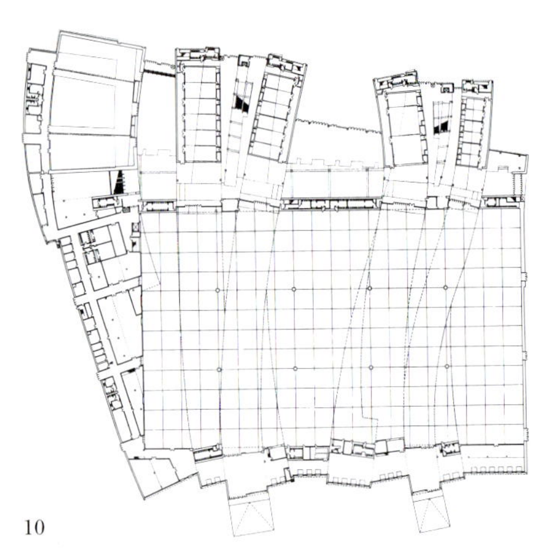
10

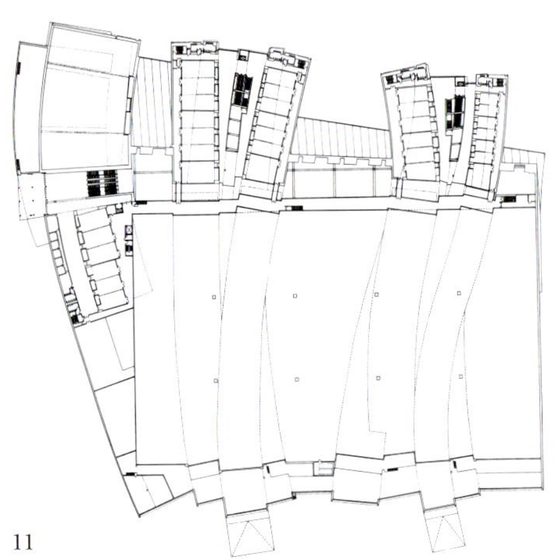
11

9 East facade, view from the north-east
10 Ground level plan
11 Mezzanine level plan
12 Aerial view from the south
13 Competition perspective, concourse
14 Competition perspective, exhibit hall

9 从东北方向看东立面
10 首层平面
11 一、二层之间夹层平面
12 南部鸟瞰
13 建成后透视，人流
14 建成后透视，展厅

12

13

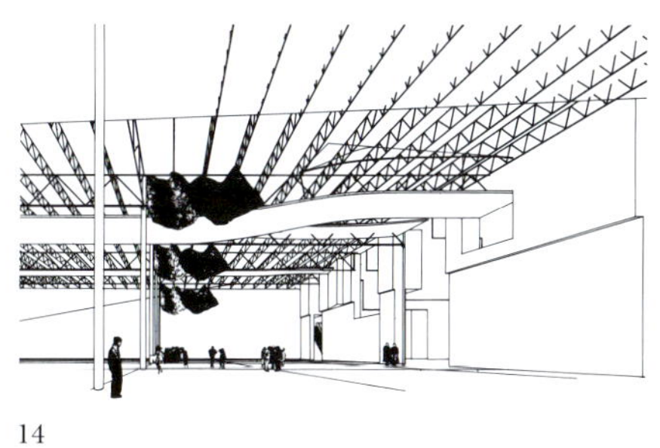

14

18

19

15

20

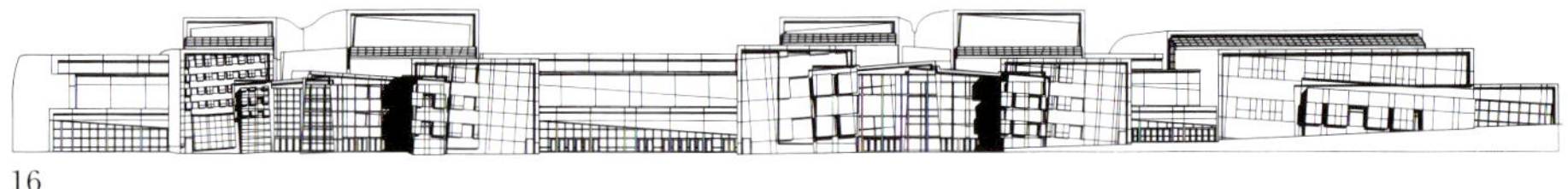

16

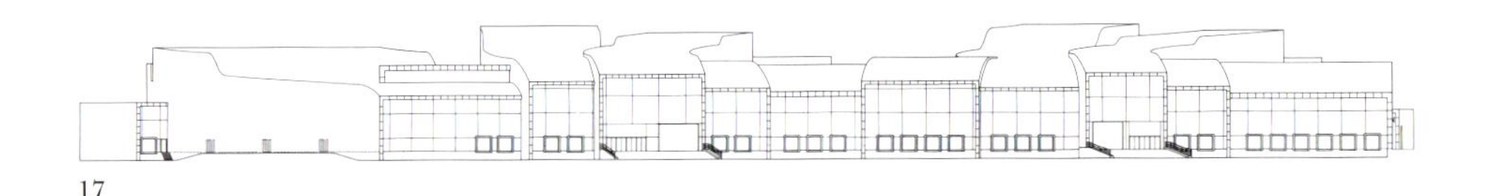

17

15 Aerial view from the north
16 West elevation
17 East elevation
18 Concourse, view from the north
19 Concourse, view from the south
20 Prefunction, view from the east
21 Competition west elevation
22 Competition east elevation
23 North elevation
24 South elevation

15 北部鸟瞰
16 西立面
17 东立面
18 从北看大厅
19 从南看大厅
20 从东看前厅
21 竞赛西立面
22 竞赛东立面
23 北立面
24 南立面

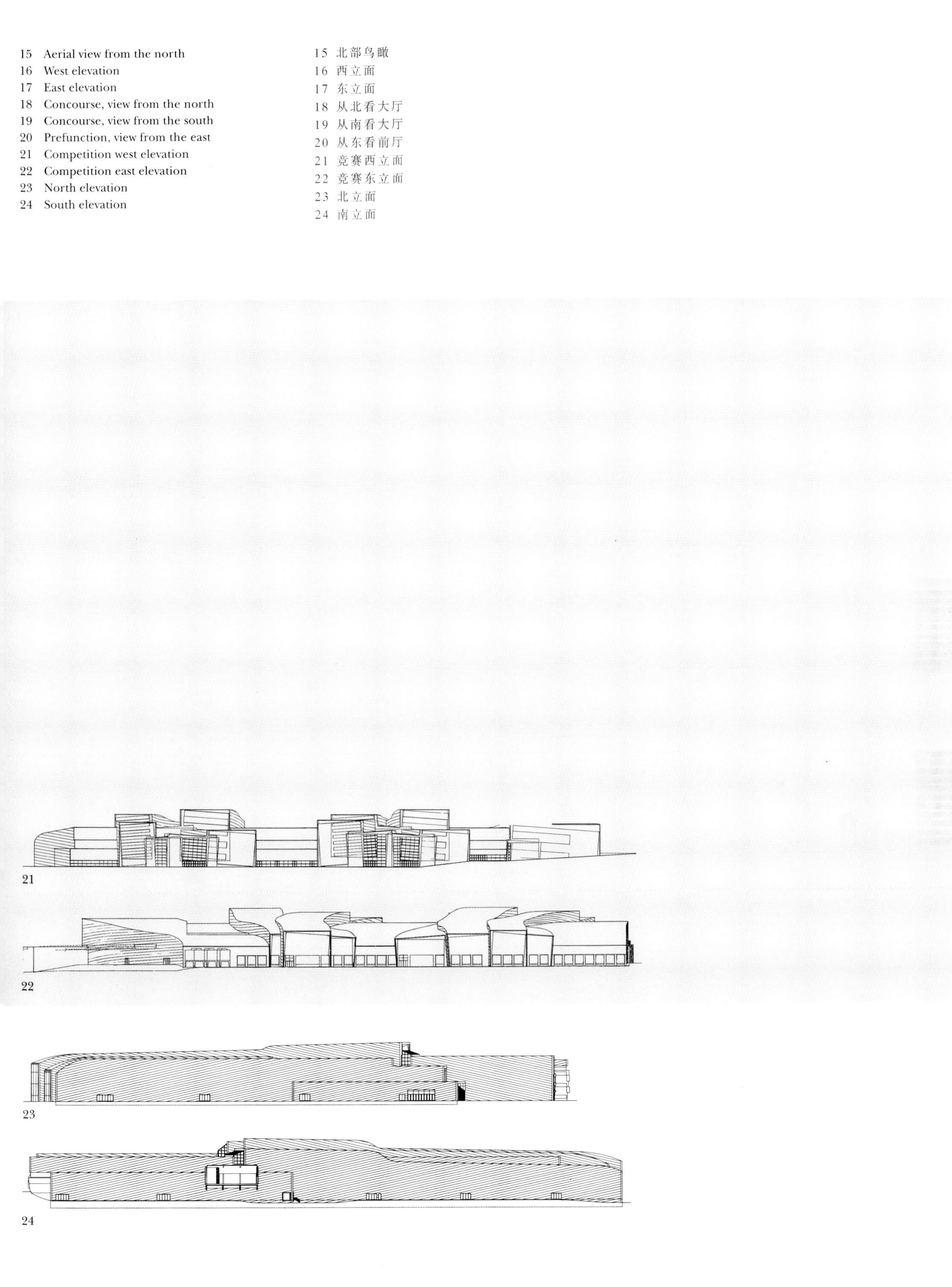

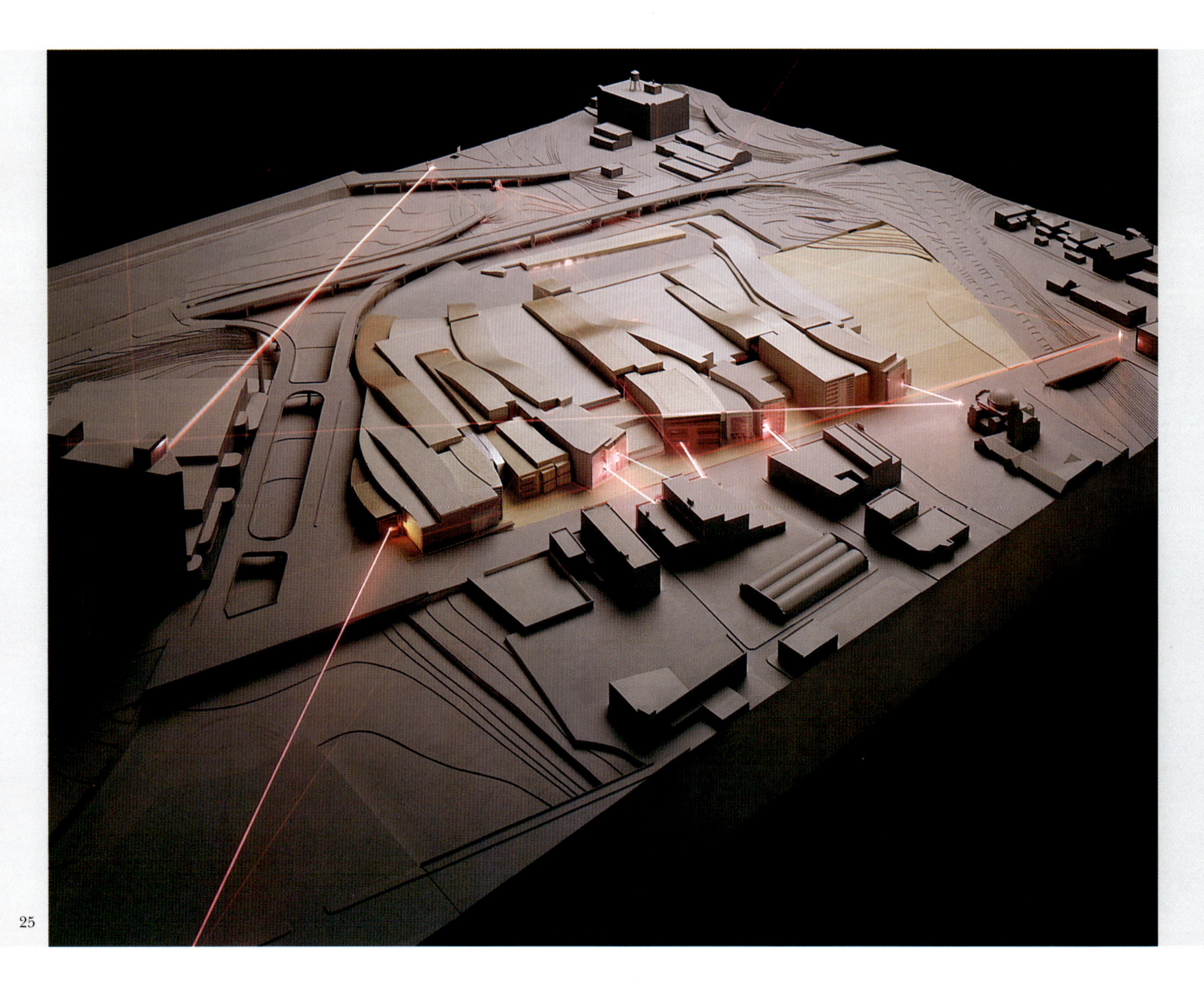
25

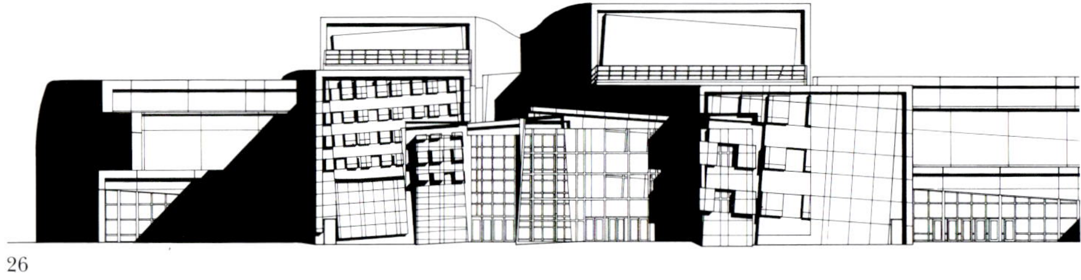
26

25 Competition model
26 West elevation, north segment
27 West elevation, south segment
28 Ballroom
29 Mezzanine prefunction, view from the west

25 竞赛模型
26 西立面，北面部分
27 西立面，南面部分
28 舞厅
29 从西看夹层前厅

28
29

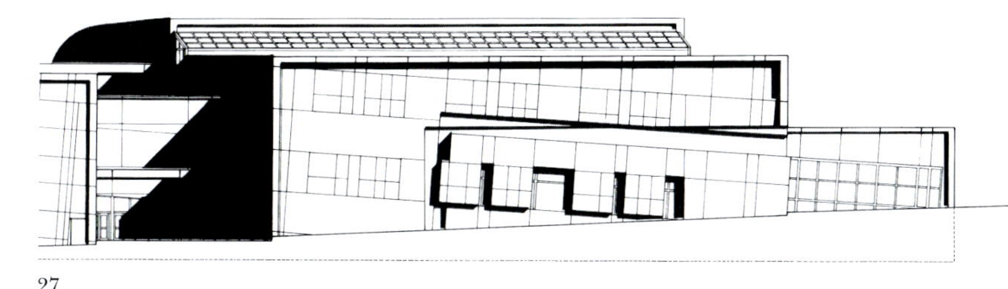
27

Banyoles Olympic Hotel

Design 1989
Banyoles, Spain
Consorci Pel Desenvolupament de la Vila Olimpica
82,000 square feet

巴塞罗那奥林匹克旅馆

设计　1989
巴塞罗那，西班牙
奥林匹克开发财团
82000平方英尺

In our project for a hotel at the site of the 1992 Olympic rowing events in Banyoles, the building is no longer a primary form—a single metaphysical enclosure. Instead, there are exponential torsions and phase shifts which characterize the line. This produces a building of richness and complexity, while at the same time preserving the simple autonomy and replication of bedroom units. It is also a building which is also part landscape.

Equally the "interior" space of the building is no longer merely the static lobby-corridor-room stacking of the traditional hotel. Instead, there is a sliding and a slipping found in the possibility of the form of the line which creates another condition of interior/exterior space.

我们的旅馆项目坐落在巴塞罗那1992年赛艇竞赛场地中，建筑不再是原始的形式——一个纯粹的简单围合。相反，这里包含幂数的扭转以及相位的转换，这些都丰富了地平线。这样就产生了一栋融丰富性和复杂性于一身的建筑，同时又保护了简单的自主以及客房单元的复制。建筑同时也成为风景的一部分。

同样地，室内空间不再仅是静态的大厅—走廊—房间的传统罗列。相反，在地平线中有一个滑行的倾斜的基础，这一基础创造出另一种室内/室外空间景象。

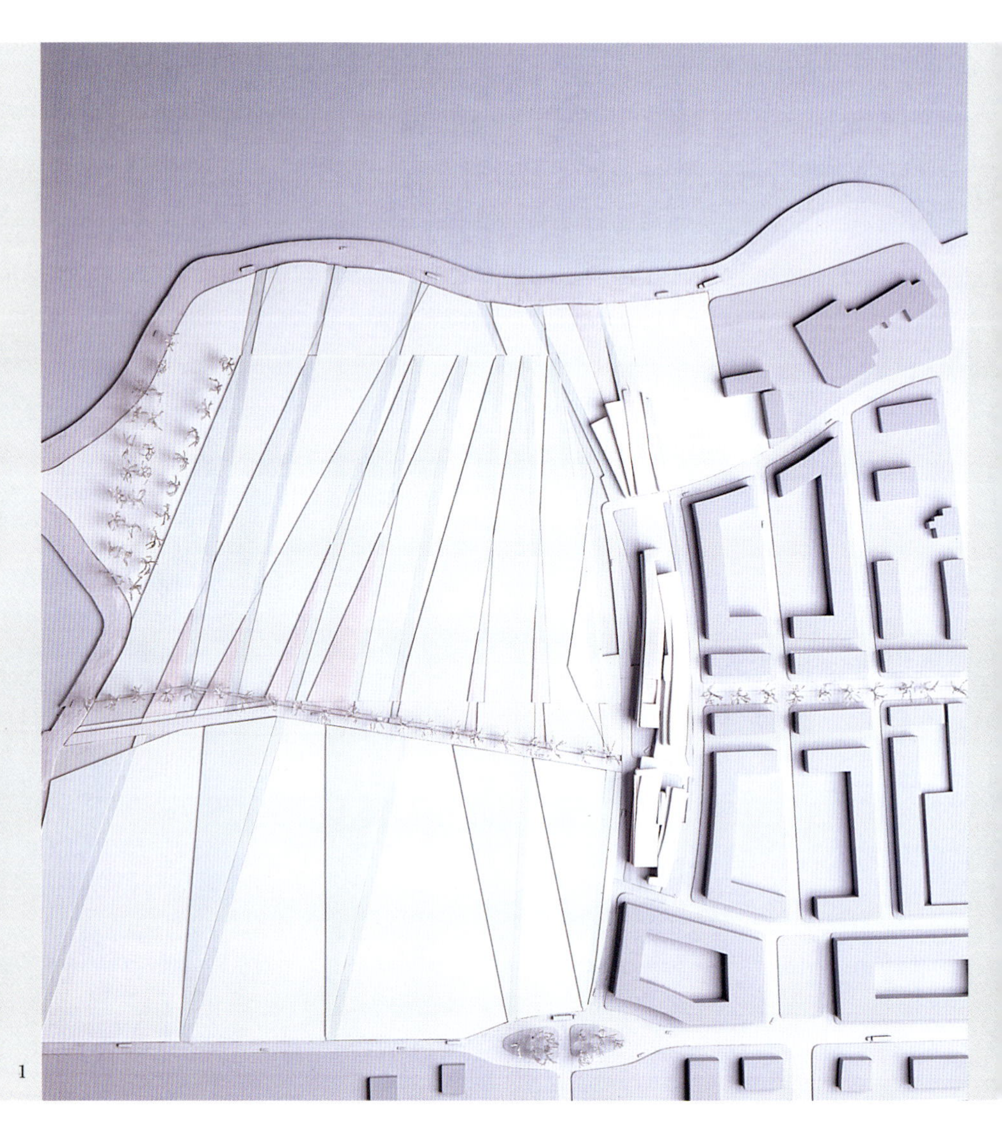
1

2

1 Presentation site model, aerial view
2 Site plan
3 Concept diagram, base bar and landscape registration
4 Concept diagram, rowing displacement and bar tilt
5 Concept diagram, imprint and overlap
6–7 Longitudinal section, partial

1 展示场地模型，鸟瞰
2 总平面
3 概念示意图，地下酒吧和景观定位
4 概念示意图，赛艇场地的覆盖以及倾斜的酒吧
5 概念示意图，痕迹和相互叠加
6－7 部分纵向剖面

3 4 5

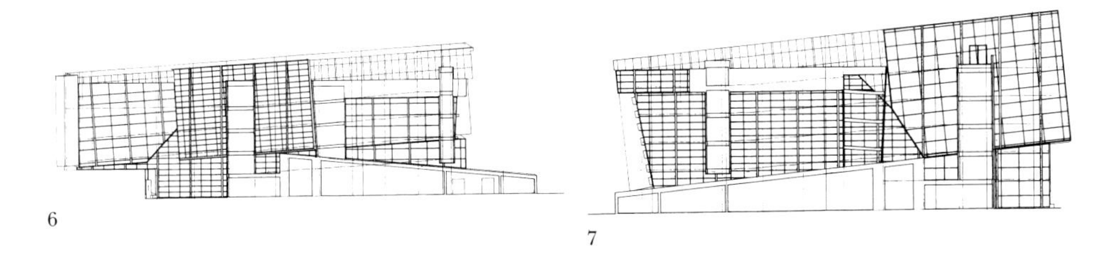

6 7

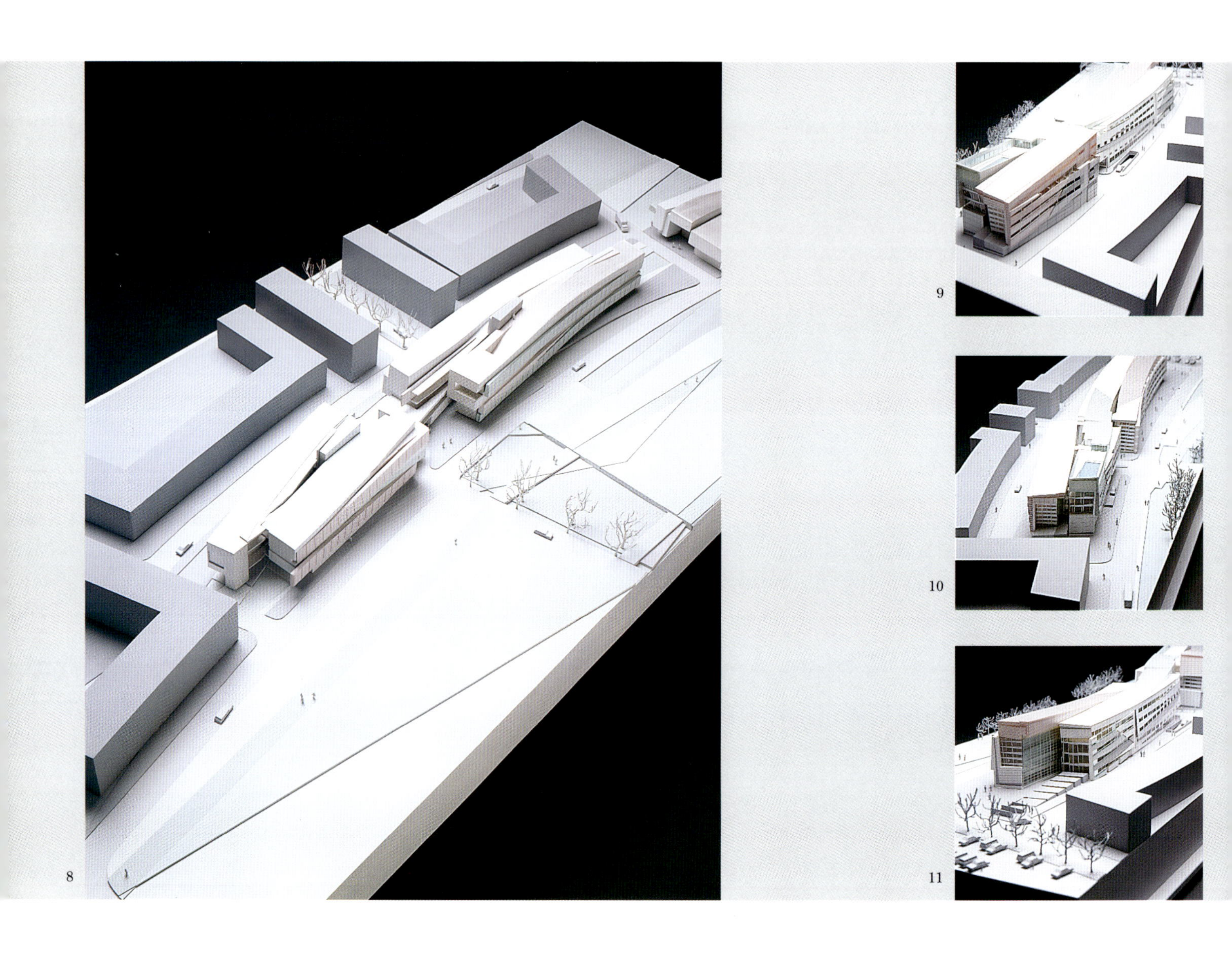

8

9

10

11

8 Presentation building model, view from the north-east
9 Presentation building model, view from the south-east
10–11 Presentation building model, view from the east
12 North elevation
13 South elevation

8 从东北方向看建筑展示模型
10-11 从东南方向看建筑展示模型
12 北立面
13 南立面

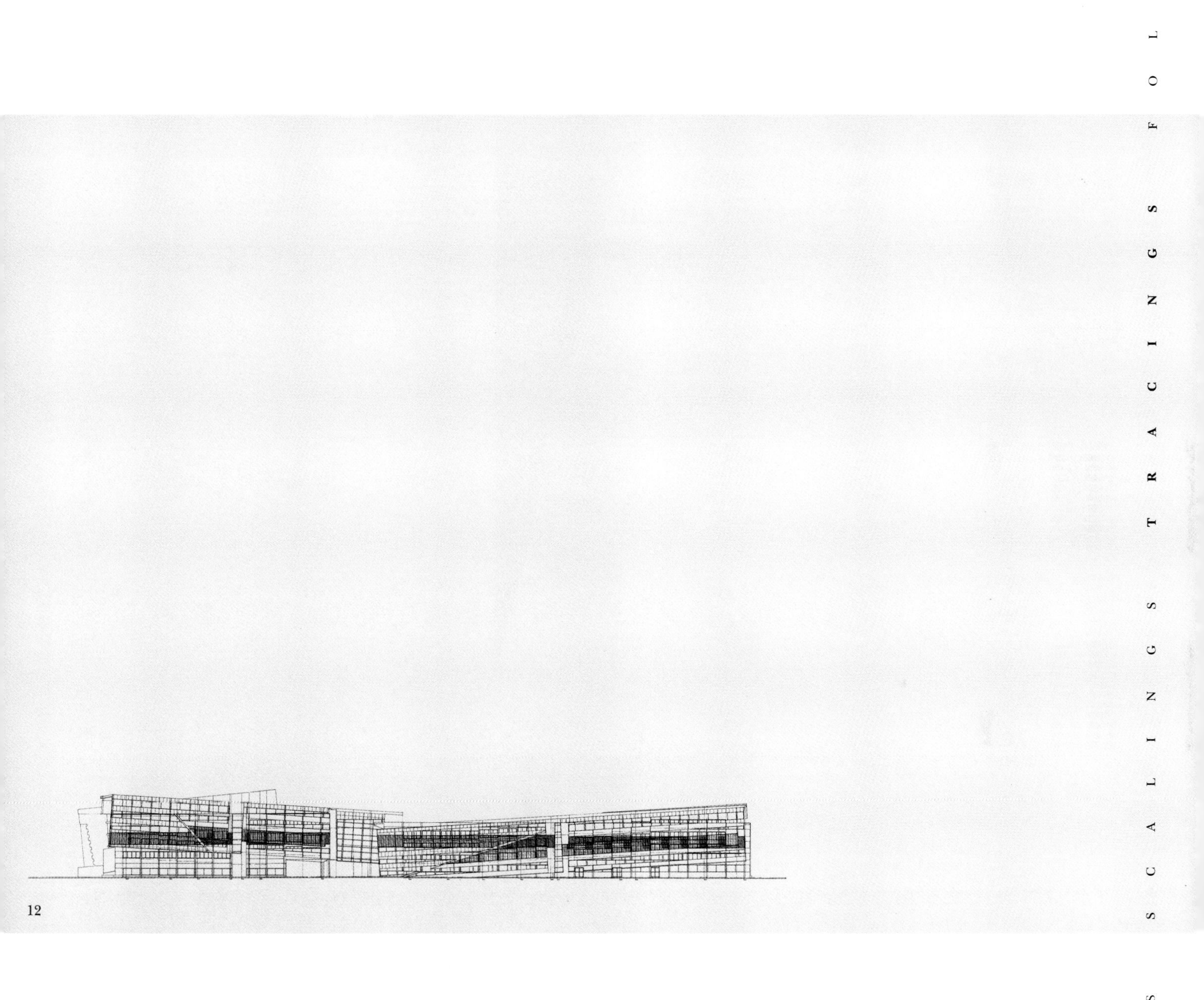

12

13

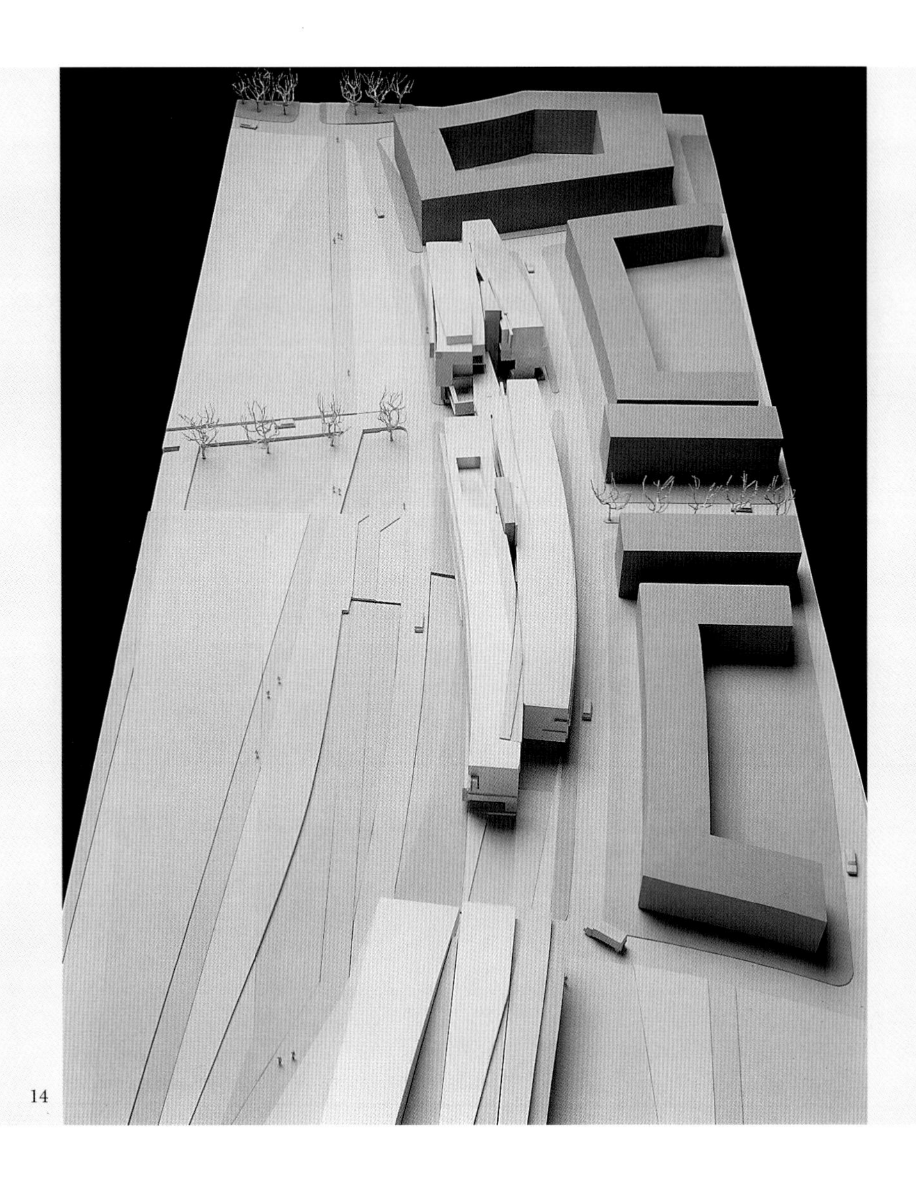
14

14 Presentation building model, view from the west
15 Second level plan
16 Fourth level plan
17 Fifth level plan
18 Roof level plan

14 从西向看建筑展示模型
15 三层平面
16 五层平面
17 六层平面
18 屋顶平面

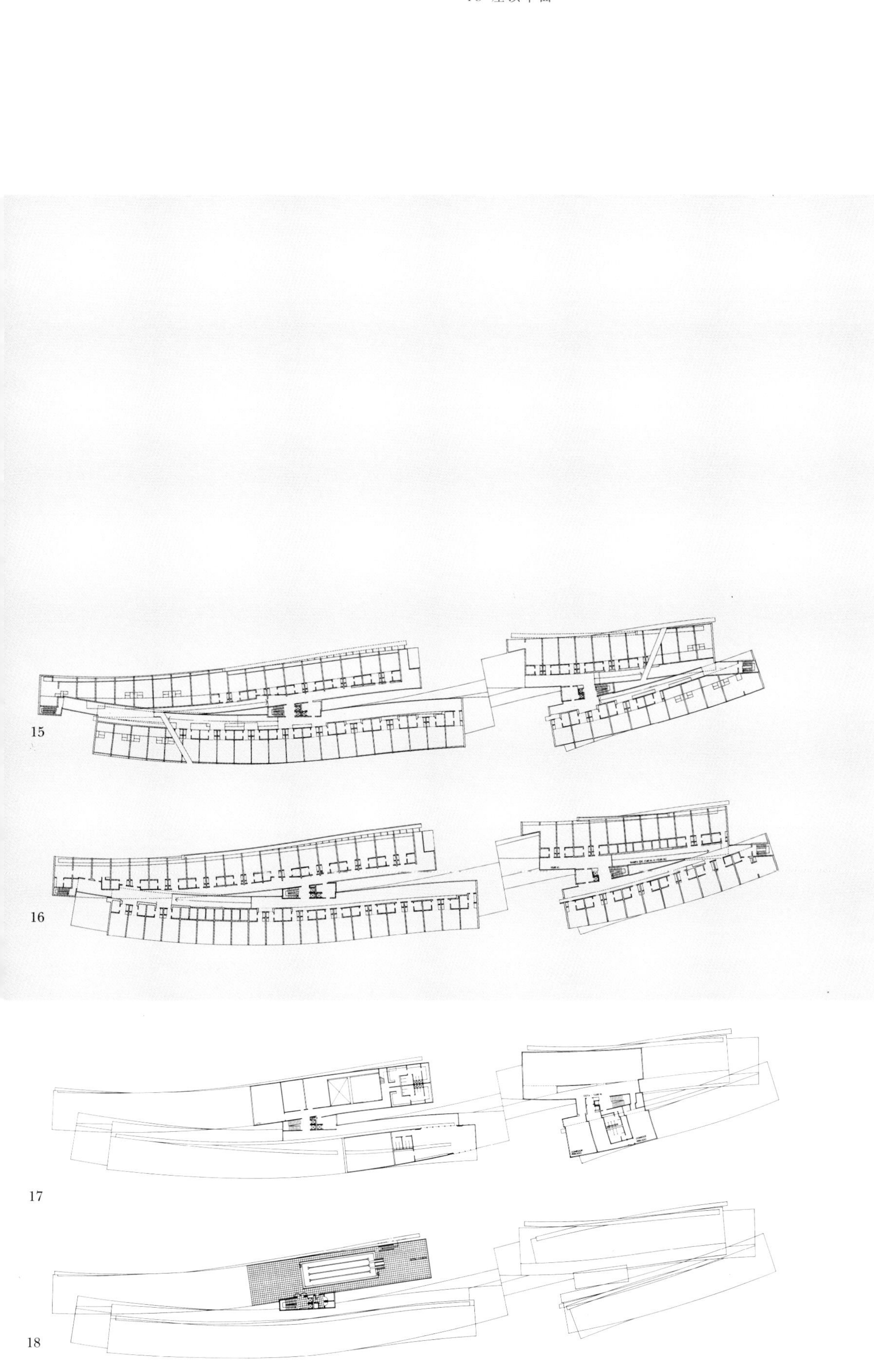

Cooper Union Student Housing

Design 1989
New York, New York
The Cooper Union
50,000 square feet

库珀学生住宅联合会

设计　　1989
纽约，纽约州
库珀联合会
50000 平方英尺

This project breaks down traditional aspects of classical monumentality and replaces them with a freer, richer, more playful massing which has no defined frame, no single axis, and no conformity of material to shape or form to function.

The Cooper Union is the "home" for students and the portal through which they venture into the life of the city. Thus, our project addresses both symbolic and functional aspects. It begins with private units for two people, then facilities for four to six people, and then loft-style duplex living areas for 16 to 32 students. The organization provides for both community and privacy, flexibility and order, breaking down the scale of a large building into recognizable human units.

该项目打破了古典主义肃穆的传统视角，取而代之的是更自由、更丰富、更有趣的混合，它没有明确的结构，没有简单的轴线，也没有形状与材料、形式与功能的统一。

库珀联合会是学生之家，同时也是他们在城市生活中冒险的开始。于是，在我们的设计中，同时强调符号与功能。它从两个人的私有单元开始，然后是 4～6 个人的房间，最后是容纳 16～32 个学生的阁楼形式的复合生活区。这种组织方式既提供私密性和公共性、自由与顺序，又将大尺度的建筑打散成可识别的人性单元。

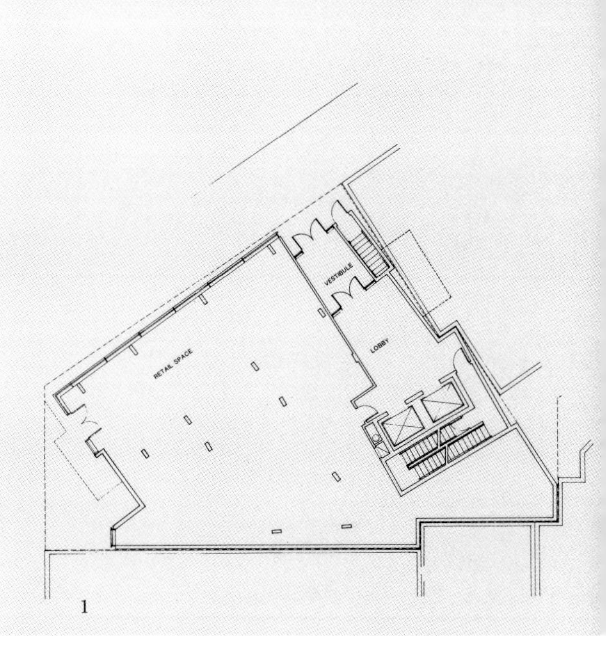

1

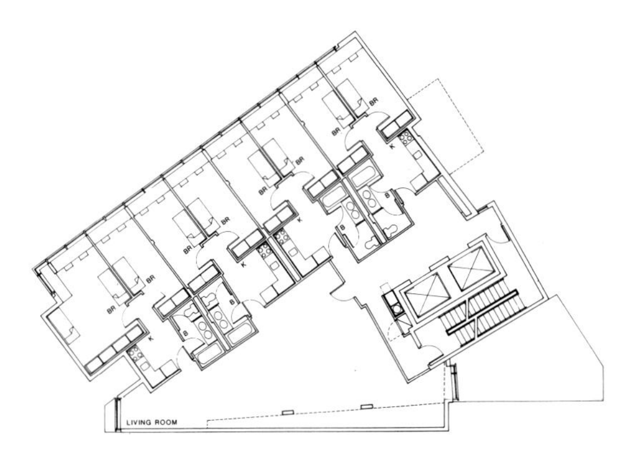

2

1 Ground level plan
2 Third level plan
3 Presentation model, view from the north
4 Eighth–tenth level plan
5 Roof level plan

1 首层平面
2 四层平面
3 从北向看展示模型
4 九至十一层平面
5 屋顶平面

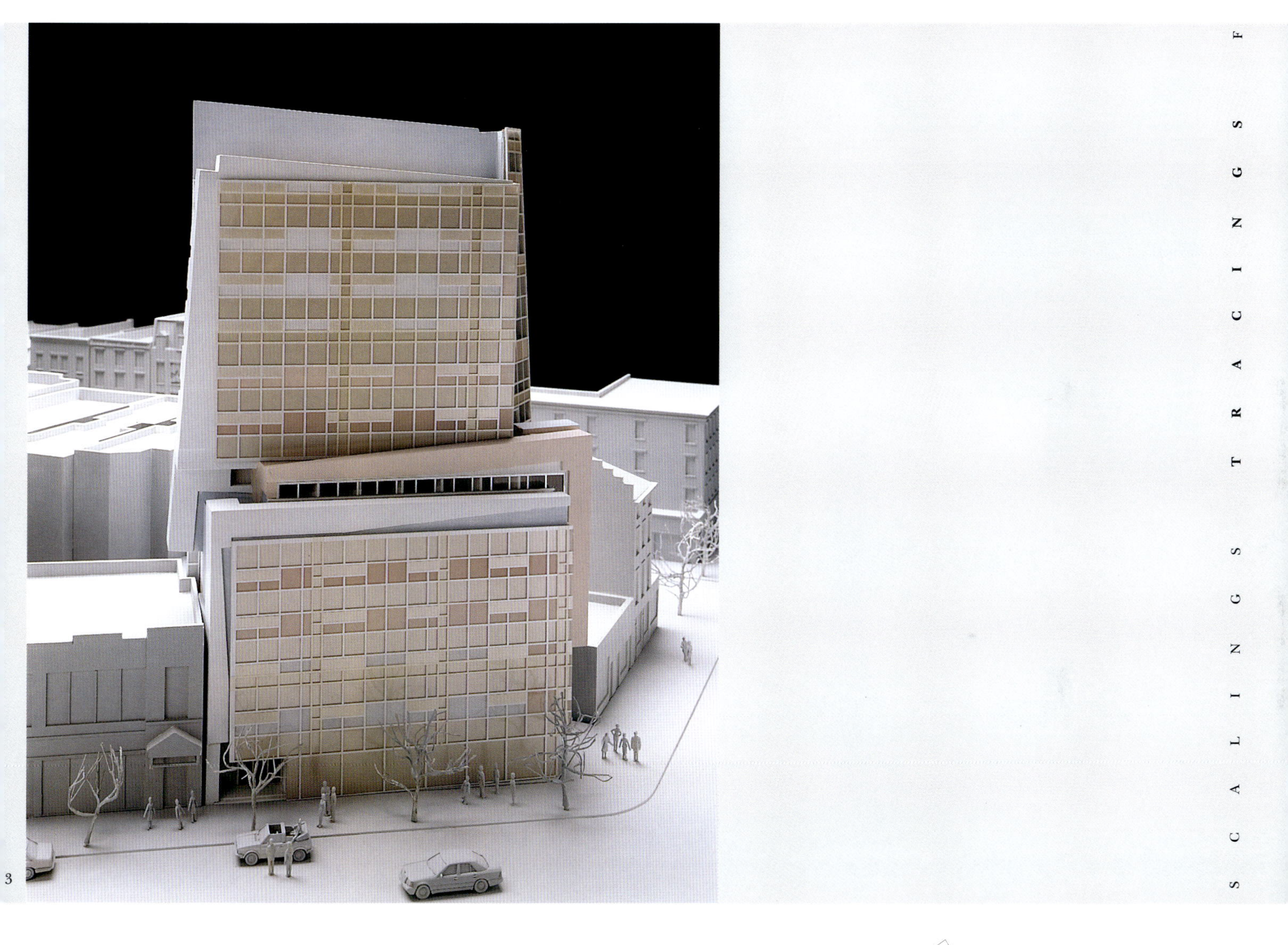

3

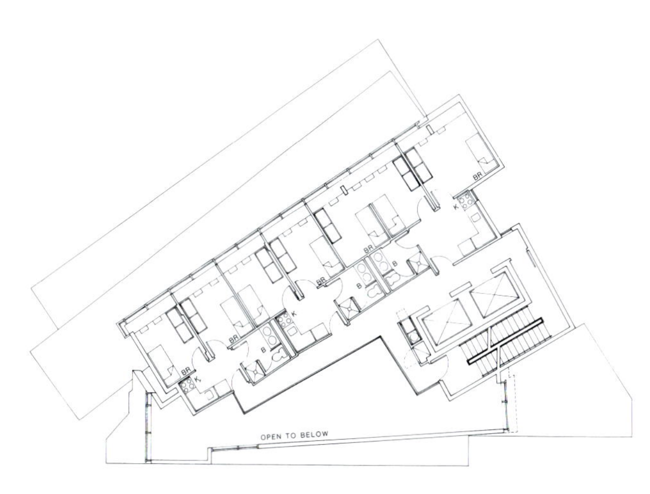

4

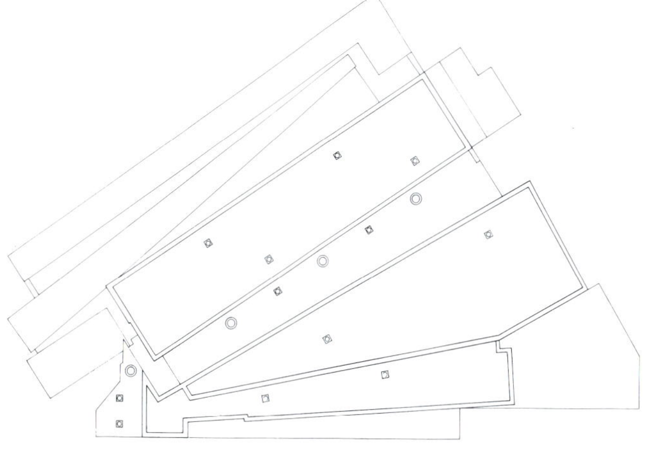

5

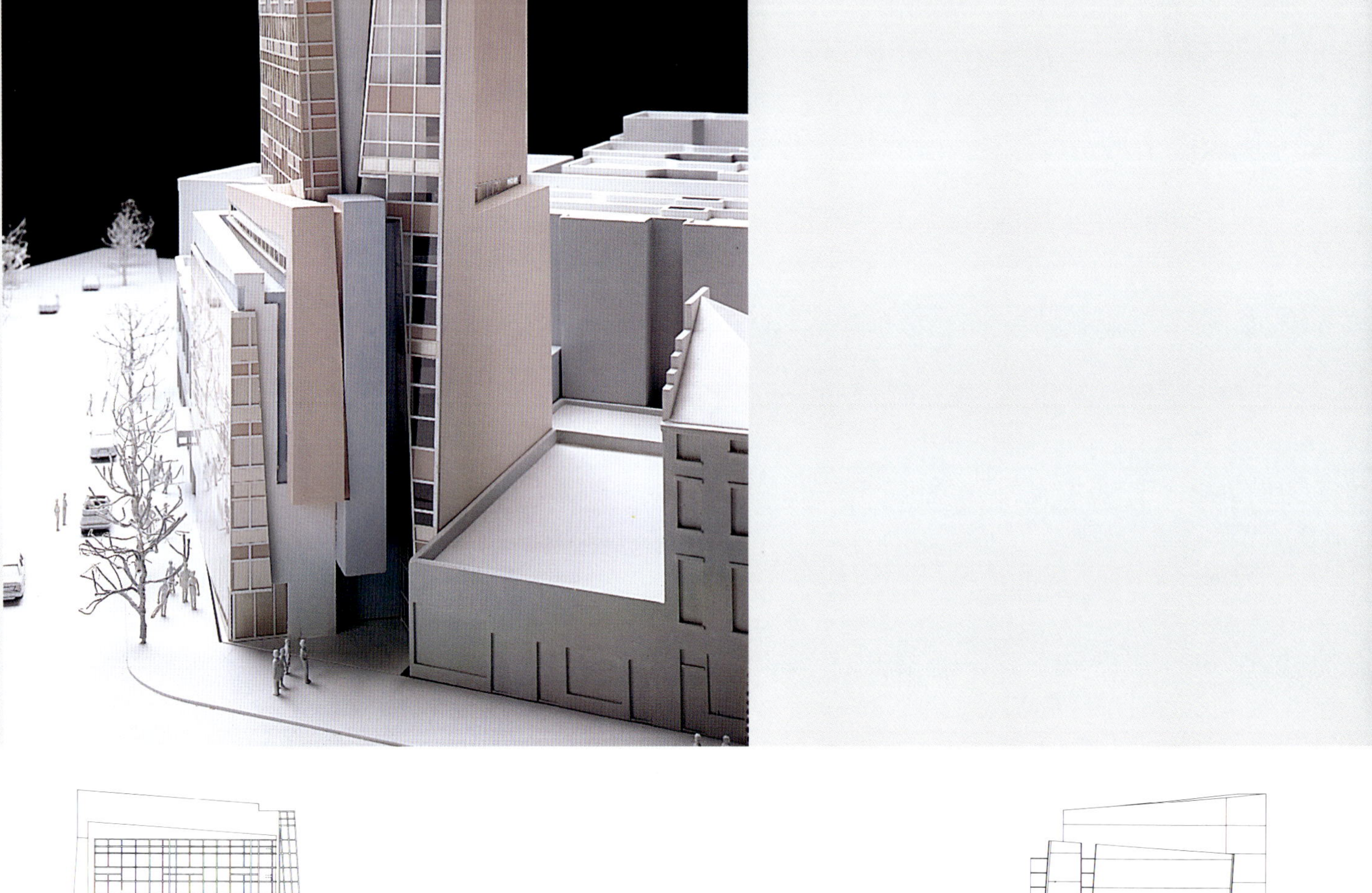

6

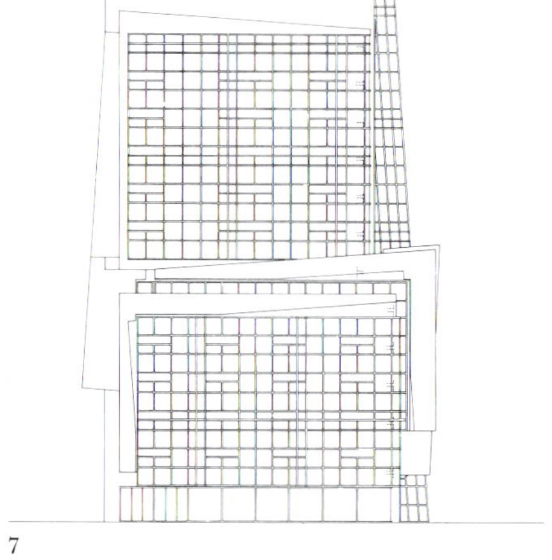
7

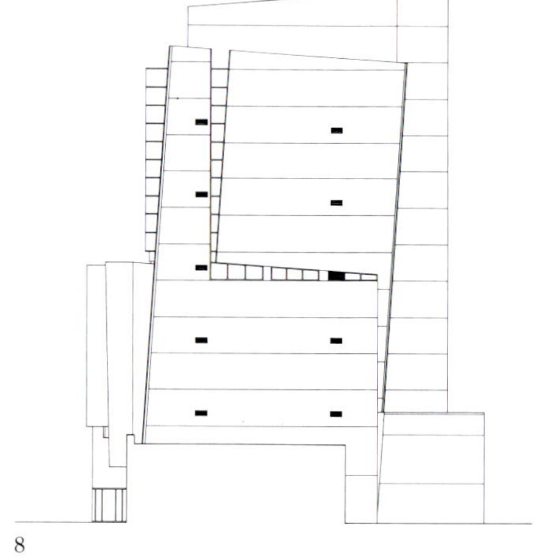
8

6 Presentation model, view from the west
7 North elevation
8 South elevation
9 Section BB
10 East elevation
11 Presentation model, view from the east
12 Presentation model, view from the south

6 从西向看展示模型
7 北立面
8 南立面
9 BB 剖面
10 东立面
11 从东向看展示模型
12 从南向看展示模型

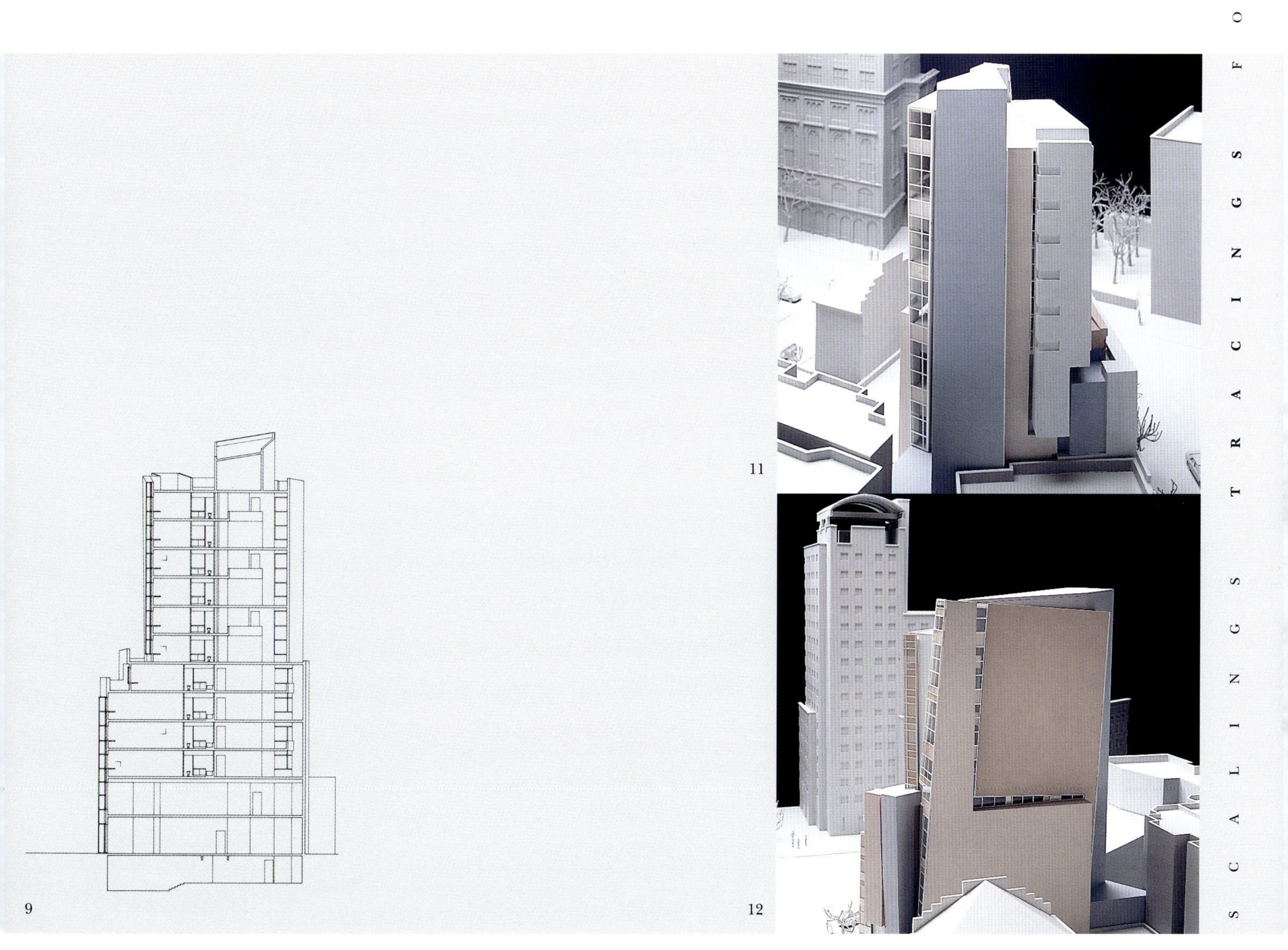

11

9

12

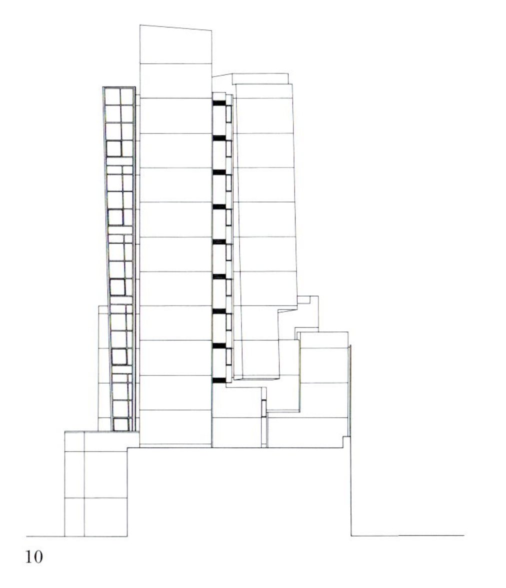

10

Groningen Music-Video Pavilion

Design/Completion 1990
Groningen, The Netherlands
Groningen City Festival
2,000 square feet

格罗宁根音乐录像馆

设计/竣工　1990
格罗宁根，荷兰
格罗宁根城市庆典
2000平方英尺

As part of the 1990 Groningen City Festival, Eisenman Architects was commissioned to build one of several satellites to the municipal museum. Our project is based on the idea that the new video technology is revolutionizing the notion of the moving image. The structure of our pavilion is based on an analysis of the way a video image is produced on a picture screen. Visitors to our pavilion follow a path which is analogous to that of a scanning beam's path on a video screen. Thus, the visitors become part of the medium itself, passing in front of viewing screens and continually crossing through images, shifting position to form images in different ways, and running interference. The project alludes to the traditional auditorium in its sloping floors.

作为1990年格罗宁根城市庆典的成员，埃森曼建筑事务所被委托设计众多博物馆中的一个。我们的设计理念是基于这样一种概念：新的录像技术使得移动影像的概念发生了革命性的变化。我们展厅的结构就是基于对录像在图像屏幕上如何产生的一种分析。到展厅来的参观者可以沿着一条路径走，这条路径与在录像屏幕扫描光柱有着相似之处。于是，参观者就成为媒体自身的一部分，在所看到的屏幕前通过，并继续从影像中穿行，以不同的方式变换位置从而形成影像，相互间干扰碰撞。用倾斜的屋顶，建筑隐喻了传统的会堂。

1

1 View from the south-east
2 Concept diagram, scanning beam
3 Concept diagram, trace
4 Concept diagram, retrace
5 Concept diagram, scanning beam
6 Concept diagram, trace
7 Concept diagram, retrace

1 从东南方向看
2 概念示意图，扫描光柱
3 概念示意图，轨迹
4 概念示意图，光影跟踪
5 概念示意图，扫描光柱
6 概念示意图，轨迹
7 概念示意图，光影跟踪

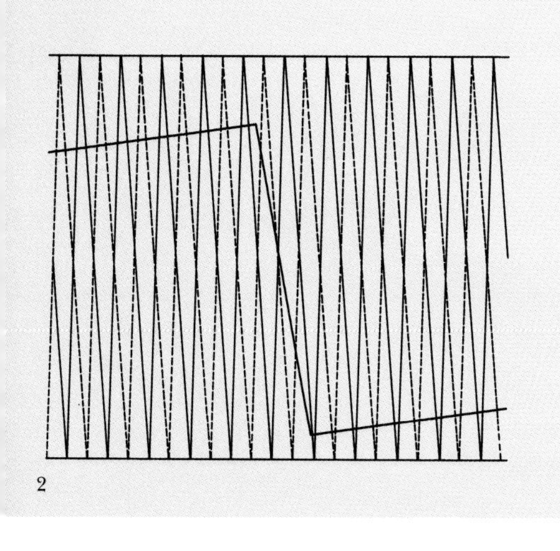

2

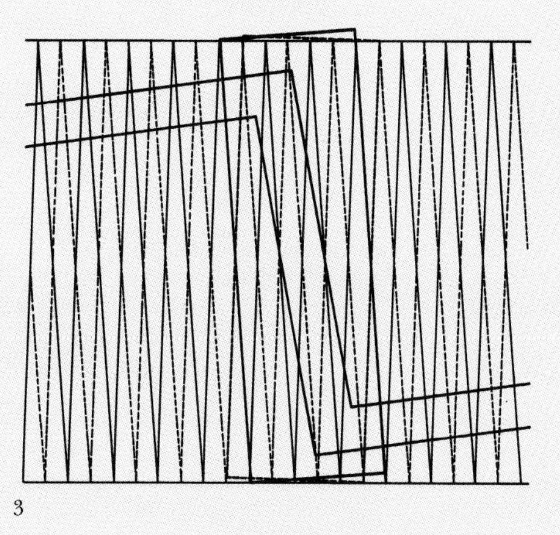

3

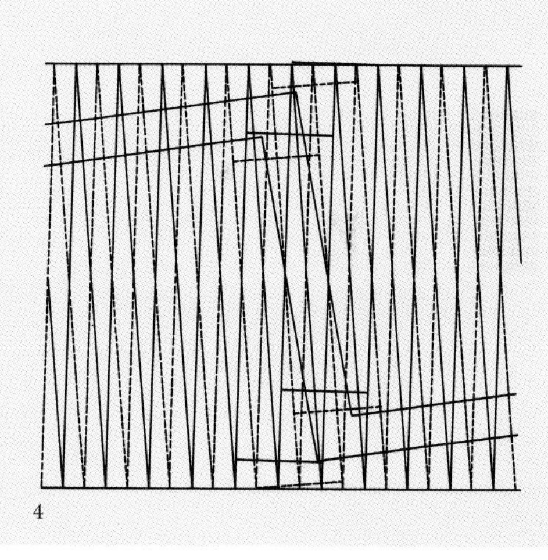

4

5

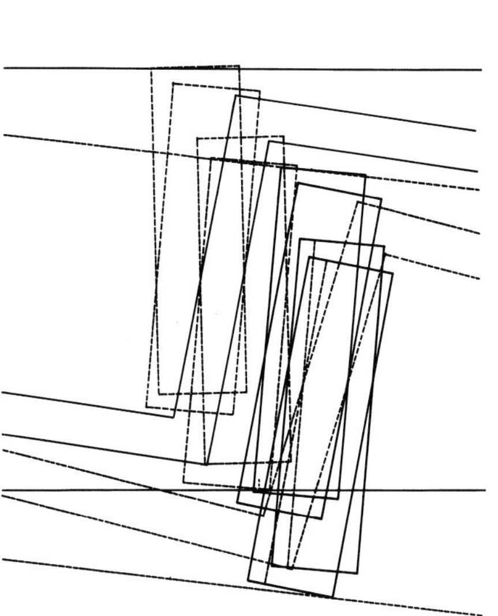

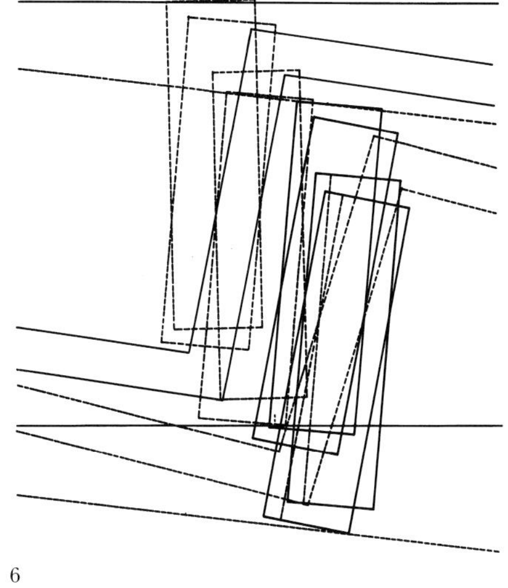

6

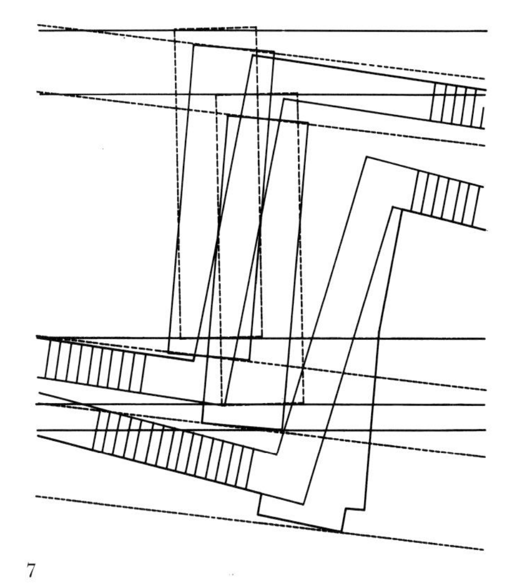

7

8

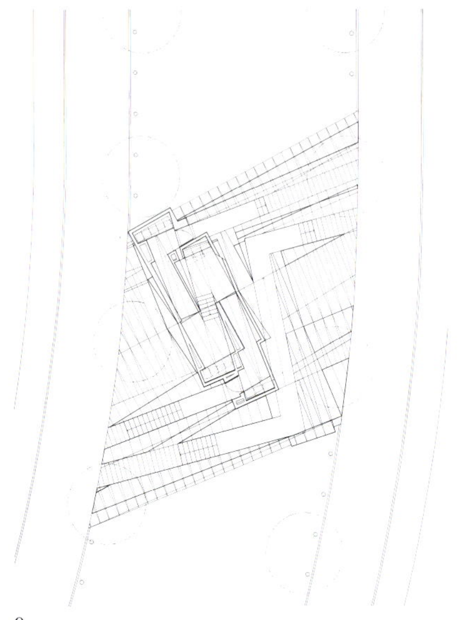
9

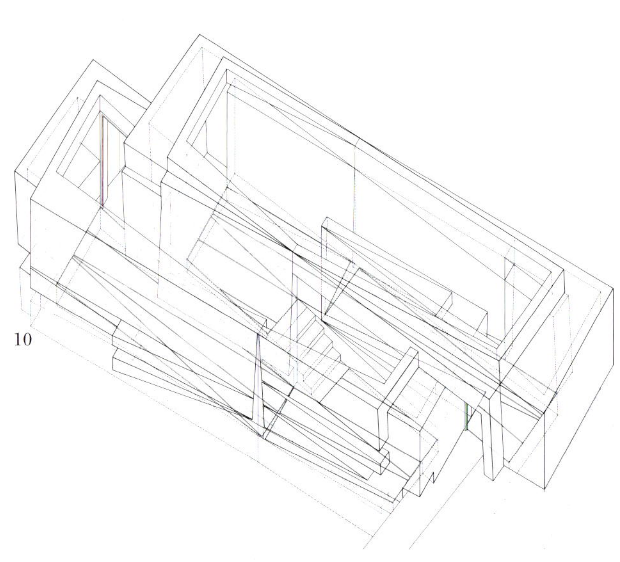
10

8 View from the south-west
9 Site plan
10 Axonometric, view from the south-east
11 South elevation
12 North elevation
13 Plan section through upper level
14–15 Interior

8 从西南方向看
9 总平面
10 从东南方向看轴测图
11 南立面
12 北立面
13 从上往下看平面剖面
14-15 室内实景

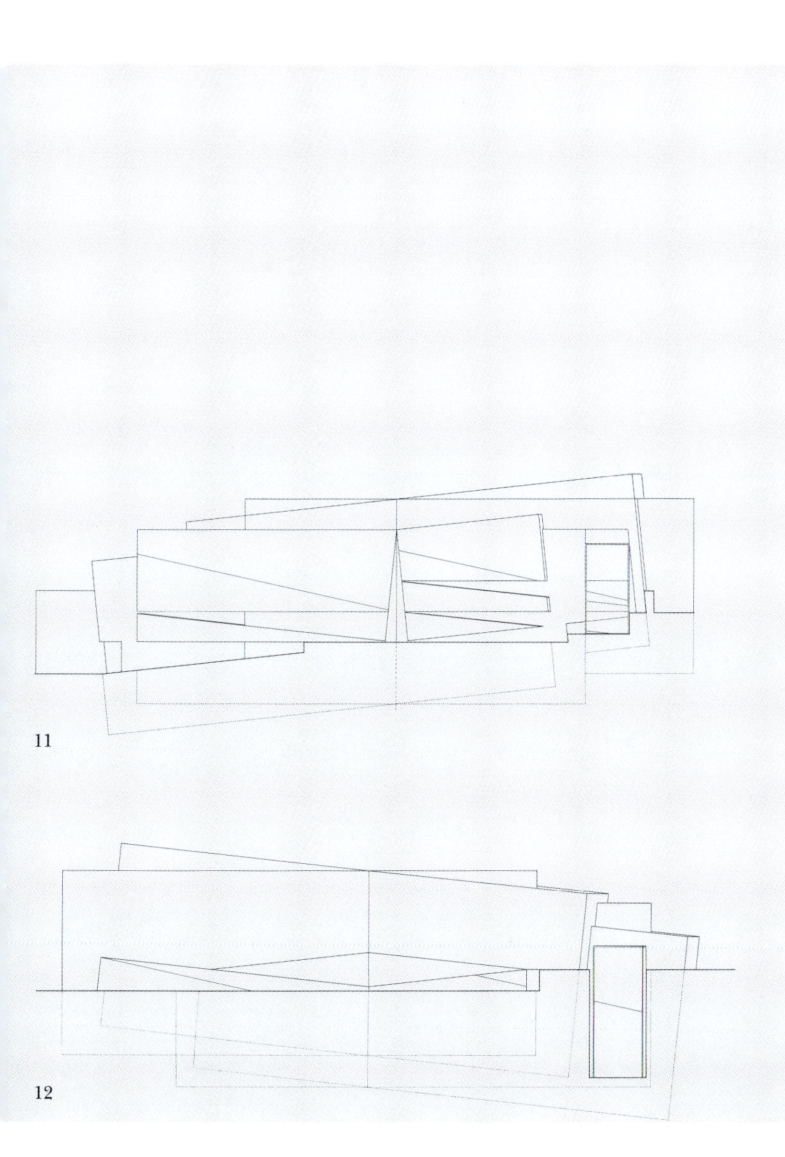

11

12

14

15

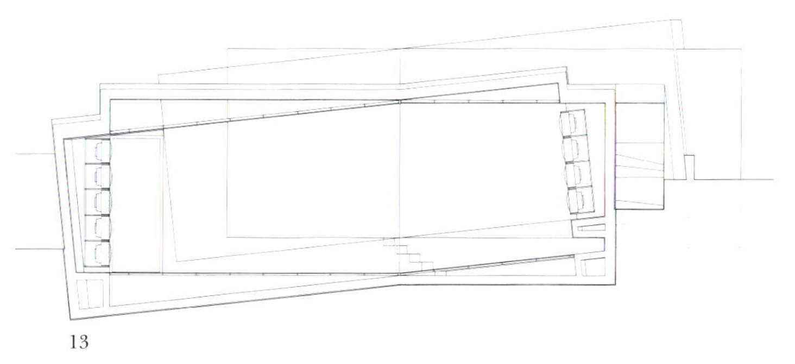

13

Nunotani Office Building

Design/Completion 1990/1992
Tokyo, Japan
Nunotani Company Ltd
40,000 square feet

布谷办公楼

设计/竣工 1990/1992
东京，日本
布谷公司
40000平方英尺

The land mass of Japan is constantly subjected to earthquake activity, and the Nunotani building is seen as a metaphoric record of these continuous waves of movement. Simultaneous to this analogy, the project represents an attempt to rethink the symbolism of the vertical office building, first by producing a building that is not metaphorically skeletal or striated, but rather made up of a shell of vertically compressed and translated plates; and second, by producing an image somewhere between an erect and a "limp" condition.

The building consists of studio and office spaces, a multimedia presentation room, library, cafeteria, CAD workrooms and traditional Japanese resting rooms.

日本的大陆地块经常发生地震活动，布谷办公楼被看作是那些连续的振动记录的隐喻。与此同时，该项目试图对垂直的办公建筑进行反思，首先创建了一栋没有骨架和线条的建筑，它是由一个壳体和被压缩、翻新的金属板建成的；其次制造了一种介于挺拔和柔软之间情形的景象。

该建筑包括：工作室、办公空间、一个多媒体演示中心、图书馆、咖啡间、CAD工作室和传统的日本休息室。

1

1 Exterior, view from the north-east
2 Site plan
3 Roof plan
4 Exterior, view from the east
5 Exterior, view from the south

1 外景，从东北方向看
2 总平面
3 屋顶平面
4 从东向看外景
5 从南向看外景

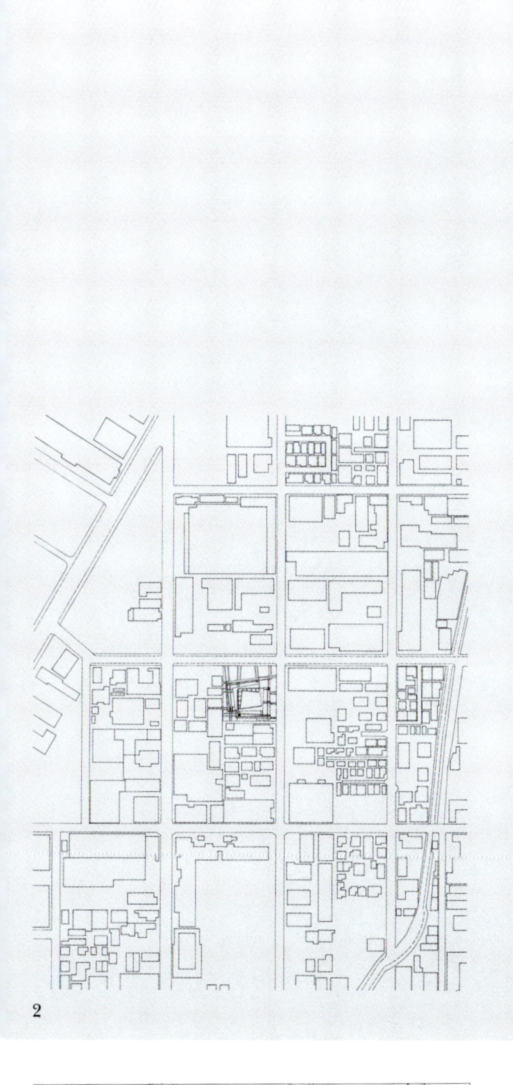

2

4

5

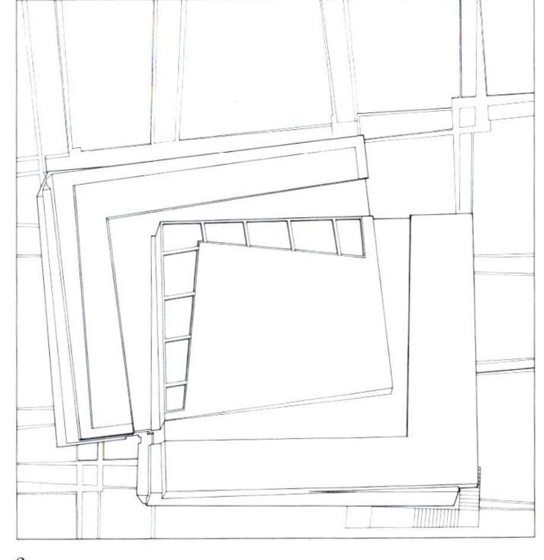

3

6

7

6 Exterior night view from the north-east
7 Section AA
8 Third level plan
9 Fifth level plan
10 Interior, view of atrium
11 Interior, upper level

6 从东北方向看夜晚外景
7 AA 剖面
8 四层平面
9 六层平面
10 从中庭看室内实景
11 从上层看室内实景

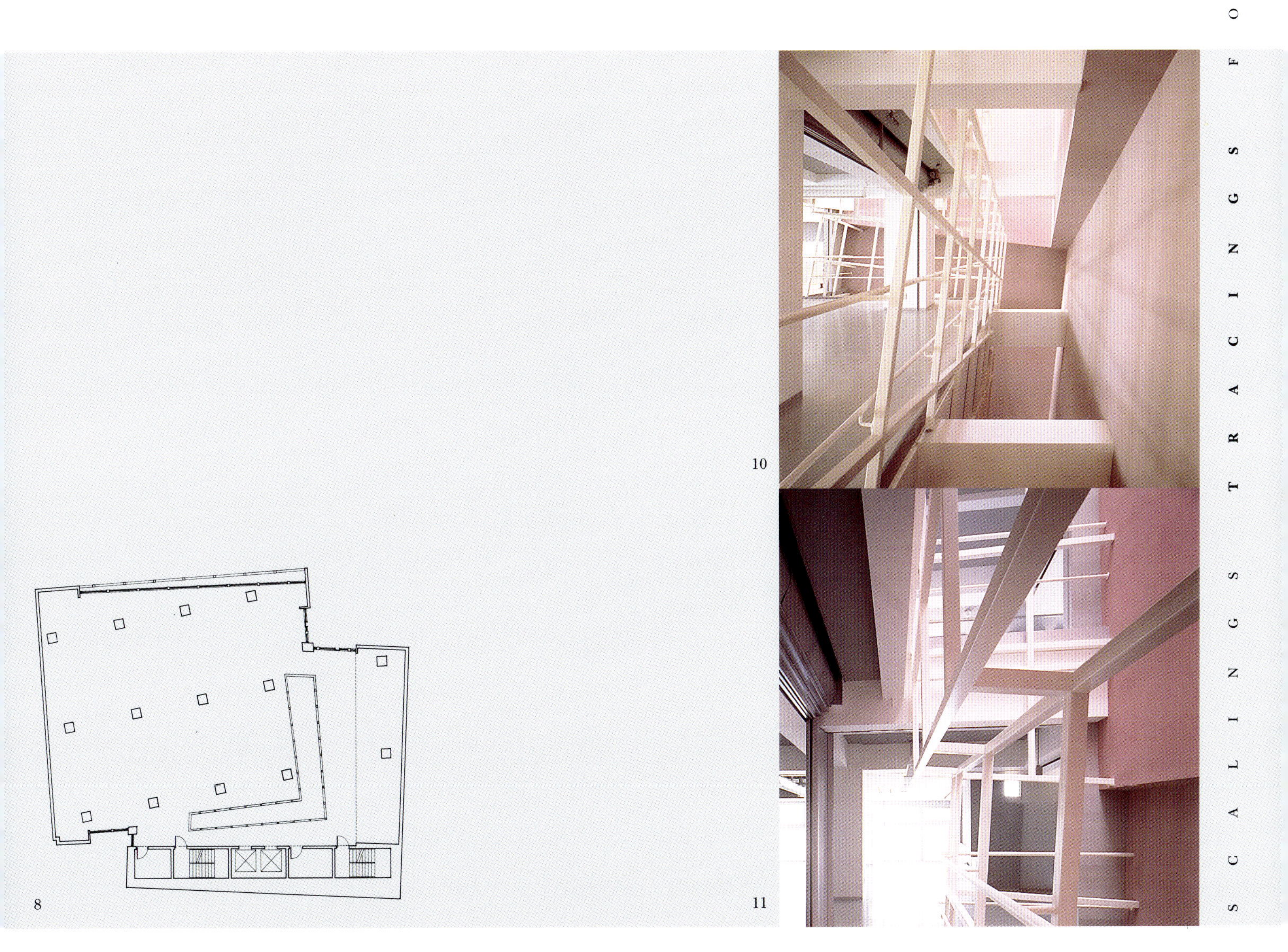

10

8

11

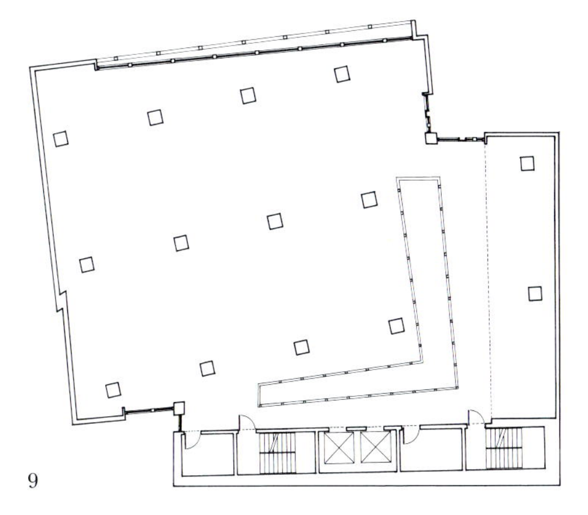

9

12

12	Exterior, view from the east	12	从西向看外景
13	North elevation	13	北立面
14	South elevation	14	南立面
15–17	Interior, entry level gallery	15–17	室内实景，入口层走廊

15

16

17

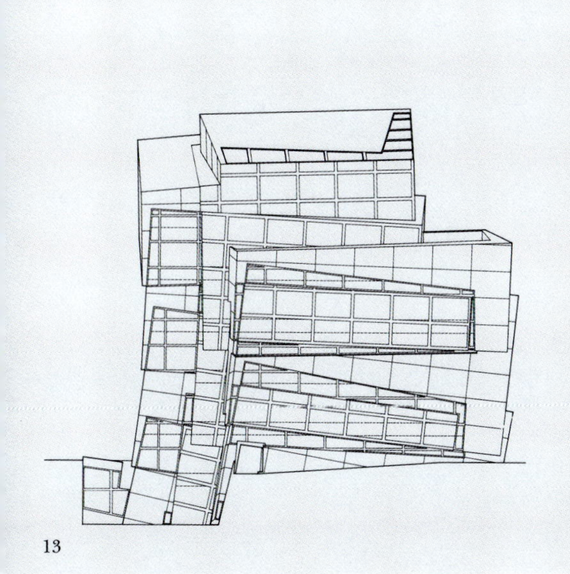

13

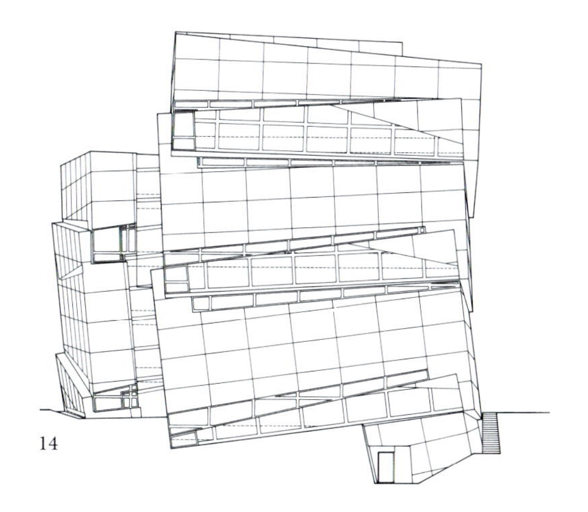

14

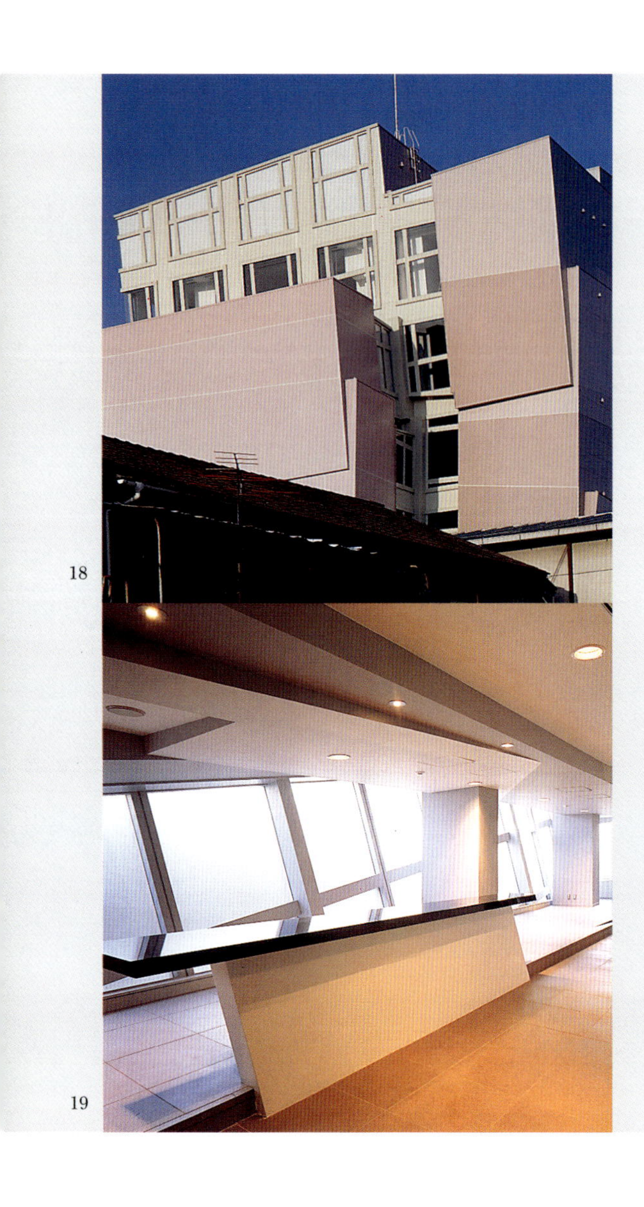

18

19

18 Exterior, view from the south-west
19 Interior, view of lobby
20 West elevation
21 East elevation
22 Exterior, view of the east facade from the north-east
23 Interior, view of atrium

18 从西南方向看外景
19 从大厅看室内实景
20 西立面
21 东立面
22 室外，从东北方向看东立面
23 室内，从中庭看

22

23

20

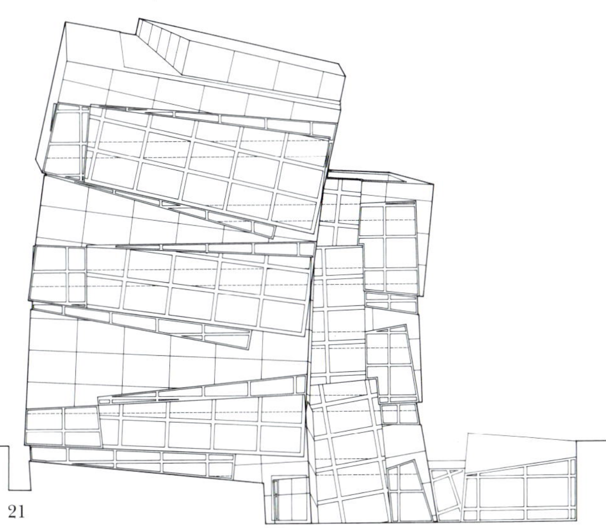

21

Atocha 123 Hotel

Design 1990
Madrid, Spain
Sociedad Belga de los Pinares de el Paular
47,000 square feet

阿托查 123 旅馆

设计　1990
马德里，西班牙
Belga de los Pinares de el Paular 协会
47000 平方英尺

This 92-room, four-star hotel was designed for the corner of Atocha and Alameda streets in downtown Madrid. With depths and heights established by zoning requirements, a series of diagrams was developed to derive the building form. First, bars were laid out parallel to Atocha, with a depth of one room, a height of two and a half floors, and separated by the width of a corridor. Second, bars of the same depth, height, and separation were repeated parallel to Alameda. Third, the bars parallel to Alameda were spread exponentially along the site until parallel to Prado, and the bars parallel to Atocha spread perpendicular to Prado. These manipulations produced a building form that responds to the richness of its urban environment.

1

这栋拥有 92 个房间的四星级宾馆位于马德里市区的阿托查和阿尔梅达大街的拐角处。由于城市规划对开间和高度的限制，一系列的设计构思图也就应运而生。首先酒吧间被放置与阿托查大街平行，并拥有一个标准间的开间，其高度是两层半，并被一个走廊分离开来。其次，拥有相同开间、层高和分隔的酒吧间也沿着阿托查大街平行重复布置。然后，与阿尔梅达大街平行布置的酒吧间一直延伸到与普拉多平行的位置，那些与阿托查大街平行布置的酒吧间沿着与普拉多垂直的方向伸展。这些重复所形成的建筑形式呼应了该城市的丰富的城市空间环境。

2

1 Interior, lobby perspective
2 Site plan
3–4 Concept diagram, Cartesian line
5–6 Concept diagram, rotation and overlay plan
7–8 Concept diagram, exponential translation plan
9 Sixth level plan
10 Ground level plan

1 室内，大厅透视
2 总平面
3-4 概念设计示意图，笛卡儿线
5-6 概念设计示意图，旋转和重叠的平面
7-8 概念设计示意图，幂数的和转换平面
9 七层平面
10 首层平面
11 展示模型，阿托查大街的立面

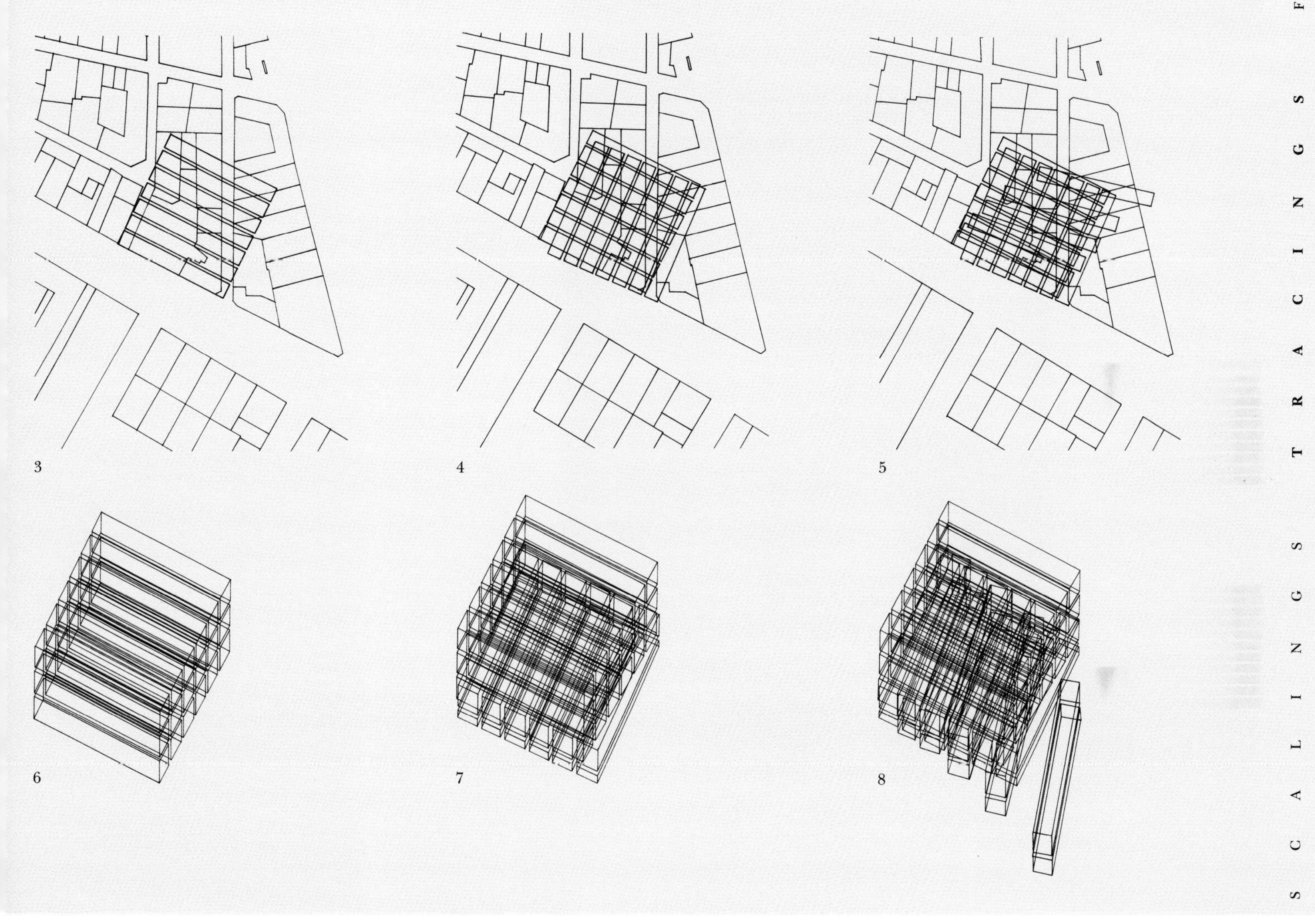
3 4 5 6 7 8

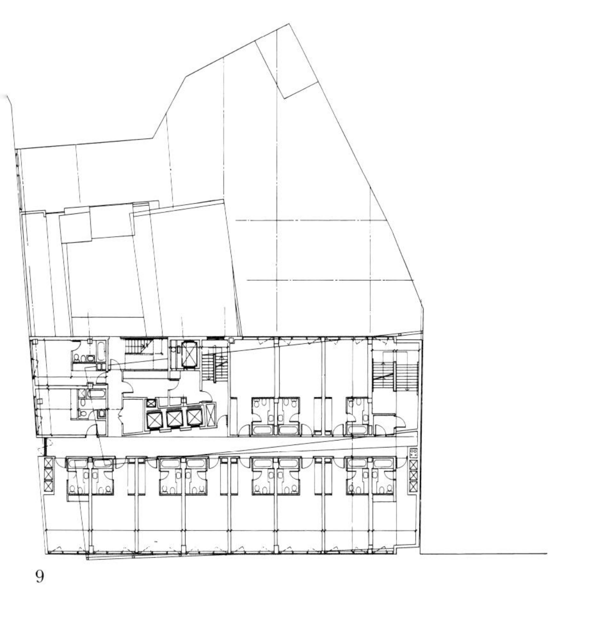
9

10

11

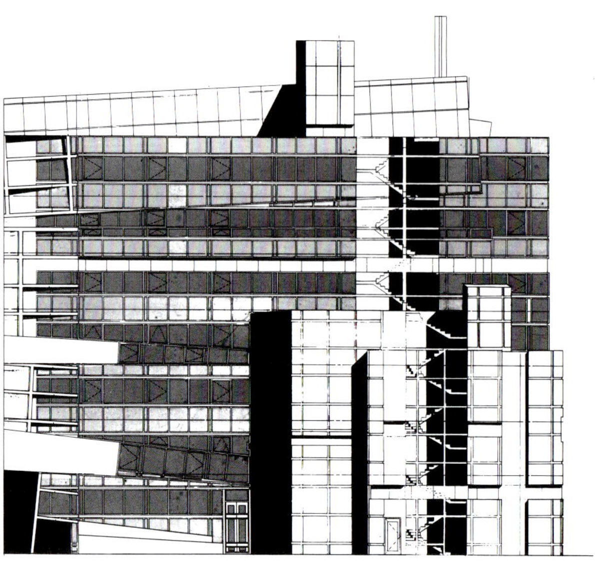

12

11 Presentation model, Atocha elevation
12 Courtyard elevation
13 Alameda (west) elevation
14 Atocha (south) elevation
15 Presentation model, view from the north-east
16 Presentation model, plan view
17 Presentation model, view from the north

12 庭院立面
13 阿尔梅达大街（西）的立面
14 阿托查大街（南）的立面
15 从东北方向看展示模型
16 从上方看展示模型
17 从北向看展示模型

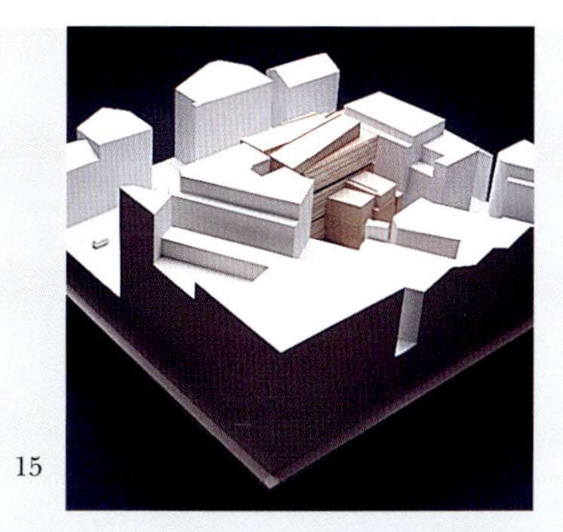
15

16

17

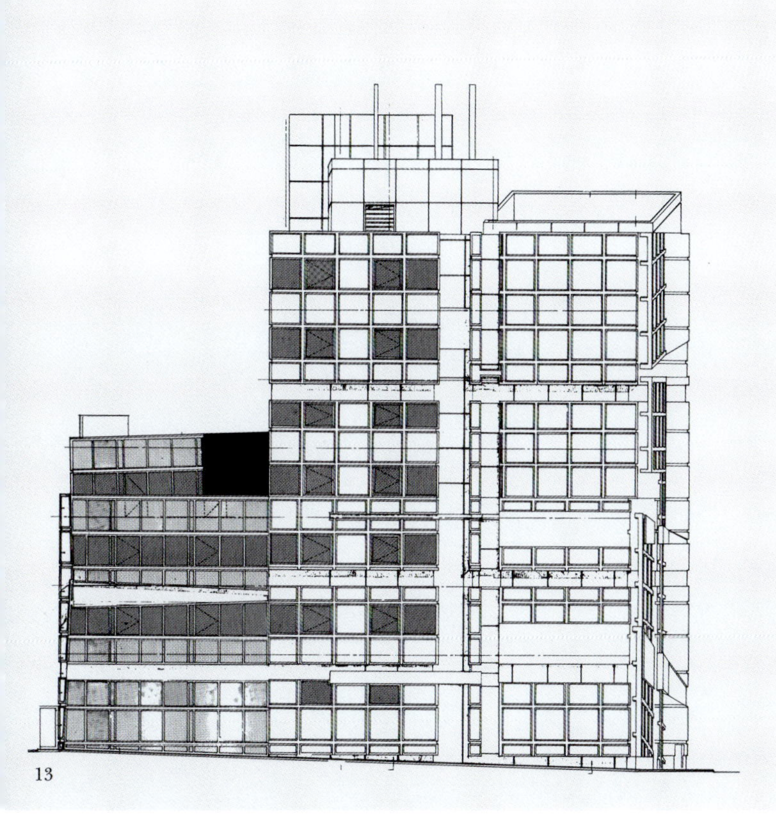
13

14

Rebstockpark Master Plan

Design 1990
Frankfurt, Germany
Advanta Management, AG/Dieter Bock
5,000,000 square feet

Rebstockpark 总体规划

设计　1990
法兰克福，德国
Advanta Management，AG/Dieter Bock 公司
5000000 平方英尺

The Rebstockpark Master Plan reassesses the idea of a static urbanism; the temporal dimension of the present becomes an important aspect of the past and the future. This reading might reveal other conditions which may have always been immanent in the urban fabric.

Framed by a segment of the Mercator grid, the Rebstockpark Master Plan floats within a rectilinear container to obscure the residual position it occupies along Frankfurt's third green belt. By compressing the large grid segment onto the site perimeter and similarly compressing the small-scale grid onto the close site, contingent readings emerge as the two site figures fold and unfold, each relative to its expanded position.

Rebstockpark 总体规划再次审视了静态的城市主义，对现状的临时衡量尺度成为了看待历史和未来的重要视角。这一事务或许将揭示城市机理中的其他固有的条件。

由一段墨卡托网格构架，Rebstockpark 规划飘浮于直线排列的集装箱之间，从而弱化了它的居住区的位置，并沿着法兰克福第三绿色地带。通过压缩大的网格片断到该场地中，同样将小比例的网格压缩到相近的场地，同样的事物——作为两个场地特征的折叠与展开这一形式出现了，每一个场地都与它另一个场地有着相互关联。

1

1 Site plan
2 Concept diagram, superposition of net
3 Concept diagram, transformation of net
4 Concept diagram, folded net
5 Concept diagram, typological fabric
6 Concept diagram, building typology
7 Concept diagram, folded typology

1 总平面
2 概念示意图，网络的重叠
3 概念示意图，网络的变形
4 概念示意图，折叠起来的网络
5 概念示意图，典型机理
6 概念示意图，建筑类型
7 概念示意图，折叠的类型

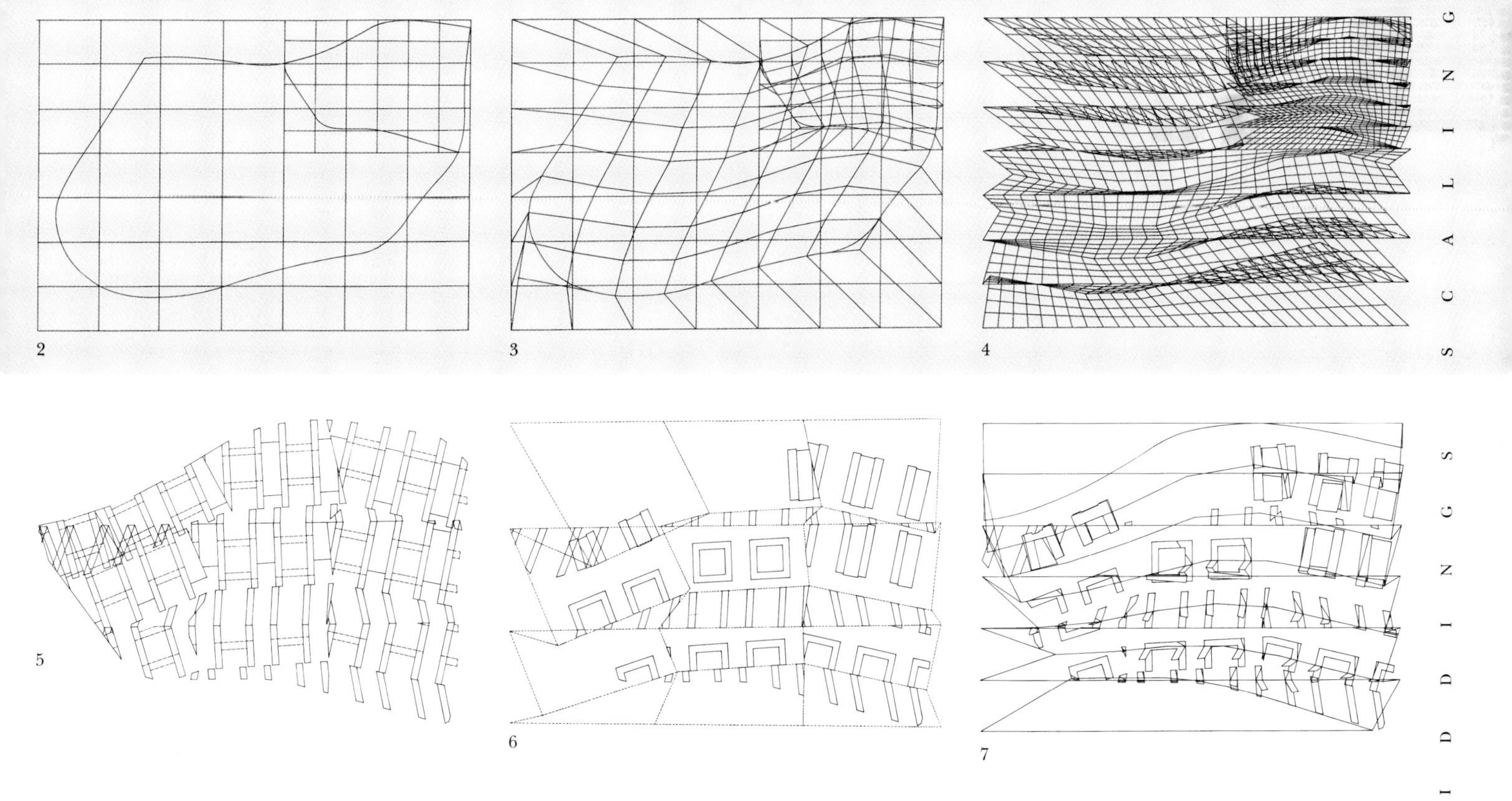

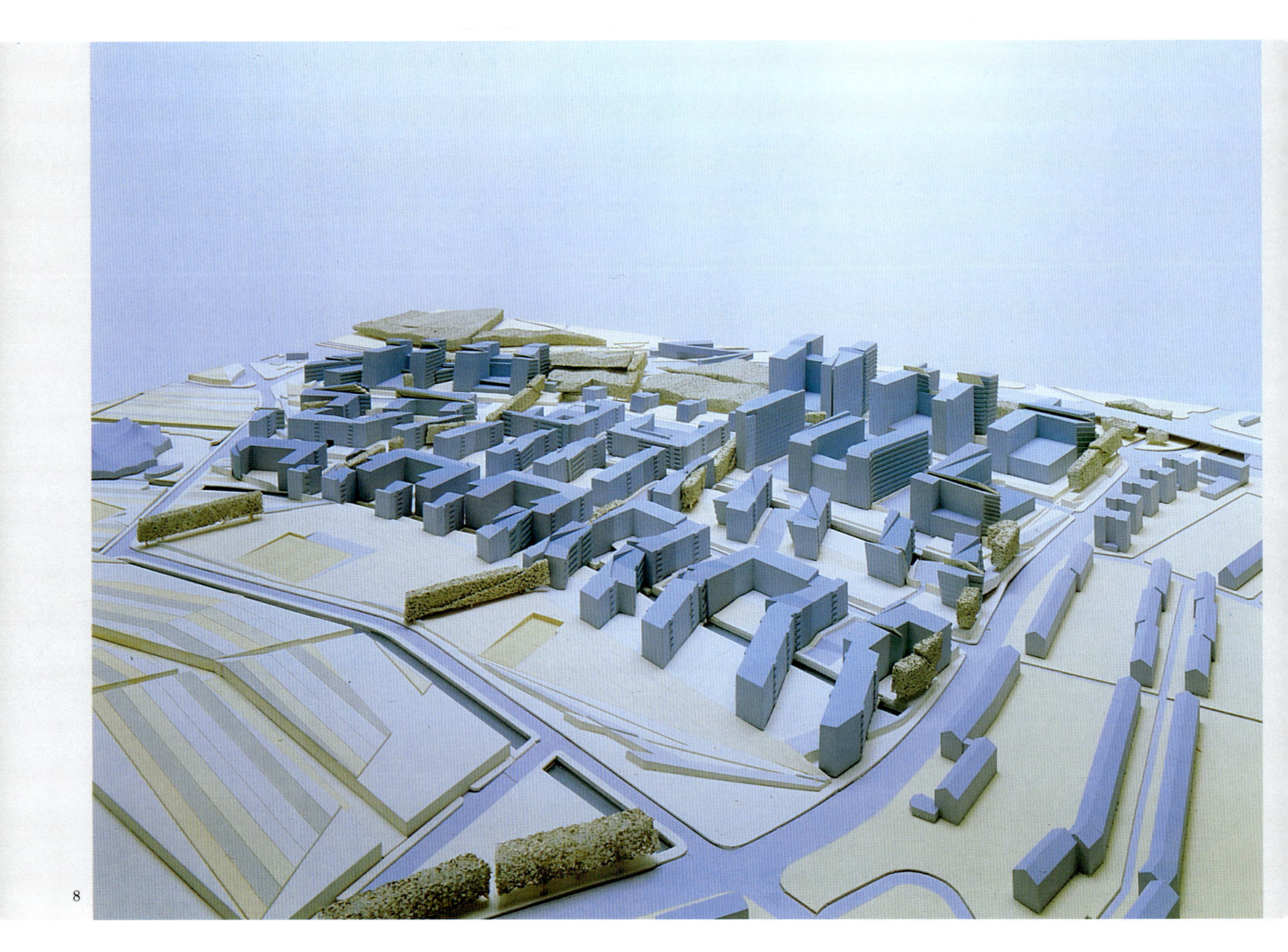

8

9

8 Presentation model, view from the south-east
9 Concept diagram, folded wire frame
10 Site plan with large-fold net overlay
11 Building plans with large-fold net overlay
12–13 Site perspectives

8 从西南方向看展示模型
9 概念设计示意图，重叠在一起的网格
10 大比例的重叠网格平面
11 重叠网格的建筑平面
12-13 场地透视

12

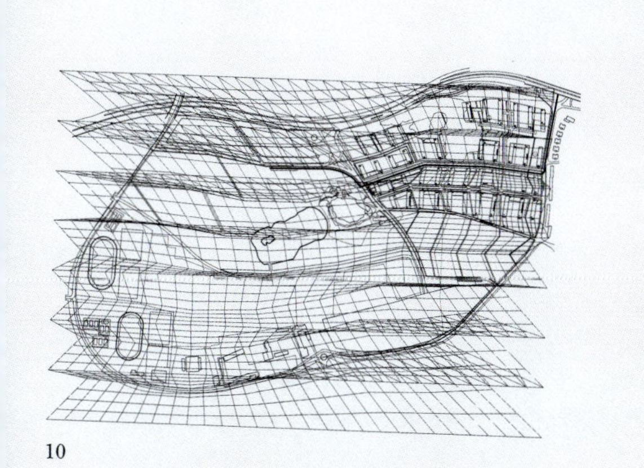

10

13

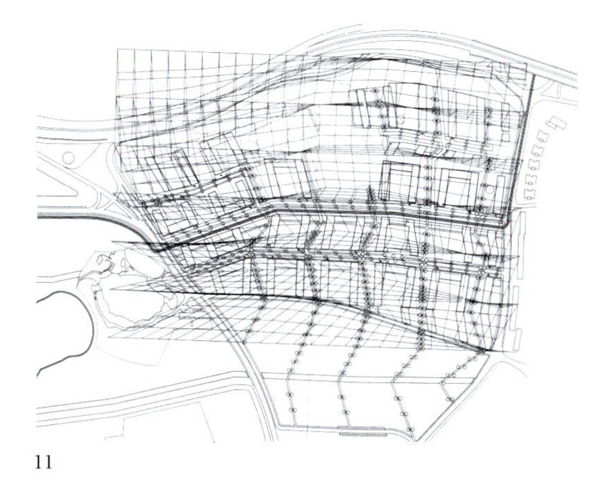

11

14

14 Presentation model, view from the east
15 Technical site plan with building footprint
16 Site plan with base and deformed grid
17–19 Diagrammatic building model

14 从东向看展示模型
15 建筑总平面定位
16 带有基础和变形网格的总平面
17–19 模型示意

17

18

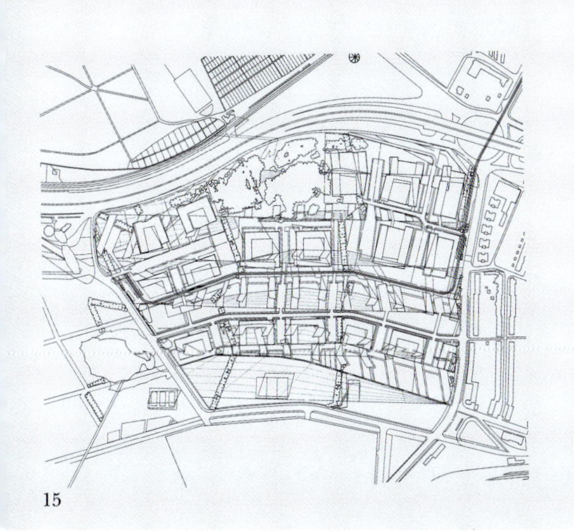

15

19

16

Alteka Office Building

Design 1991
Tokyo, Japan
Alteka Corporation
30,000 square feet

Alteka 办公楼

设计　1990/1992
东京，日本
Alteka 公司
30000 平方英尺

A paradigmatic city of accumulation, juxtaposition, and compression, Tokyo is an index of contingent, tentative relations and new, complex urban realities. Our project suggests another relationship to the city. Caught between the traditional city fabric and the Jigamae, the site suggests a building defined by fluctuation, where the object takes place in a continuum of variation. Thus, the building does not correspond to a spatial mold, but to a temporal modulation that implies a continual variation of the matter and a perpetual development of the form.
The typological el frees its folds from their subordination to the finite body, emerging from the context to fold and unfold.

作为一座人口聚集、周边城市毗邻、拥挤的典型城市，东京是偶然的、临时的关系和新的复杂的城市实体之标志。我们的项目却暗示了建筑与城市的另一种关系。场地受限于其所处环境中的传统城市肌理与 Jigamae 之间，从而创建了一栋蕴育着波动的建筑，在该建筑中，事物在一系列连续的波动中发生。这样，建筑就不再与空间模式相呼应，而是与一个临时的调节相呼应，这一临时调节暗示了事物的连续变化和形式的不断发展。该类型的建筑将它的折叠从附属的位置释放到具体的实体中，这一实体从折叠与伸展的文脉中产生。

1

1 Presentation model, view from the south-east
2 Concept diagram, infolding section
3 Concept diagram, infolding section
4 Concept diagram, unfolding section
5 Concept diagram, unfolding plan
6 Concept diagram, envelope plan
7 Concept diagram, envelope plan

1 从西南方向看展示模型
2 概念示意图，折叠的剖面
3 概念示意图，折叠的剖面
4 概念示意图，打开的剖面
5 概念示意图，打开的平面
6 概念示意图，包围的平面
7 概念示意图，包围的平面

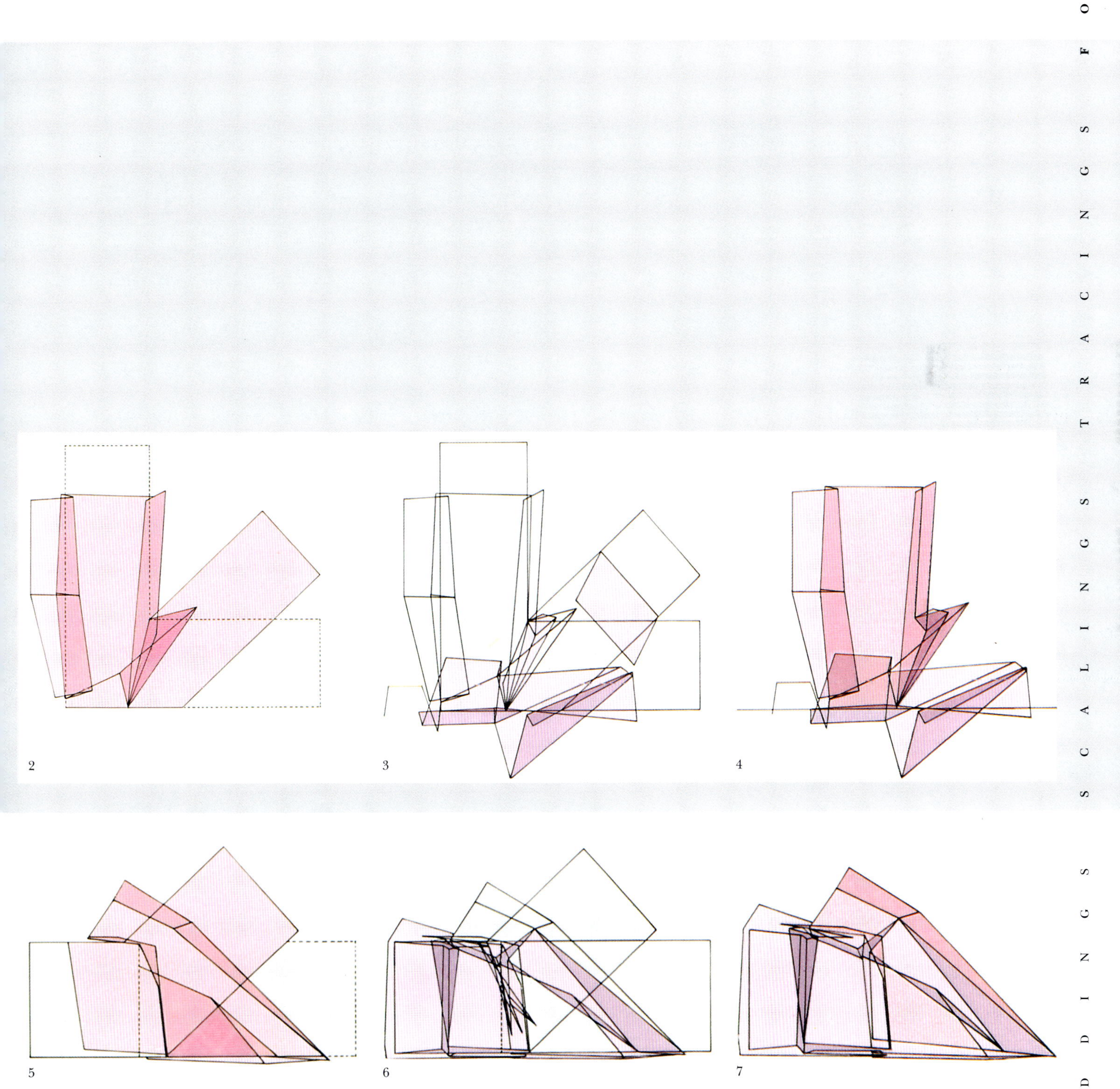

8

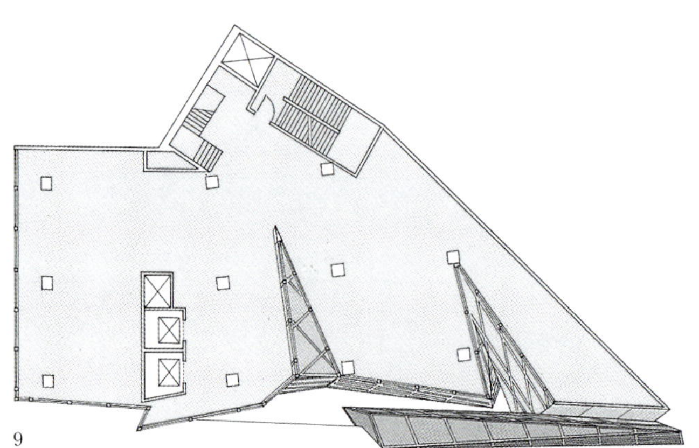

9

8 Perspective, view from the south-east
9 Second level plan
10 Fourth level plan
11 Eighth–tenth level plan
12 Presentation model, view from the north
13 Presentation model, view from the west

8 从东南方向看透视图
9 三层平面
10 五层平面
11 九至十一层平面
12 从北看展示模型
13 从西看展示模型

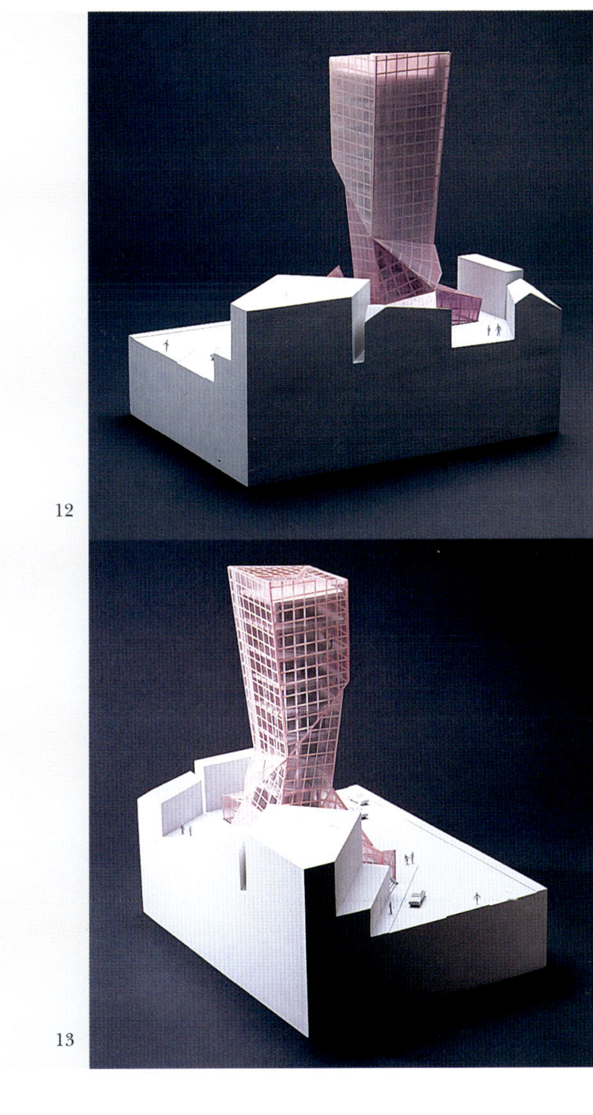

12

13

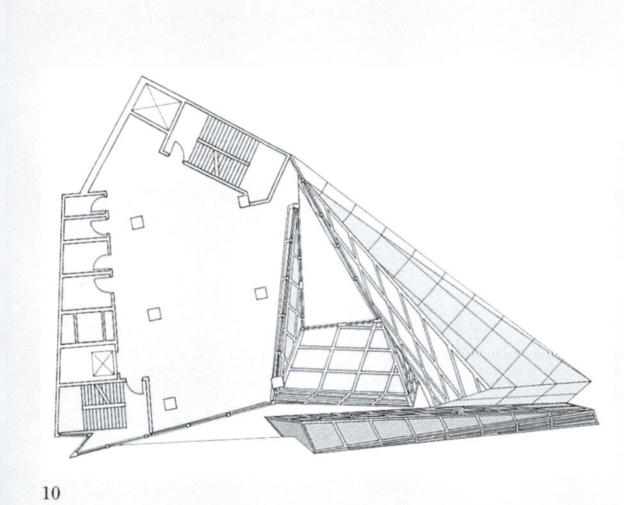

10

11

Emory Center for the Arts

Design 1991
Atlanta, Georgia
Emory University
160,000 square feet

埃默里艺术中心

设计　　1991
亚特兰大，佐治亚州
埃默里大学
160000 平方英尺

The Center for the Arts at Emory University accommodates four major performance spaces (a 1,100-seat music hall, and a recital hall, studio theater, and cinema each seating 200), and is designed to be a national and international center for scholarship and performance in the fields of theater, music, and film.

The four performance halls are linked by an expansive, multi-level lobby traversing the length of the building and functioning as a link between the campus boundary and a new open-air amphitheater. Academic spaces are located to the east of the lobby over the parking garage, and rehearsal and support space is provided adjacent to the performance halls.

埃默里大学的艺术中心包括四个主要的表演空间（一个 1100 人的音乐厅，以及一个独奏厅、一个演播室剧场、一个电影院，每个都是 200 座位），它被设计成一个国家甚至是国际学者、各种艺术表演家的中心。

四个表演厅由一个开敞的、多层的大厅连接，该大厅横穿建筑，起到连接校园边界与一个室外剧场的功能。学术区在大厅的东边，位于停车场上方，同时在表演厅周围还有排练区和附属用房。

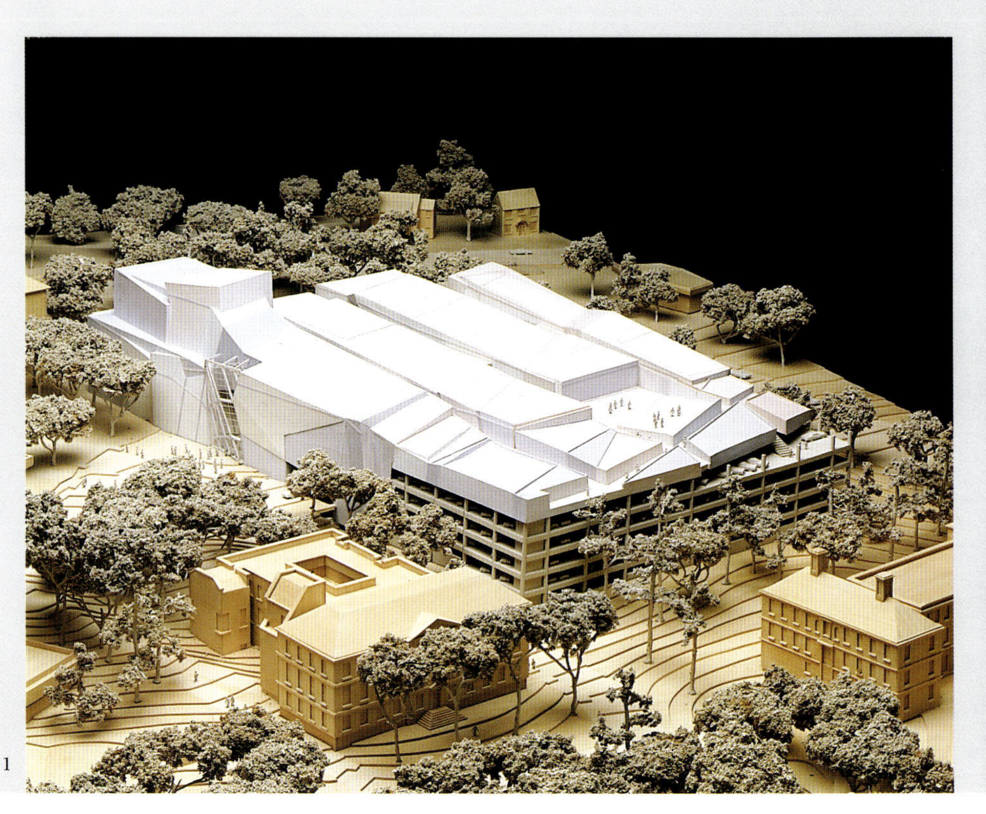

1

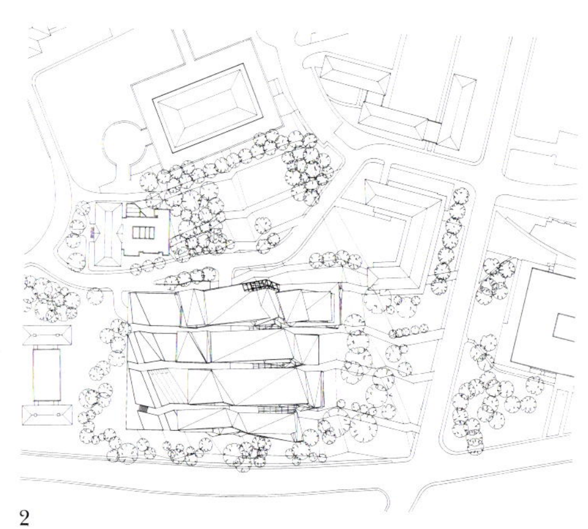

2

1 Presentation model, view from the north-west
2 Site plan
3 Concept diagram
4 945 level plan
5 977 level plan
6 964 level plan

1 从西北方向看展示模型
2 总平面
3 概念示意图
4 945 标高平面
5 977 标高平面
6 964 标高平面

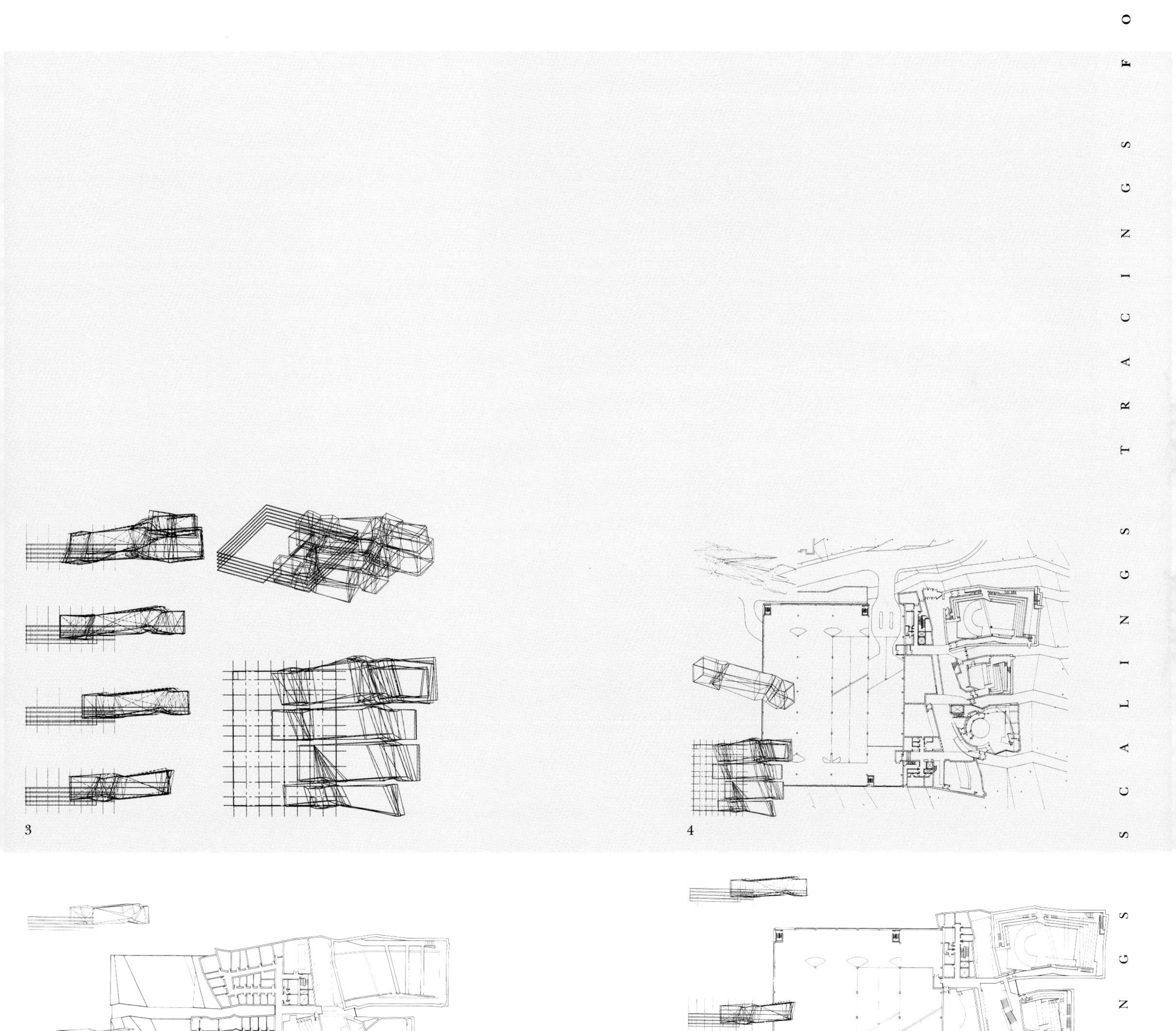

3

4

5

6

7

7 Presentation model, view from the east
8 Presentation model, view from the south-east
9 Presentation model, view from the north-east
10 Concept diagrams

7 从东看展示模型
8 从东南方向看展示模型
9 从东北方向看展示模型
10 概念示意图

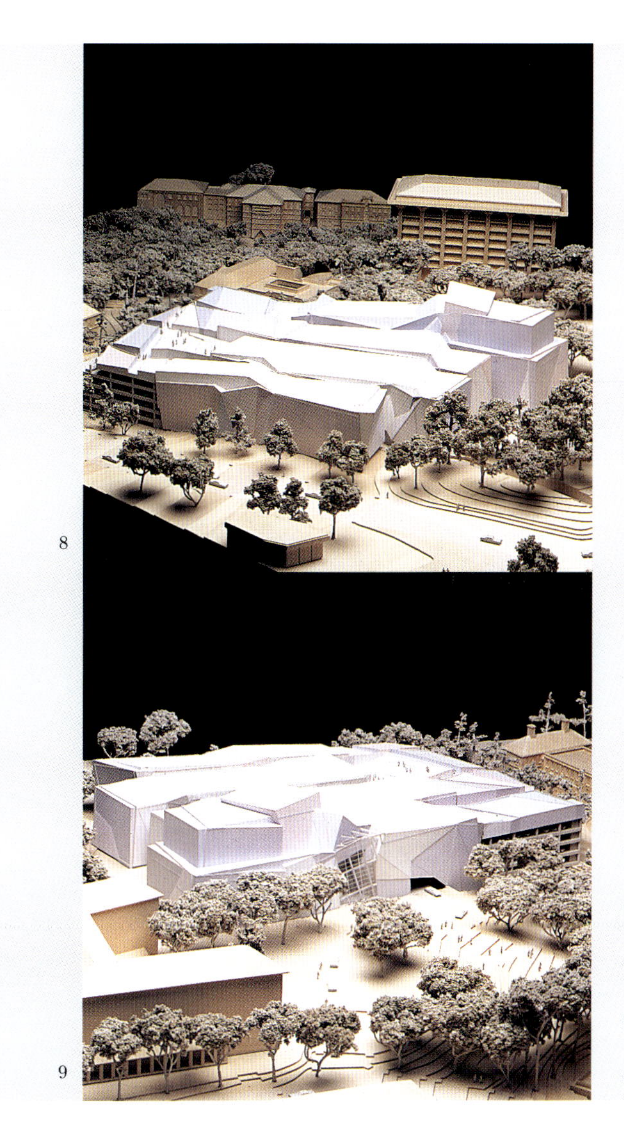

8

9

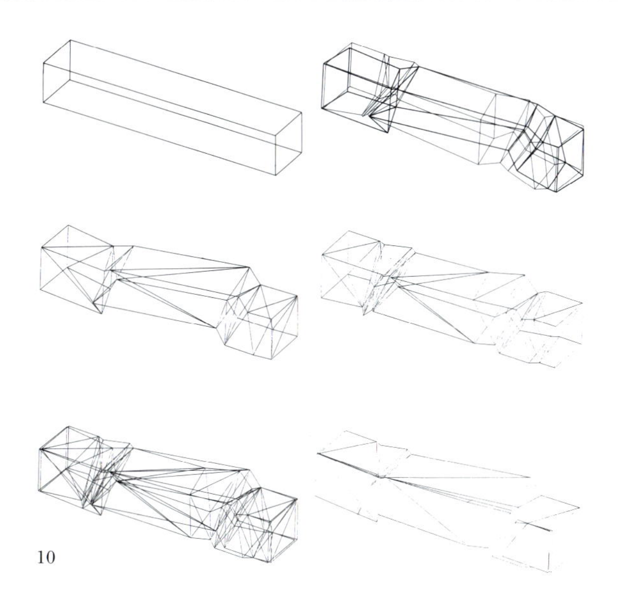

10

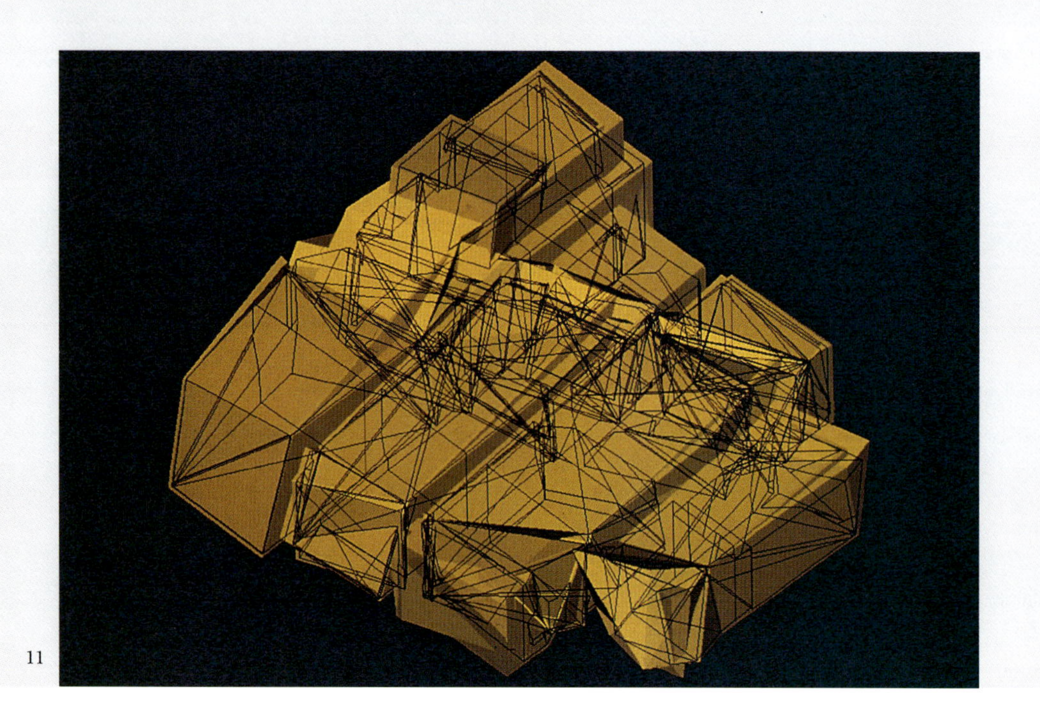

11

11 Computer-generated study model
12 Bar X section, view from the south
13 Bar W section, view from the south
14 Lobby section, view from the west
15 Performance spaces section, view from the west
16 Bar W study model
17 Computer-generated Bar W study model

11 计算机绘制的研究模型
12 从南看酒吧 X 剖面
13 从南看酒吧 W 剖面
14 从西看大厅剖面
15 从西看表演空间剖面
16 酒吧 W 研究模型
17 计算机绘制的酒吧 W 研究模型

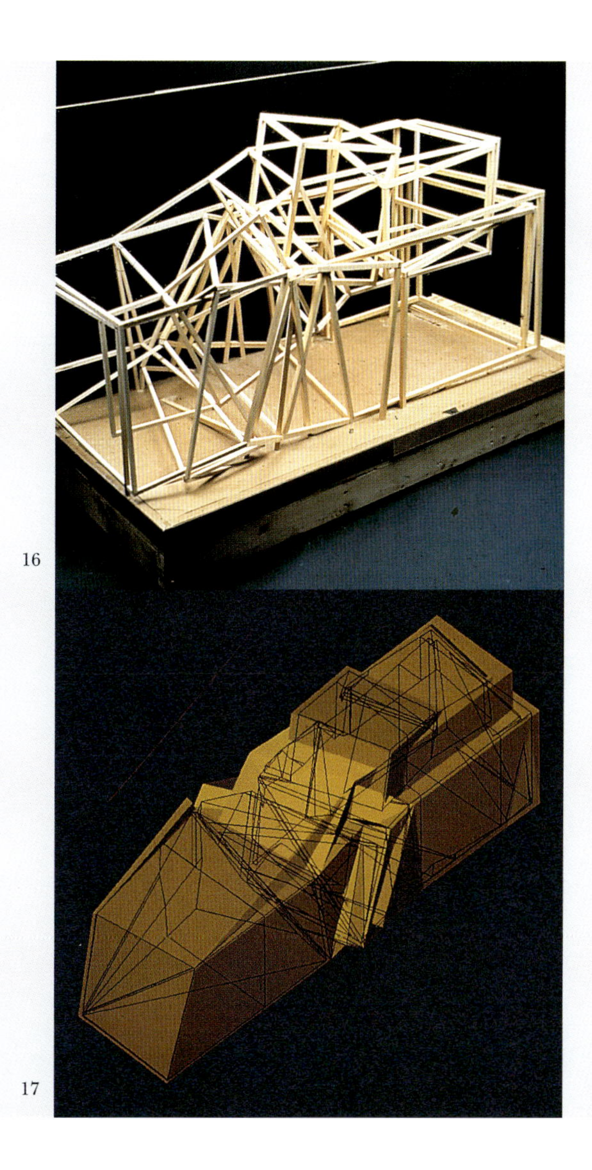

16

17

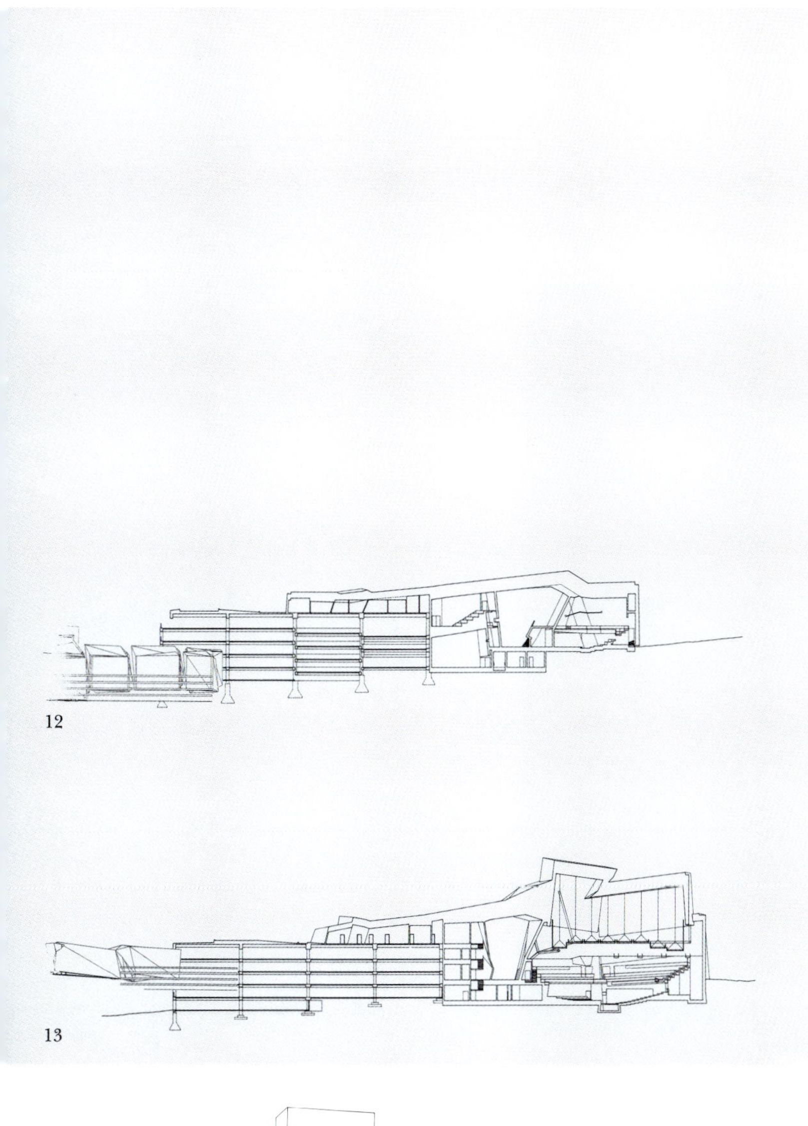

12

13

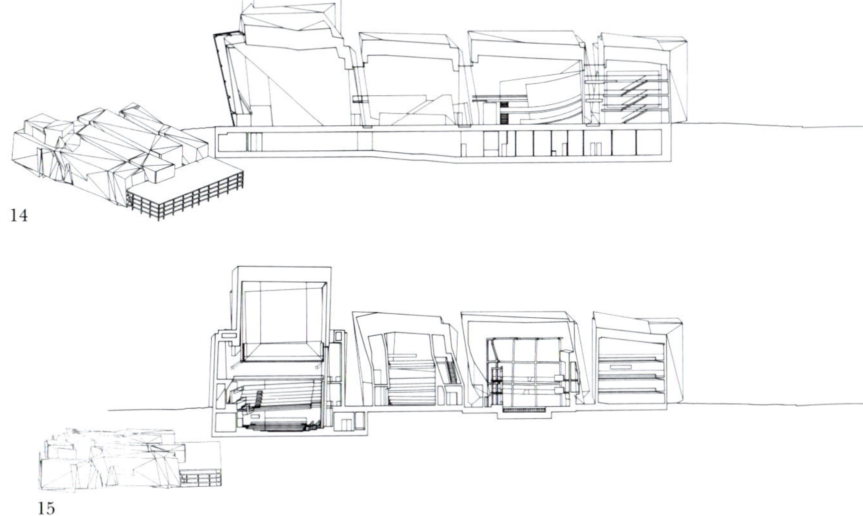

14

15

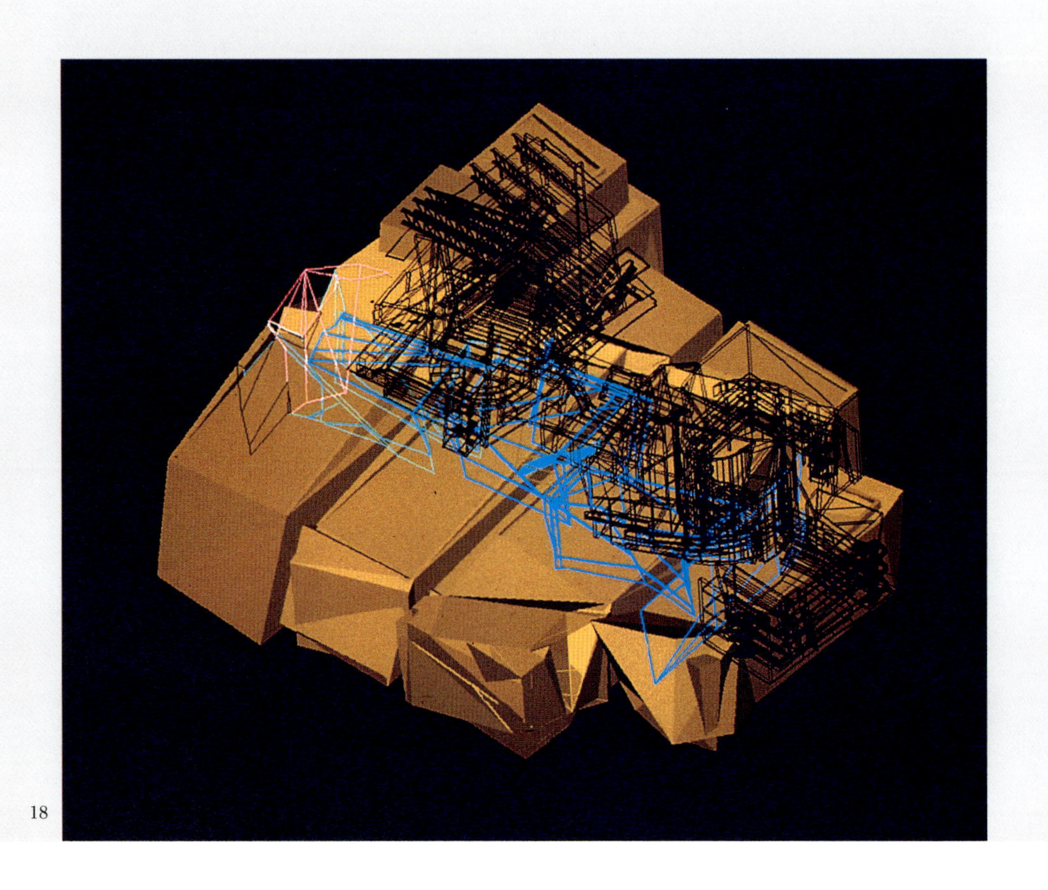

18

18 Computer-generated study model
19 Theater section, view from the south
20 Cinema section, view from the south
21 Isometric of music hall structural framing
22 Computer-generated study models

18 计算机绘制的研究模型
19 从南看剧场剖面
20 从南看电影院剖面
21 音乐大厅的桁架结构
22 计算机绘制的研究模型

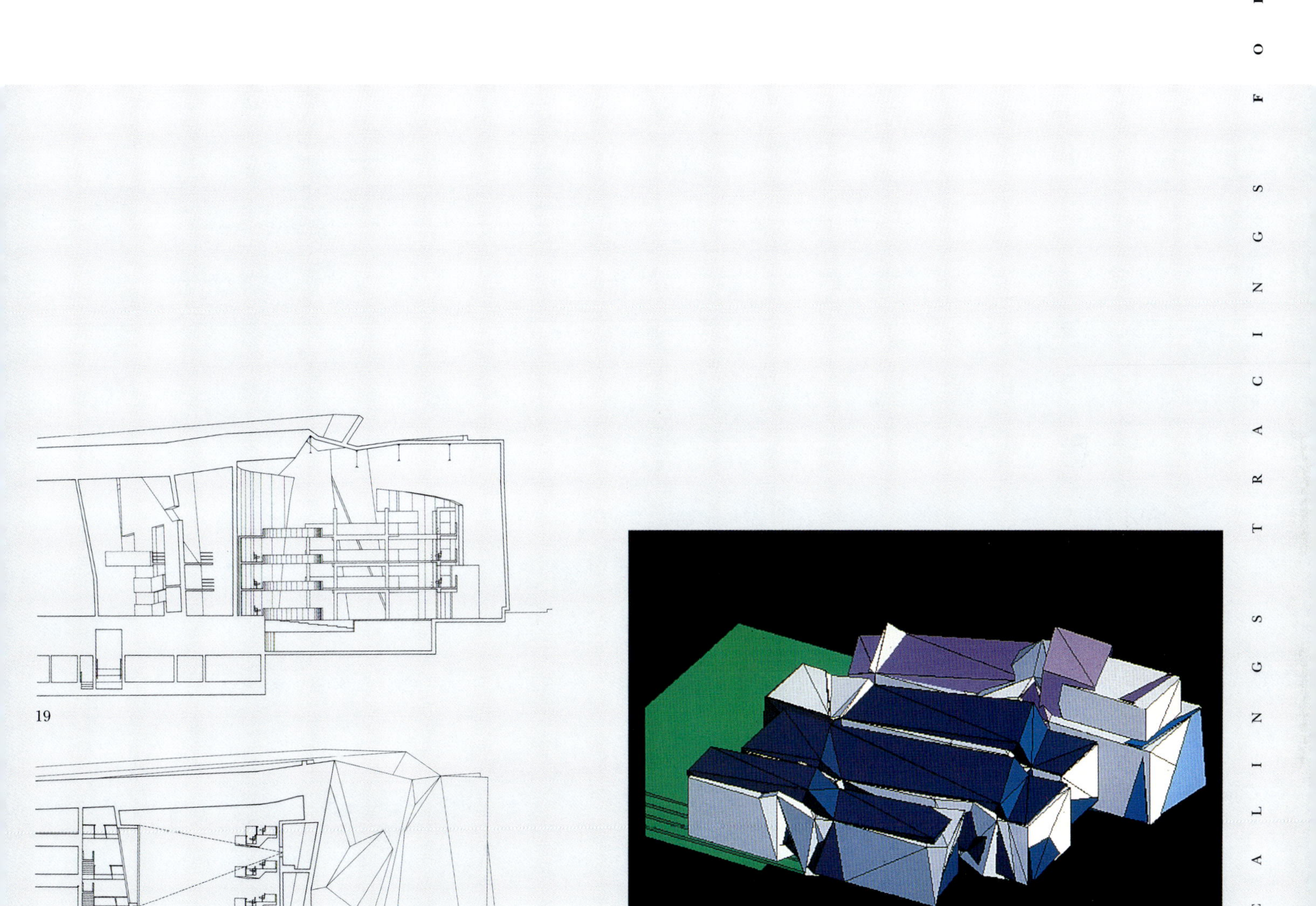
19

20

22

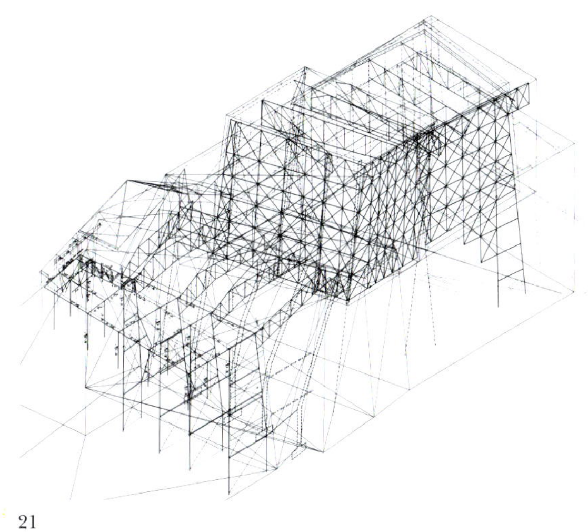
21

Max Reinhardt Haus

Design 1992
Berlin, Germany
Advanta Management, AG/Dieter Bock
1,000,000 square feet

Max 赖恩哈茨住宅

设计　　1992
柏林，德国
Advanta Management，AG/Dieter Bock 公司
1000000 平方英尺

The dominant character of the Max Reinhardt Haus is both symbolic and recreational. Named for the famous German theatrical entrepreneur, it occupies the site of his former *schauspielhaus.* Its symbolism is intended to be forward rather than backward-looking, combining the best of what is German with a symbolic vision of the future. Its program is representative of Reinhardt's energy and vision: a present-day media center. Almost by definition, the building has to assume a "prismatic" character; that is to say, it needs to fold into itself—but also open itself out to—an infinite, always fragmentary, and constantly changing array of metropolitan references and relationships.

Max 赖恩哈茨住宅的突出特征就是兼有象征性和娱乐性。由德国著名戏剧企业家得名，Max 赖恩哈茨住宅占有的场地从前是舒奥斯佩尔住宅，它的象征性试图展望未来而不是回顾历史，同时采纳德国精华之带有对未来展望的象征派眼光。它的设计意图是重现赖恩哈茨的活力和景象：一个当代的媒体中心。近乎是顾名思义，该建筑必须承担一个“三棱镜”的特征；也就是说它需要将自己混入——有时又要跳出——一个无限的、经常是断断续续并经常变幻的阵列中，大都市内及其相互关系中的阵列。

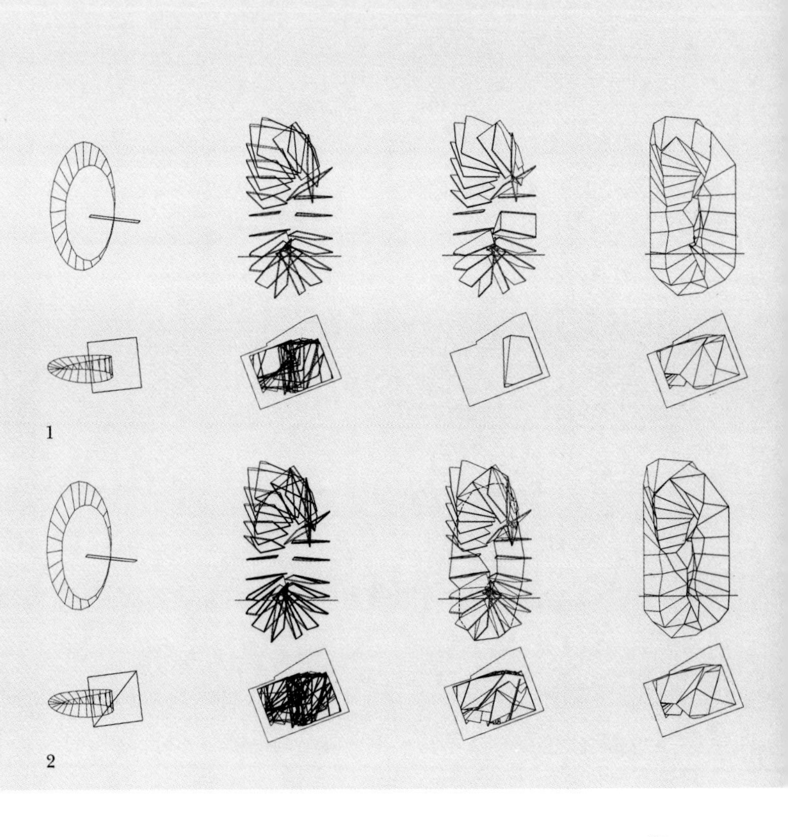

1

2

3

1–3 Concept diagrams
4 Presentation rendering, view from the west

1－3 概念示意图
4 从西看渲染图

4

5

6

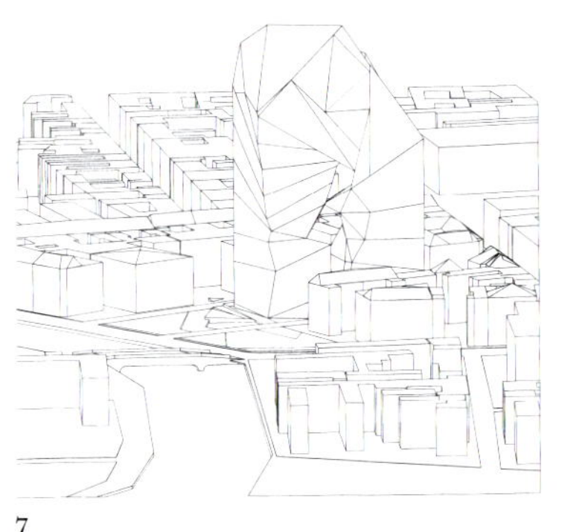

7

5 Presentation rendering, view from the east
6 Site plan
7 Axonometric, view from the east
8 Ground level plan
9 Second basement level plan
10 Section A
11 East elevation

5 从东看渲染图
6 总平面
7 从东看轴测图
8 首层平面
9 地下二层平面
10 剖面 A
11 东立面

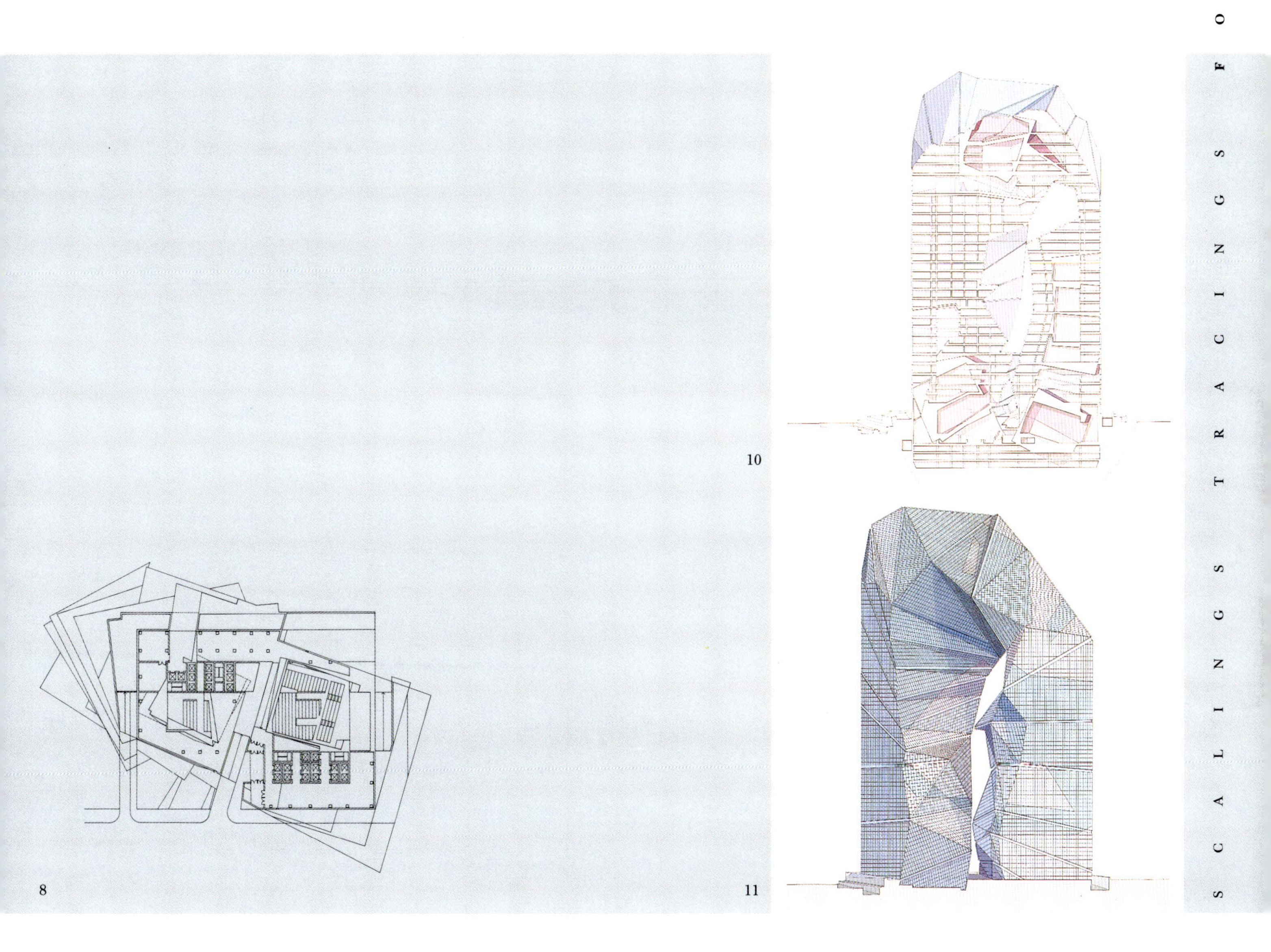

10

8

11

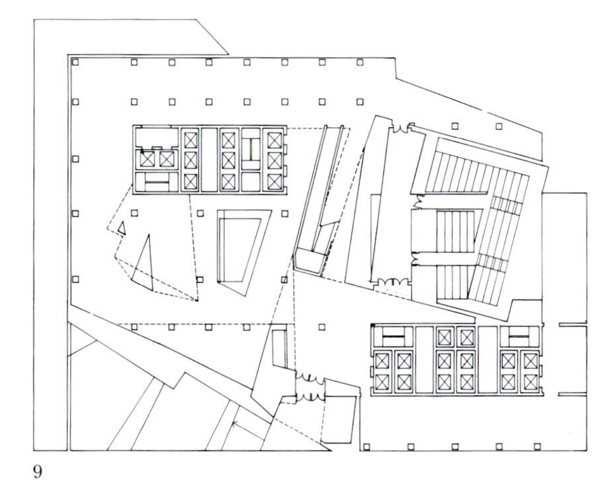

9

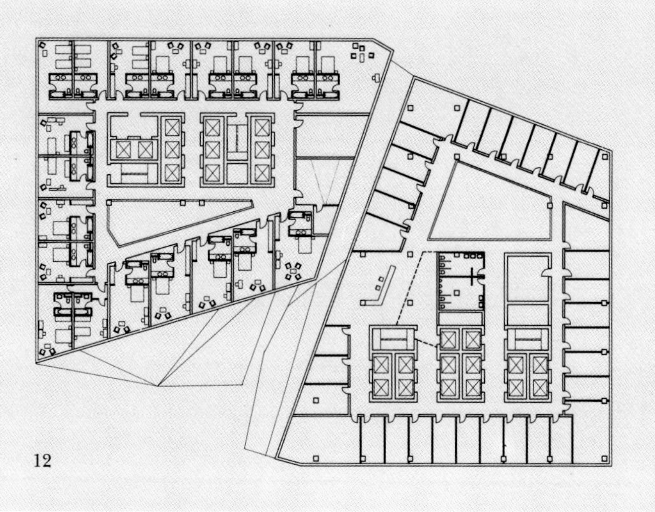
12

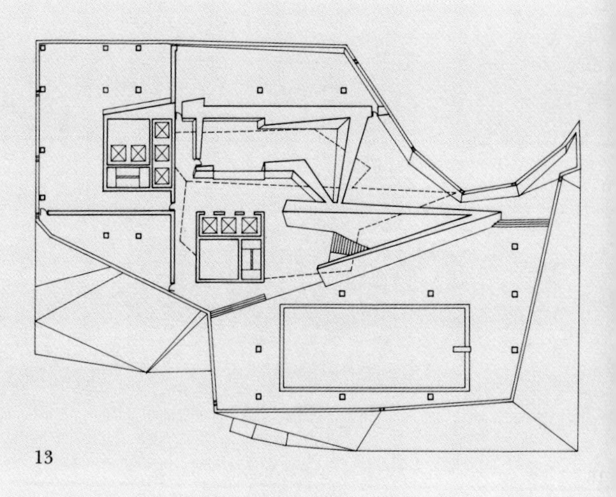
13

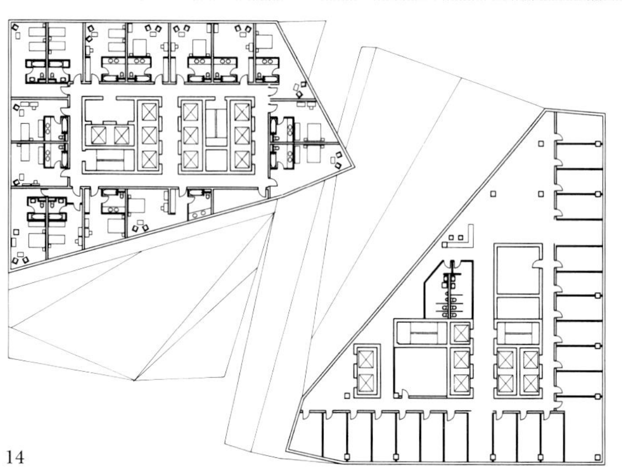
14

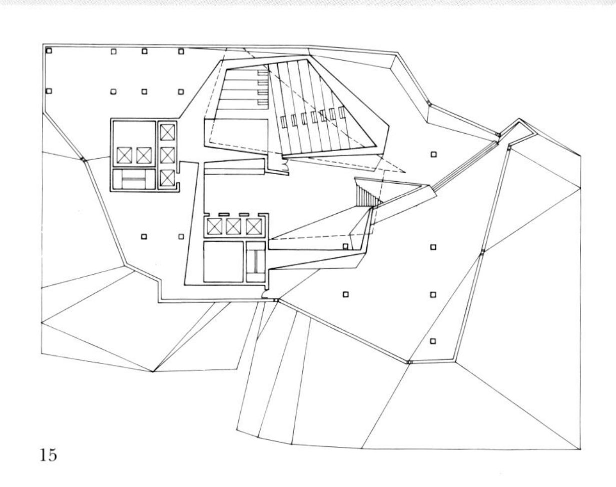
15

12 Eighth level plan
13 Twenty-sixth level plan
14 Thirteenth level plan
15 Twenty-ninth level plan
16 Site model, view from the west
17 Section B

12 第九层平面
13 第二十七层平面
14 第十四层平面
15 第三十层平面
16 从东看场地模型
17 剖面 B

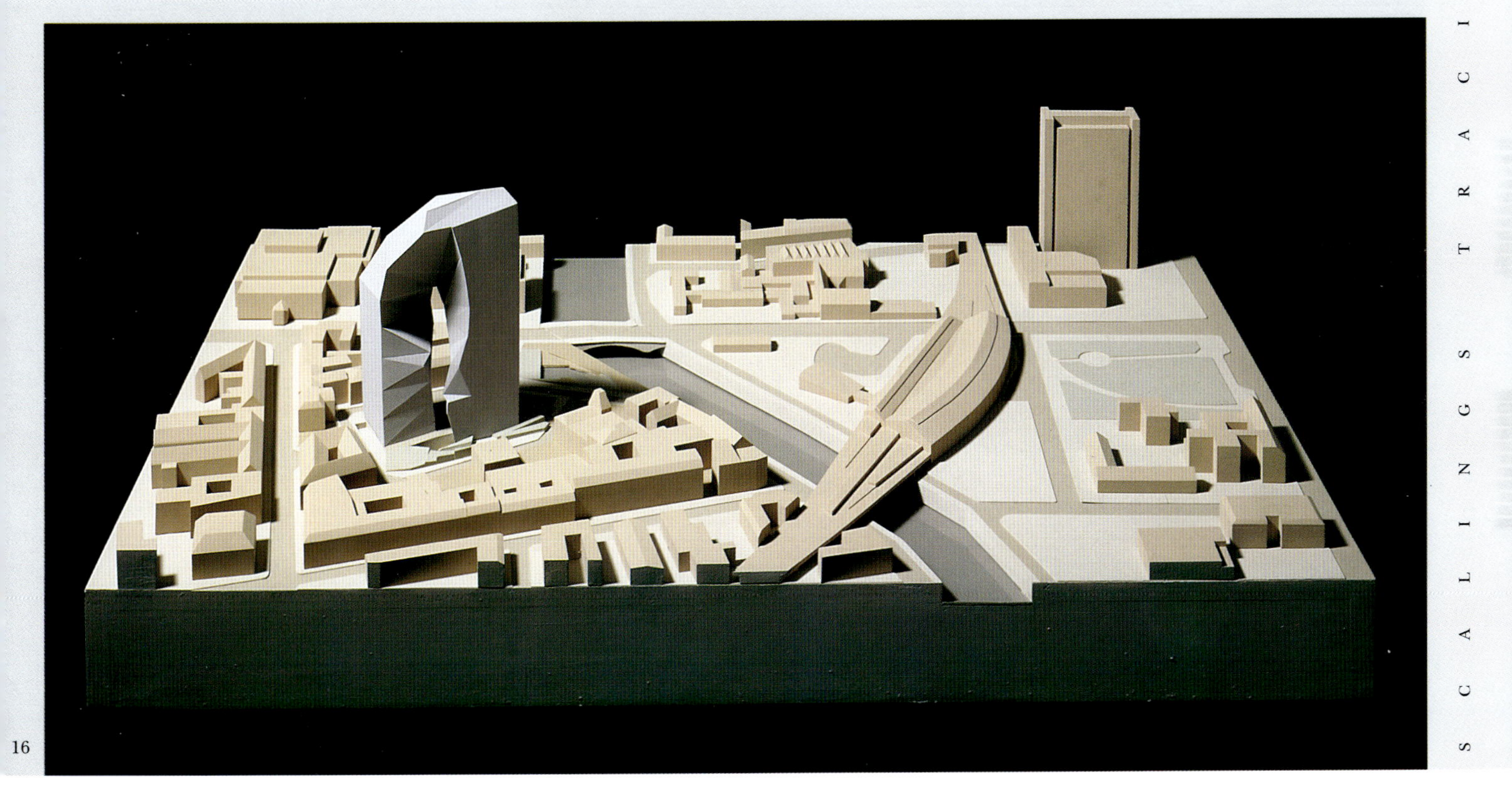

16

17

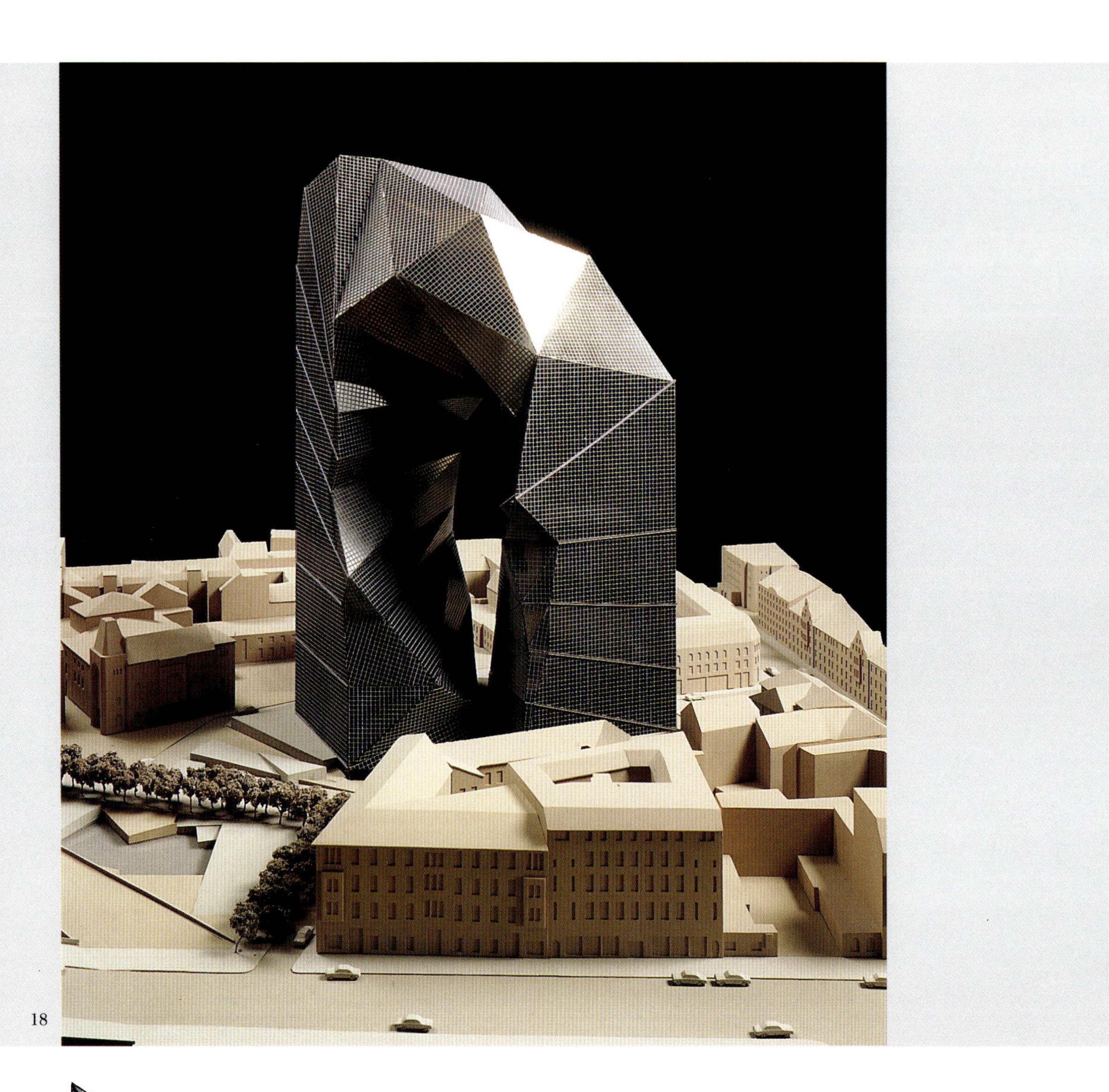

18

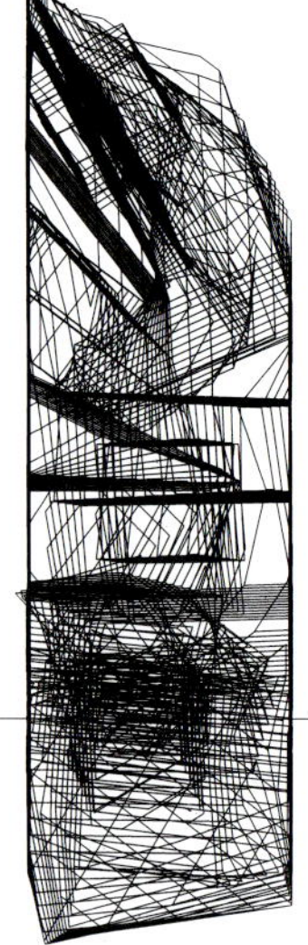

19

18	Presentation model, view from the east	18	从东看展示模型
19	Wire frame diagram	19	网格设计示意图
20–21	Wire frame diagrams	20–21	网格设计示意图
22	Presentation model, view from the west	22	从西看展示模型

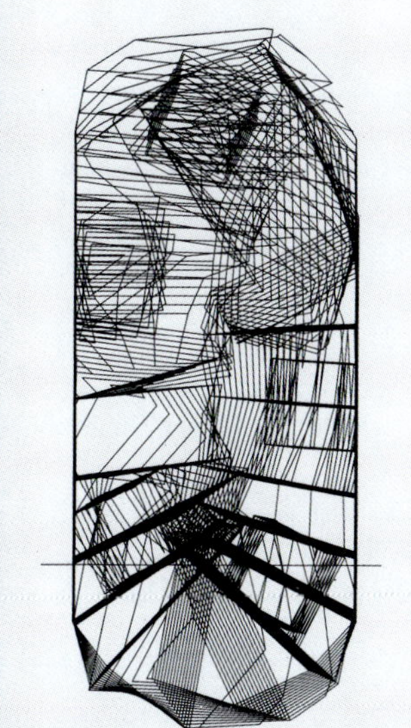

20

22

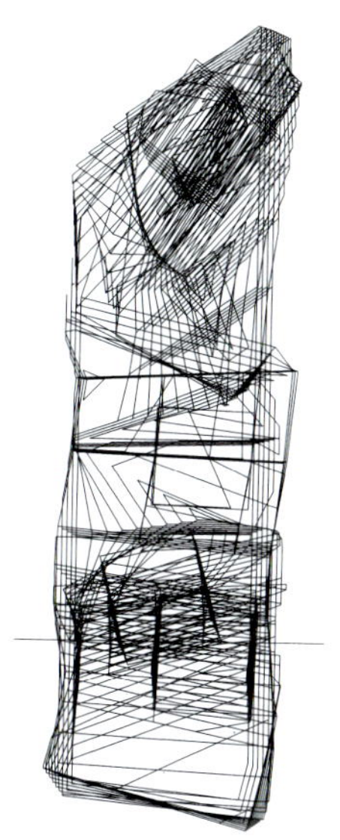

21

Nordliches Derendorf Master Plan

Design 1992
Dusseldorf, Germany
City of Dusseldorf Planning Department
5,800,000 square feet

诺德里奇·德伦多夫总体规划

设计　　1992
杜塞尔多夫，德国
杜塞尔多夫城市规划委员会
5800000平方英尺

Our proposal recognizes the fact that we are living in an electronic era, which has replaced the mechanical one. In the movement from the era of utility to the era of information, electronic information systems become one of the new limitations to urban growth.

In Dusseldorf, one of these new limits is the system of radar and radio. The proximity of the airport's flight path causes certain height restrictions to be mapped onto this project. This mapping derives from the intersection of the radar and radio patterns, which produces an interference pattern that becomes the form-generator on the site.

我们认识到这样的一个事实，也就是我们生活在电子时代，电子时代已经取代了机械工业时代。从效用时代到信息时代的转换中，电子信息系统已经成为城市发展的限制因素之一。

在杜塞尔多夫地区，新的限制因素之一就是雷达和无线电波系统。飞机飞行路径的接近引起了对其领空下辐射到的该项目的高度的限制。辐射来源于雷达和无线电波模式的介入，它们的介入导致了一种干涉模式，从而也就产生了在该场地中的建筑形式。

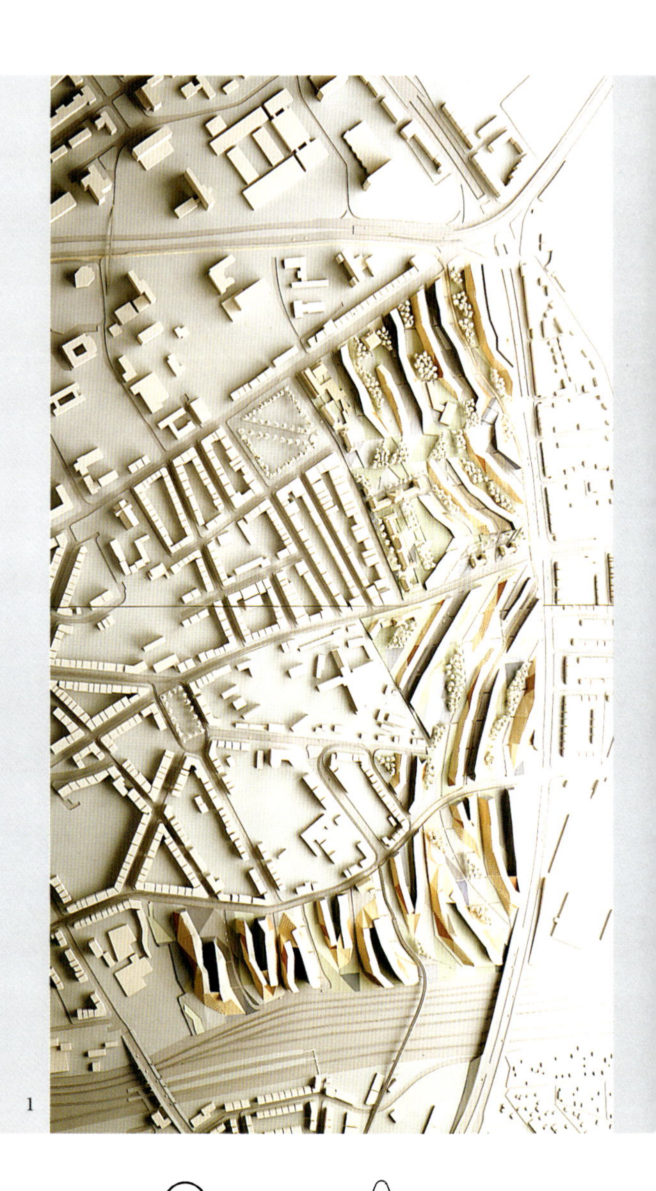

1

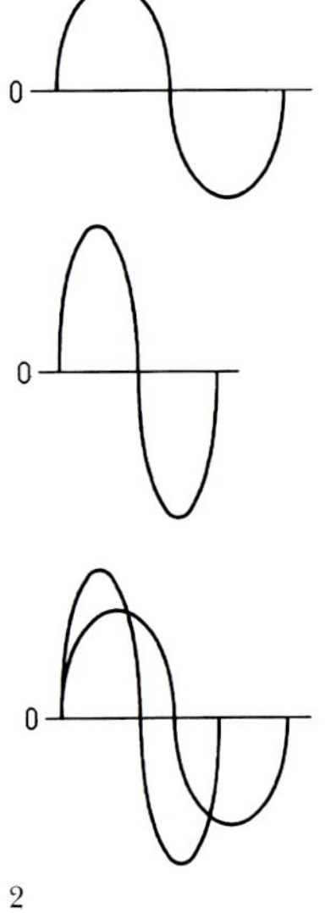

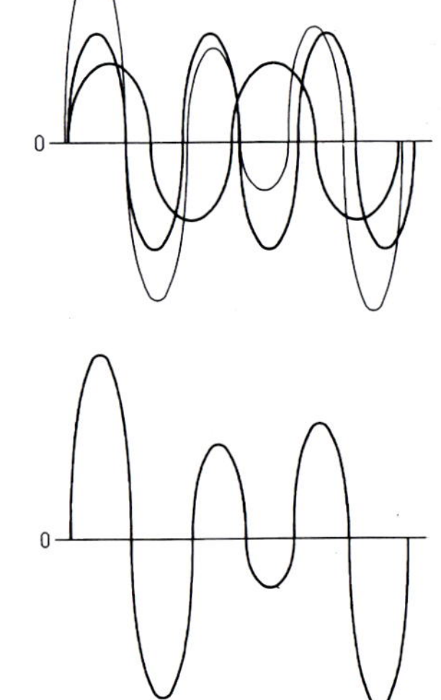

2

1 Presentation model, aerial view
2 Concept diagram, wave and interference
3 Concept diagram, vertical topographical section
4 Concept diagram, interference
5 Concept diagram, superposition of radar

1 展示模型，鸟瞰
2 概念示意图，波及干扰
3 概念示意图，地形垂直剖面
4 概念示意图，干扰
5 概念示意图，雷达信息的重叠

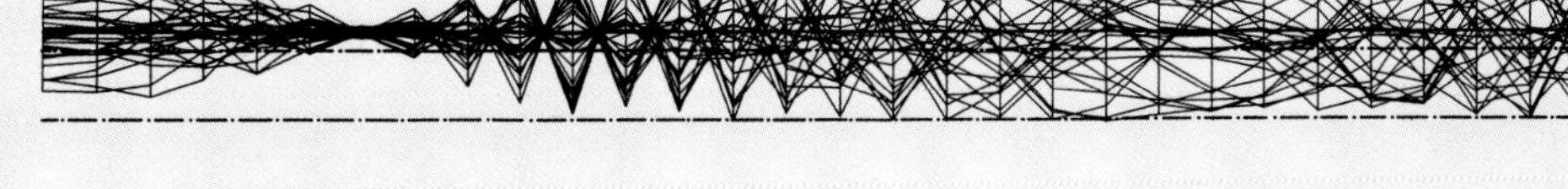

3

4

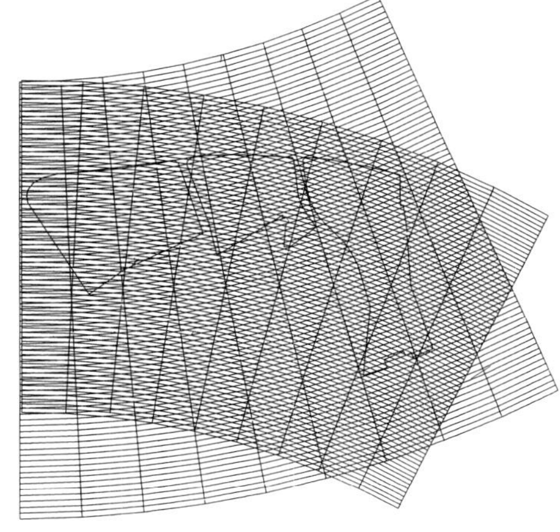

5

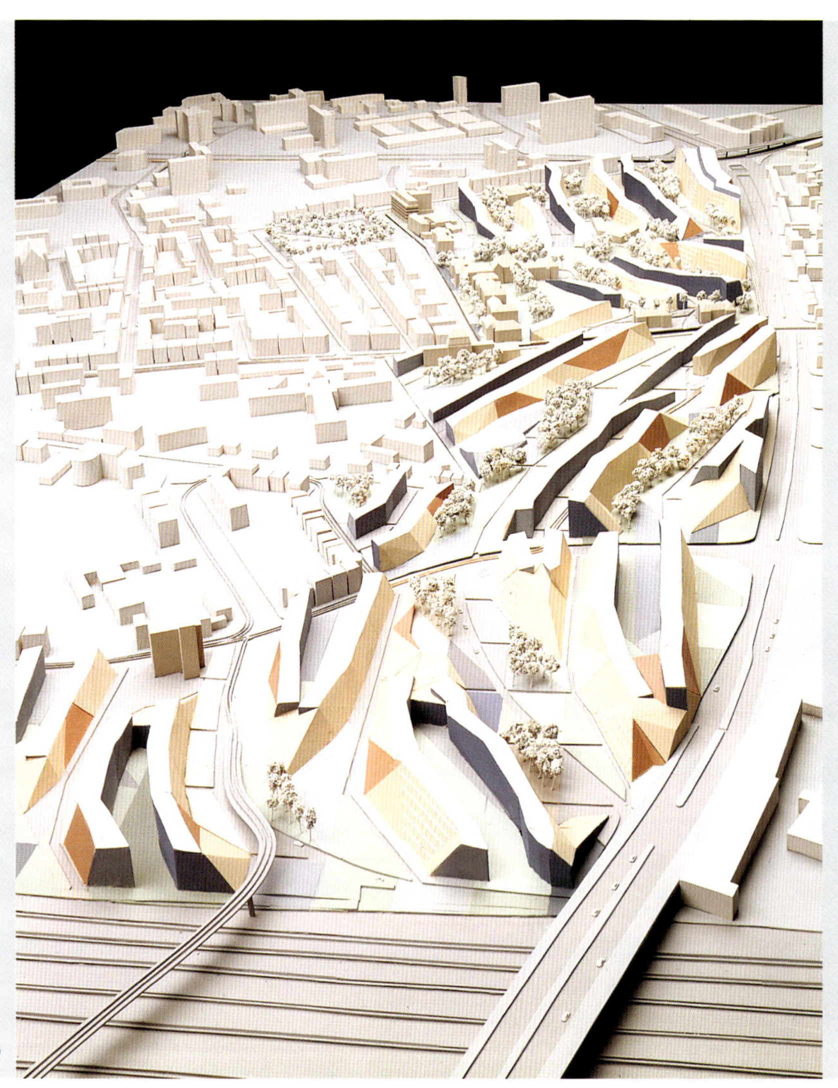
6

7

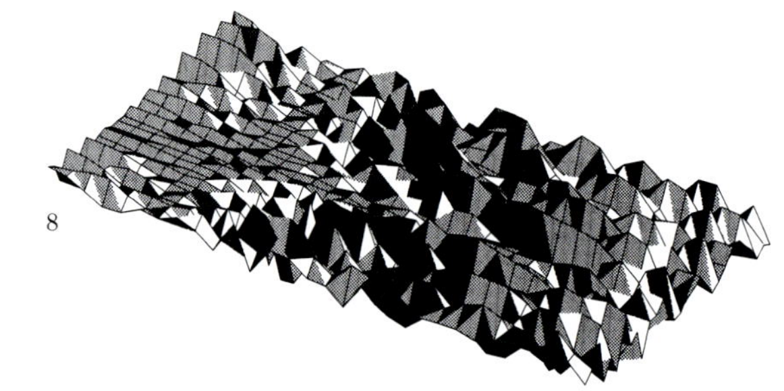
8

6 Presentation model, view from the east
7 Concept diagram, interference field
8 Concept diagram, topological interference
9 Concept diagram, wave formation
10 Concept diagram, isometric of interference
11 Concept diagram, overlap of wave and interference
12 Site plan
13 Presentation model, view from the east
14 Presentation model, view from the south

6 展示模型，从正东方向看
7 概念示意图，干扰区域
8 概念示意图，地形干扰
9 概念示意图，波的构成
10 概念示意图，等容干扰
11 概念示意图，波及干扰的重叠
12 总平面
13 从东看展示模型
14 从南看展示模型

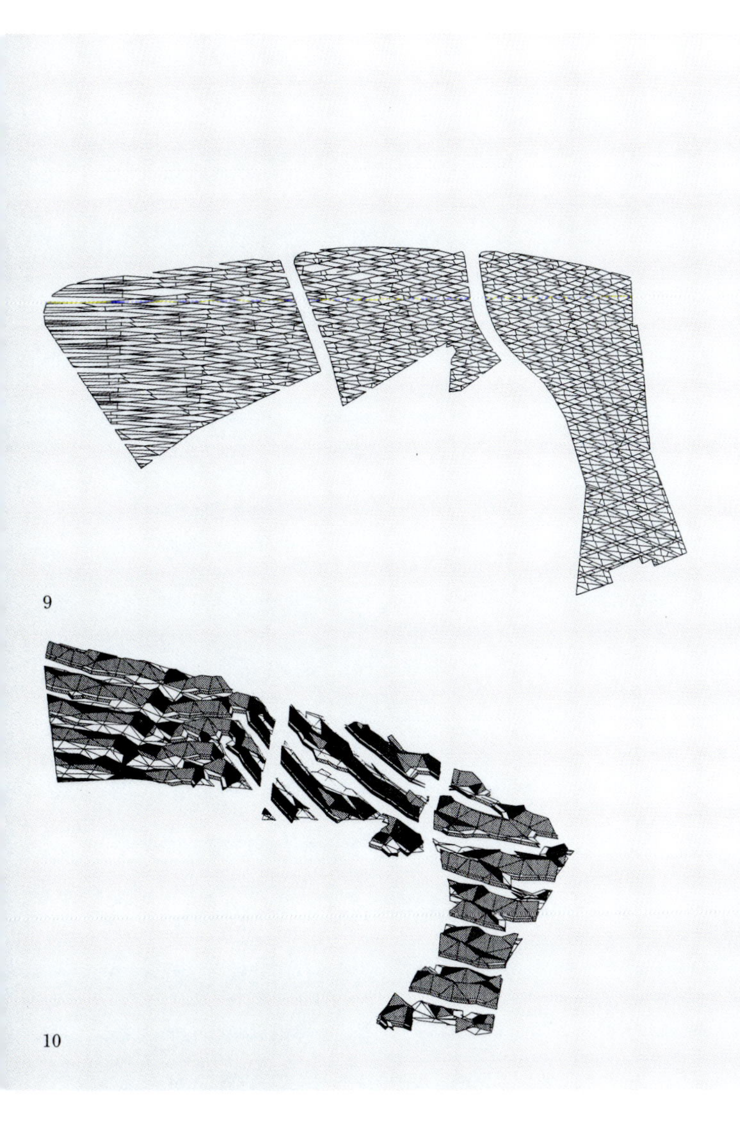
9

10

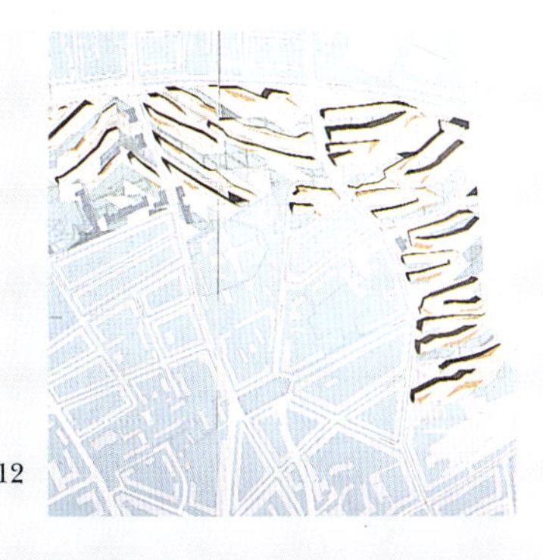
12

13

14

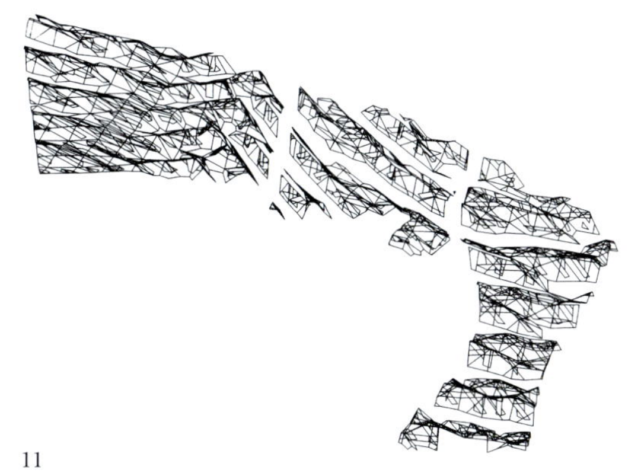
11

Haus Immendorff

Design 1993
Dusseldorf, Germany
Professor Jorg Immendorff
13,300 square feet

伊门多夫住宅

设计　1993
杜塞尔多夫，德国
约尔格·伊门多夫教授
13300平方英尺

This project for a cafe, private club, studio and office space for a painter is located on Dusseldorf's waterfront. The building's twisting form derives from an analysis of soliton waves which form non-linear interactions. Solitons undergo constant change and generate singular aqueous forms that alternately dissipate and regenerate as they move through the water. Haus Immendorff is composed of inner and outer volumes whose oblique surfaces intersect each other as they twist vertically, forming a vortex-like cone rising to the top of the building. The exterior volume is a stepped glass "skin" of bands of glass windows alternating with louvers set back at various widths from the glass. The inner volume is a solid wall with glazed cuts, to be used as a paint surface.

该项目是为一位画家设计的包含咖啡厅、私人俱乐部、画室和办公用房的综合体，坐落在杜塞尔多夫濒水地区。该建筑的弯曲的形式来源于对孤波的分析，它形成非直线的交互作用。孤波总是不断变化并产生单一的水的形式，当它们在水中穿行的时候交替地驱散、重新产生。伊门多夫住宅既有室内空间又包含室外空间，它倾斜的表面沿着垂直方向发生弯曲的时候交叉在一起，形成了一个好像是有顶点的圆锥形体，该圆锥体直冲建筑顶端。室外是阶梯装的玻璃皮肤，在各层之间窗户的宽度不断变化。室内是一个实体的墙，有着光滑的割纹，它是被作为画板使用的。

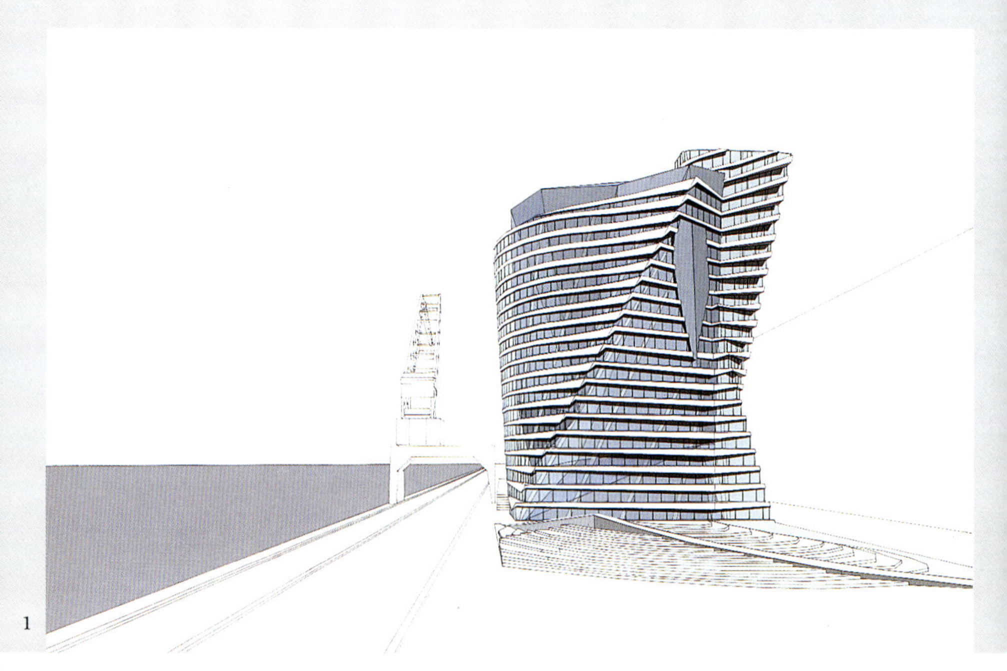

1

1 Perspective, view from the west
2–9 Concept diagrams
10 Study model
11 Computer-generated study

1 从西看透视图
2－9 概念示意图
10 研究模型
11 计算机绘图研究

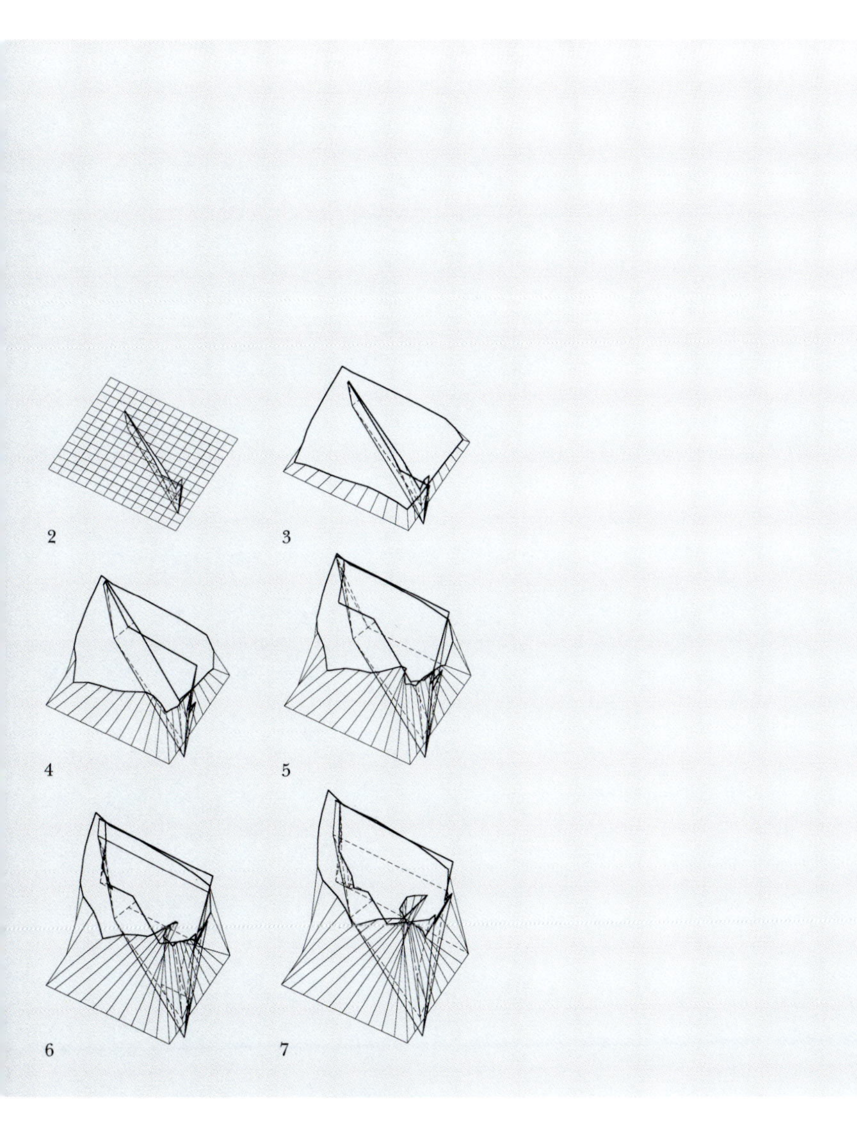

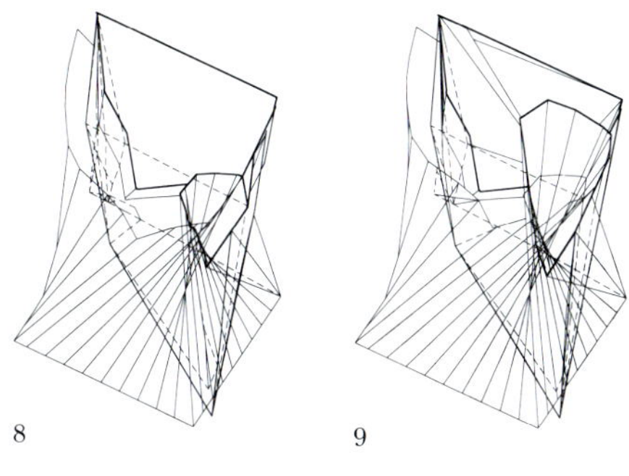

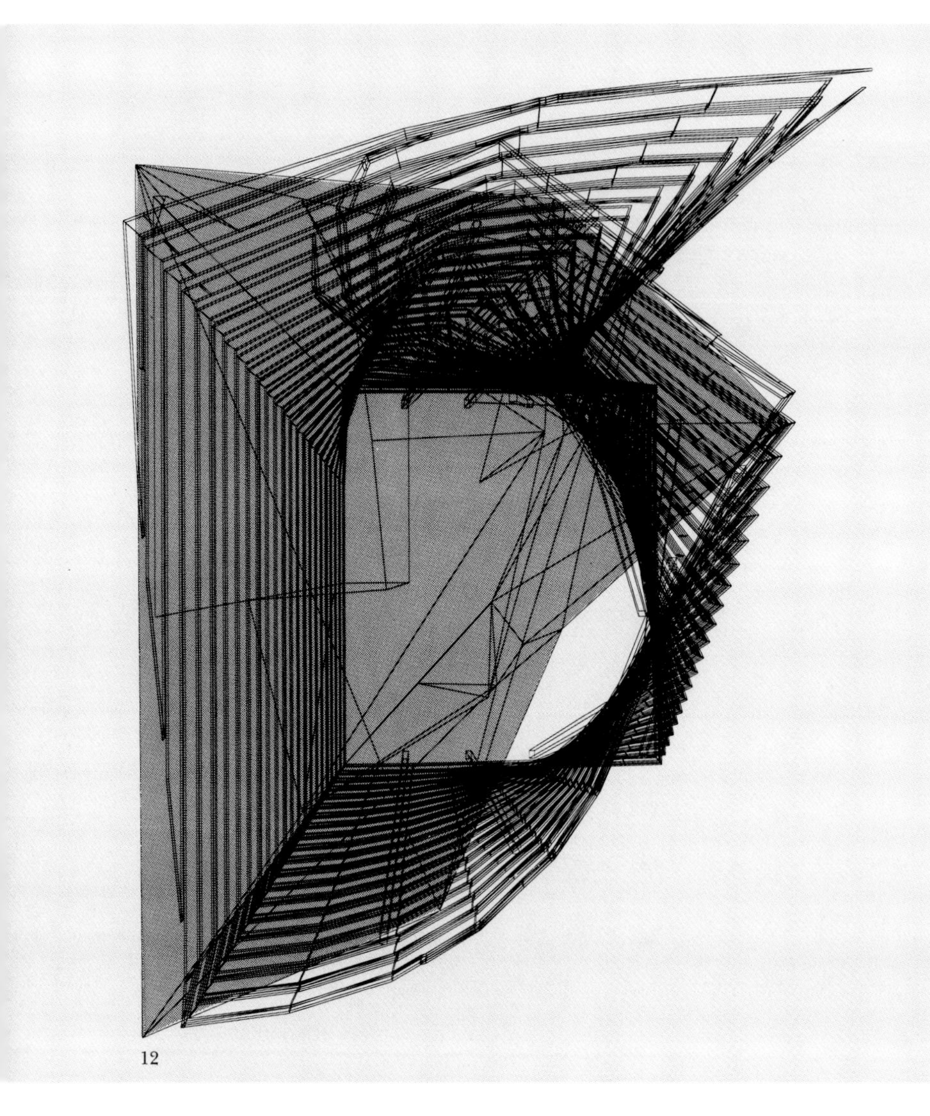
12

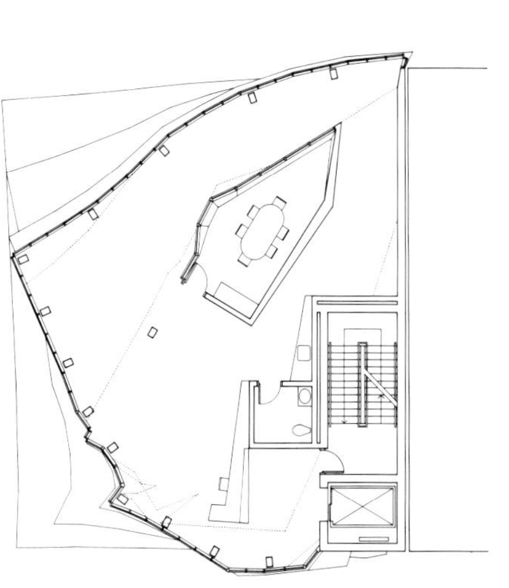
13

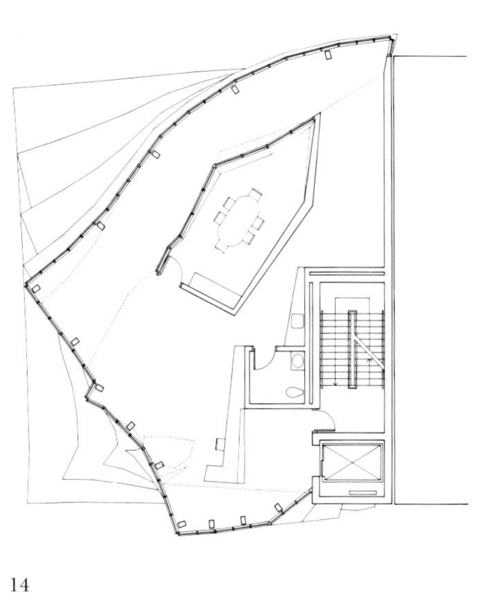
14

12 Wire frame diagram, view from above
13 Second level plan
14 Third level plan
15 Wire frame diagram, axonometric view
16 Fourth level plan
17 Fifth level plan

12 从上看网架示意图
13 三层平面
14 四层平面
15 网架示意图，轴测图
16 五层平面
17 六层平面

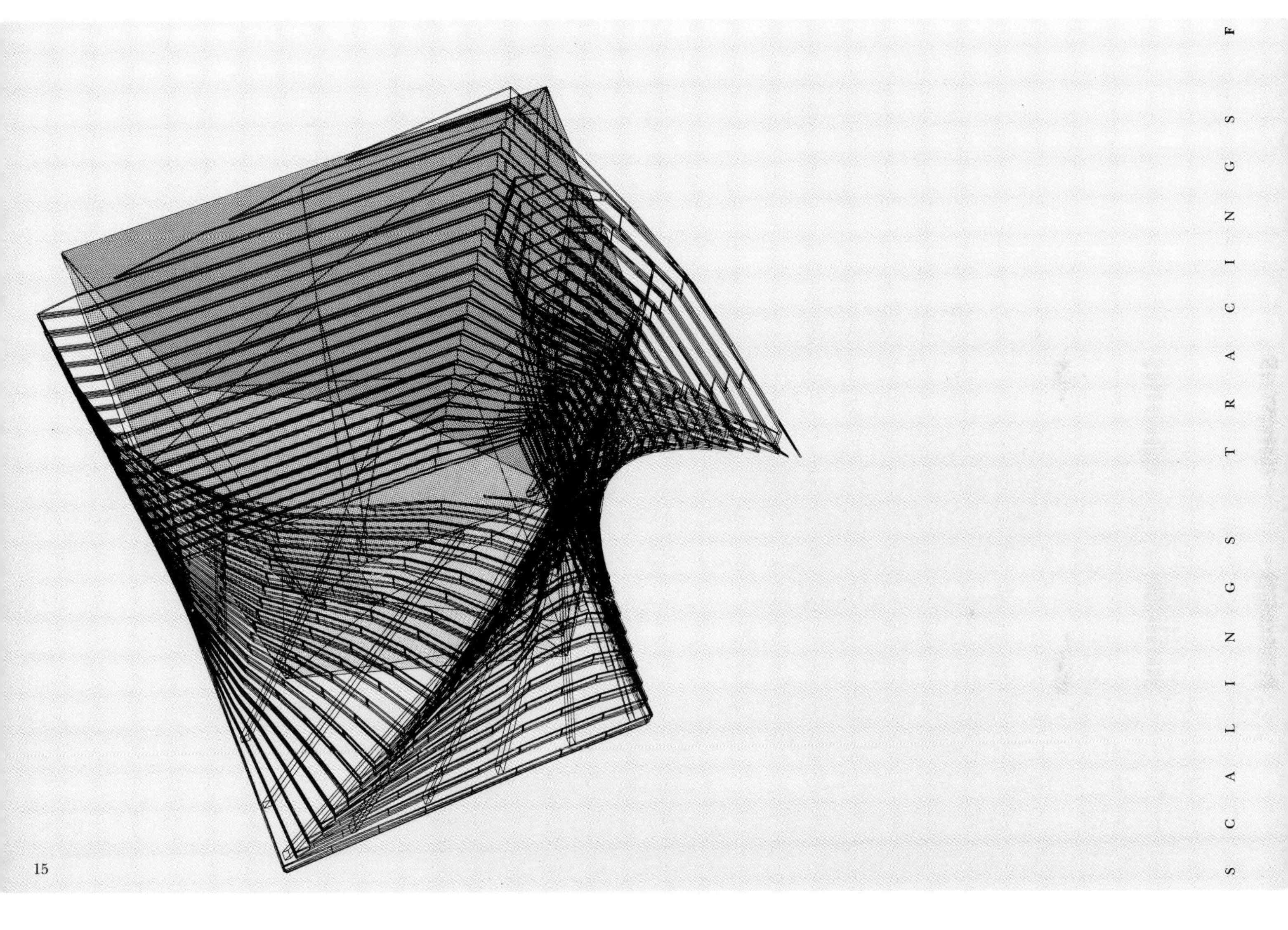

15

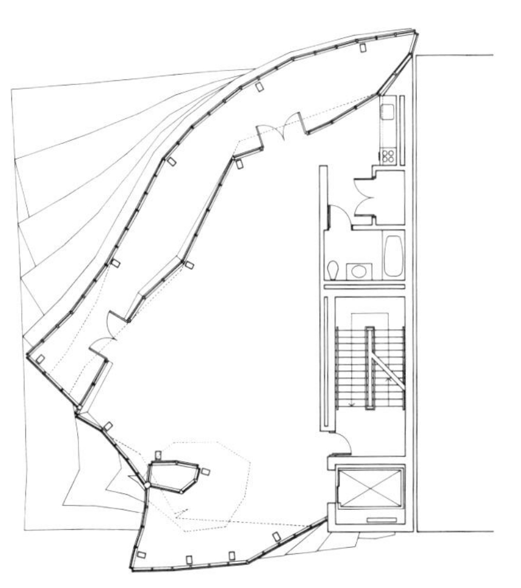

16

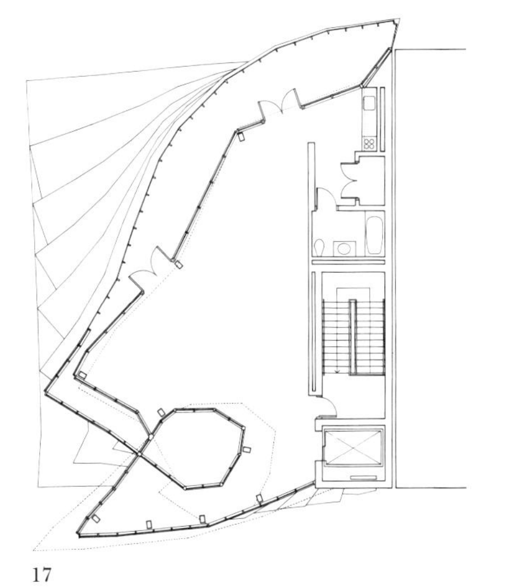

17

18

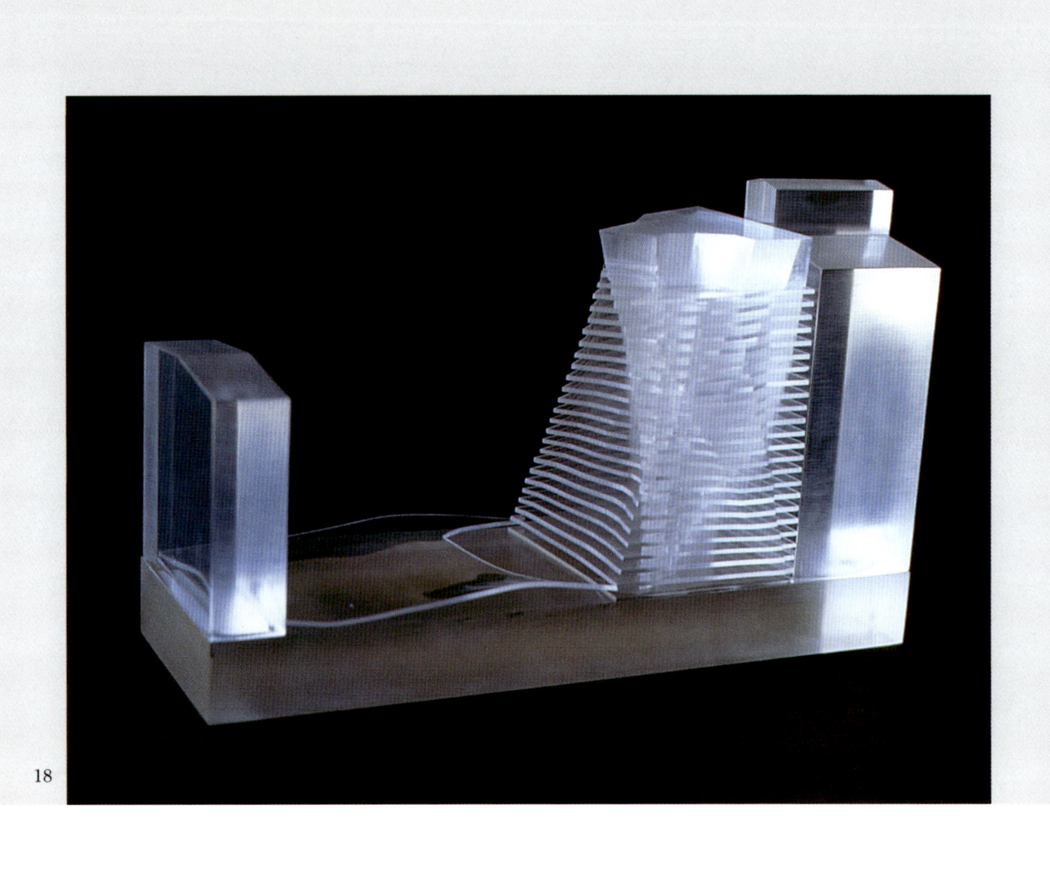

18 Study model
19 Section BB
20 Section AA
21 Study model
22 Study model, view from the west
23 Study model, view of the gallery wall

18 研究模型
19 BB 剖面
20 AA 剖面
21 研究模型
22 从西看研究模型
23 研究模型，走廊墙

21

22

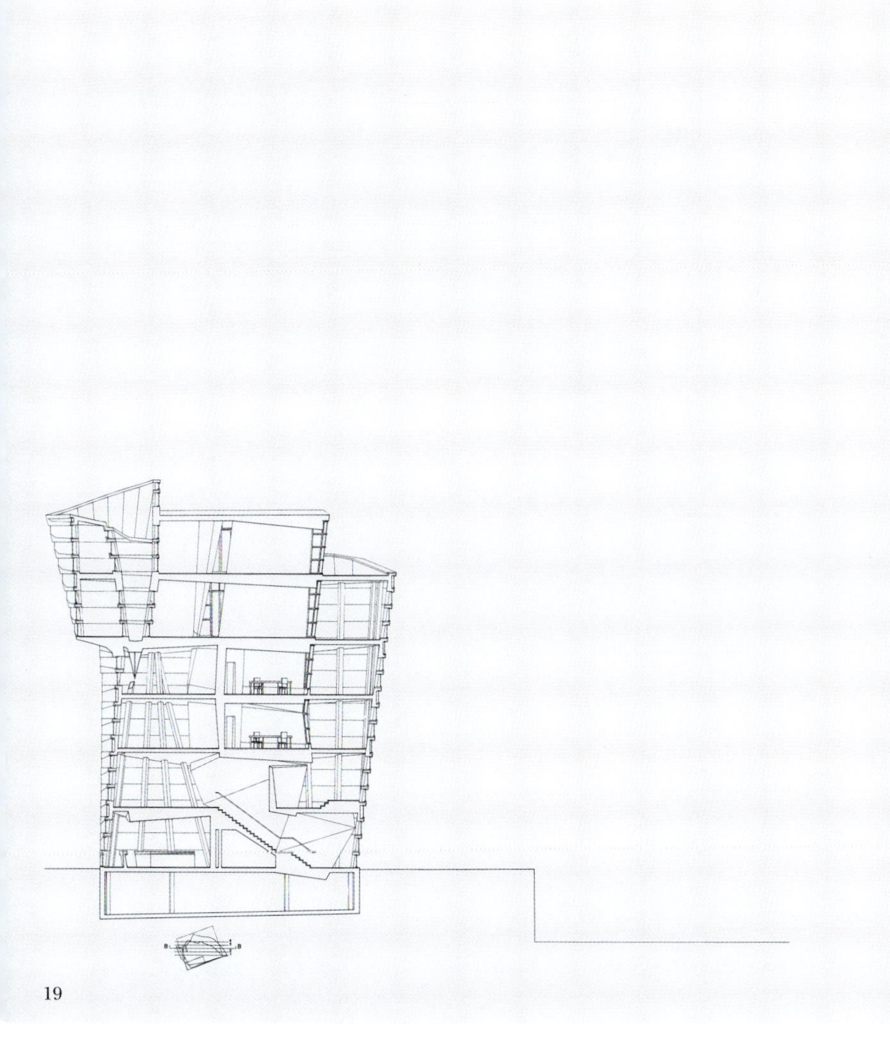

19

23

20

Biographies　个人简历

Peter Eisenman, FAIA

Peter Eisenman is an architect and educator. In 1980, after many years of teaching, writing, and producing respected theoretical work, he established his professional practice to focus exclusively on building. He has designed a wide range of prototypical projects including large-scale housing and urban design projects, innovative facilities for educational institutions, and a series of inventive private houses.

Among his built projects, the Wexner Center for the Visual Arts and Fine Arts Library at the Ohio State University in Columbus, completed in 1989, met with international acclaim, and received a 1993 National Honor Award from the American Institute of Architects. His project for social housing at Checkpoint Charlie at the Berlin Wall was honored by the West German Government, being featured on a postage stamp commemorating the 750th anniversary of the City of Berlin. He has built two office buildings in Tokyo: the Nunotani Corporation building, and the Koizumi Sangyo Corporation headquarters building, which received a 1991 National Honor Award from the American Institute of Architects.

In March 1993, opening ceremonies were held for the Greater Columbus Convention Center in Ohio, and construction had begun on the Aronoff Center for Design and Art at the University of Cincinnati. At present Peter Eisenman is working on the Center for the Arts at Emory University in Atlanta, Georgia; the master plan for Rebstockpark in Frankfurt, Germany; the high-rise Max Reinhardt Haus in Berlin; and an artist's cafe-studio in Dusseldorf.

In 1985, Peter Eisenman received a Stone Lion (First Prize) for his Romeo and Juliet Castles project at the Third International Architectural Biennale in Venice. He was one of two architects to represent the United States at the Fifth International Exhibition of Architecture of the Venice Biennale in 1991, and his projects are exhibited at museums and galleries around the world. Eisenman is the founder and former director of the Institute for Architecture and Urban Studies, an international think-tank for architectural criticism.

He has received numerous awards, including a Guggenheim Fellowship, the Brunner Award of the American Academy of Arts and Letters, and a grant from the National Endowment for the Arts.

彼得·埃森曼，美国建筑师学会资深会员（FIFA）

彼得·埃森曼是一位建筑师和教育家。1980年，在经历了许多年的教书、写作生涯，以及出版了学术上十分受尊崇的著作之后，他开始了在建筑方面的职业实践。他设计了类型广泛的建筑项目，包括大范围的住宅区和城市设计项目，具有创新性的教育建筑，以及一系列的具有创造性的私人住所。

在他的作品中，位于俄亥俄州哥伦布市的俄亥俄州立大学韦克斯纳视觉艺术中心以及艺术图书馆，竣工于1989年，获得了国际上的认可，并获得美国建筑师学会（AIA）1993年度奖章。位于德国柏林墙查理检查站的他的社会住宅项目受到前西德政府的表彰，并被印在纪念柏林750周年的纪念邮票中。他在日本东京设计了两栋办公建筑：布谷总部建筑和小泉三洋总部大楼，后者为他赢得了1991美国建筑师学会颁发的奖章。

1993年3月，俄亥俄州的大哥伦布会议中心举行了开工典礼，辛辛那提大学的阿诺夫设计和艺术中心也开始动工。目前埃森曼正忙于佐治亚州亚特兰大市的埃默里大学的艺术中心项目；德国法兰克福的Rebstokpark规划项目；柏林的超高层的Max莱恩哈茨住宅项目；以及在杜塞尔多夫的一个艺术家的咖啡厅兼创作室的项目。

1985年，彼得·埃森曼的罗密欧与朱丽叶城堡项目，获得在意大利威尼斯举办的第三届国际建筑双年展的石狮奖（一等奖）。在1991年的双年展中，他是代表美国参展的两位建筑师之一，他的作品在全世界的博物馆和展廊中展出。埃森曼是建筑与城市研究学会的创建者和成员之一，该学会是对建筑评论的一个国际思想宝库。

他已经获得许多奖项，包括古根海姆基金，美国艺术、文字学术颁发的布鲁纳奖章，以及国家艺术捐赠的批准。

他的学术事务包括在剑桥大学、普林斯顿大学、耶鲁大学以及俄亥俄州立大学任教。1982年至1985年期间，他是哈佛大学建筑系教授，1993年秋季，他是哈佛的访问学者（Eliot Noyes）。目前，他是纽约库珀协会的建筑系教授（Irwin S. Chanin）。

彼得·埃森曼撰写了许多书，包括《住宅X》(Rizzoli)，《*Find'Ou T HouS*》（建筑协会），《*Moving Arrows，Eros and Other Errors*》（建筑协会），《纸牌房子》（*House of Cards*）（牛津大学出版社）以及《韦克斯纳视觉艺术中心》（Rizzoli）。并且他还是《*Oppositions*》期刊和书籍的编辑，并已经在国际上一些期刊杂志上发表了若干他的建筑理论的文章和论文。

彼得·埃森曼曾获得康乃尔大学建筑学学士学位、哥伦比亚大学建筑学硕士学位、剑桥大学硕士和博士学位，以及芝加哥伊利诺伊大学美术学名誉博士。

George Kewin, AIA

George Kewin has recently led project teams for numerous international projects and competitions in Germany, including the Rebstockpark Master Plan, a five million square foot housing and commercial development in Frankfurt, Germany, which was the winning entry in an international competition in 1990. He was the Associate Principal-in-Charge for Eisenman Architects' entries in the Bahnhofsbereich Friedrichstrasse competition in Berlin, and the Nordliches Derendorf competition for a large-scale urban development in Dusseldorf. In addition, he was the Associate Principal-in-Charge for the Nunotani Office Building in Tokyo and the four-star Atocha 123 Hotel in Madrid. He is currently directing completion of comprehensive design and legal guidelines for the Rebstockpark Master Plan, and overseeing design development for a six-story cafe-bar and artists' studio space on the Dusseldorf Harbor in Germany.

Before joining Eisenman Architects in 1984, he was associated with Richard Meier and Partners, where he was project architect for the Des Moines Art Center Museum; and with the firm of Edward L. Barnes, where he was the project designer for the IBM Gallery of Art and Science and Equitable Tower in Manhattan.

George Kewin received his Master of Architecture degree from Harvard University, and his BA from the University of California at Berkeley. He has taught at the Graduate School of Architecture at the Ohio State University and served as a visiting critic at various other schools.

Richard N. Rosson, AIA

Richard Rosson is currently directing work on the Emory University Arts Center in Atlanta, Georgia, a $36 million, 126,000 square foot instructional and performance facility for the Department of Theater and Film Studies and the Department of Music. In addition, he is coordinating the work of Eisenman Architects' office on the Aronoff Center for Design and Art at the University of Cincinnati as the project enters the construction administration phase.

He was Project Manager for the fast-track design and construction of the $65 million, 530,000 square foot Convention Center in Columbus, Ohio, completed in March 1993, and also for a master plan of the Pittsburgh Technology Center, a 500,000 square foot laboratory and office park on the Monongahela River, for Carnegie Mellon University. He also oversaw the design of two buildings at the Center.

Since joining Eisenman Architects in 1985, he has been involved in various aspects of many projects, including the design and construction of a 350,000 square foot headquarters office building in Washington, DC and the renovation of the Harvard Club in New York City.

Formerly with Gresham Smith and Partners in Nashville, Richard Rosson served as Project Architect for hospitals and office buildings in the south-eastern United States and in Saudi Arabia.

乔治·凯文（George Kewin），美国建筑师学会会员（AIA）

乔治·凯文最近为了许多国际项目和在德国的竞赛成立了一个项目小组，德国的竞赛项目包括 Recstockpark 的规划，位于法兰克福的 5000000 平方英尺的住区规划和商业开发，该项目是 1990 年国际竞赛的获奖方案，他是埃森曼事务所参与柏林 Bahnhofsbereich Friedrichstrasse 竞赛的主要合伙负责人。而且，他还是东京布谷办公建筑和马德里的四星级阿托查 123 旅馆项目的主要合伙负责人。他目前正在负责 Recstockpark 的规划项目的全方位设计和合法方针，以及监查德国杜塞尔多夫港口的艺术家咖啡厅兼画室的项目。

在 1984 年加入埃森曼事务所之前，他曾经是理查德·迈耶的合伙人，在迈耶事务所，他曾经是德梅因艺术博物馆的项目建筑师；在爱德华·L·巴恩斯（Edward L. Barnes）公司，他曾经是 IBM 科学和艺术展廊以及曼哈顿的公平大厦（Equitable Tower）的设计者。

乔治·凯文获得哈佛大学建筑学硕士学位，伯克利加利福尼亚大学的学士学位。他曾经在俄亥俄州立大学建筑研究院任教，并且是很多大学的访问评论家。

理查德·罗森（N. Rosson）美国建筑师学会会员

理查德·罗森目前正在负责佐治亚州亚特兰大市的埃默里大学的艺术中心项目，该项目耗资 3600 万美元，建筑面积 126000 平方英尺，该项目是戏剧和电影研究系以及音乐系的教学和表演用房。而且对于辛辛那提大学阿诺夫设计和艺术中心项目，当工程进入施工管理阶段，他便开始协助项目进行。

他是俄亥俄州哥伦布会议中心 fast-track 设计和施工的项目经理，该项目耗资 6500 万美元，建筑面积 530000 平方英尺，于 1993 年竣工，同时他还是卡内基·梅隆大学的匹兹堡技艺中心总体规划的项目经理，该项目建筑面积 500000 平方英尺，包括实验室和办公园地，坐落在莫农格希拉河畔。他同时负责监督该中心的两栋建筑的设计。

自从 1985 年加入埃森曼建筑师事务所以来，他已经参与了许多项目的各个方面，包括华盛顿 350000 平方英尺的办公大楼的设计和施工，以及纽约城的哈佛俱乐部的改造。

之前在纳什维尔的格雷舍姆·史密斯及合伙人公司工作时，理查德·罗森是美国东南部以及沙特阿拉伯的医院和办公建筑的项目建筑师。

Project Credits 项目荣誉

House I
Architect: Peter Eisenman
Design Assistants: Russell Swanson, Robinson O. Brown
Drawings: Russell Swanson, Thomas Pritchard, Gregory A. Gale
Contractor: Bard Construction Co.

House II
Architect: Peter Eisenman
Design Assistants: Gregory A. Gale, Robinson O. Brown
Drawings: Gregory A. Gale, Judith Turner, Christopher Chimera
Structural Engineer: Geiger-Berger
Contractor: Dutton Smith

House III
Architect: Peter Eisenman
Structural Engineer: Geiger-Berger
Mechanical Engineer: George Langer
Contractor: Joseph Maloney

House IV
Architect: Peter Eisenman
Design Assistant: Rodney Knox
Drawings: Ellen Cheng Koutsoftas

House VI
Architect: Peter Eisenman
Design Assistants: Randall Korman, Rodney Knox
Drawings: Read Furguson, Caroline Sidnam, Wliliam Jackson
Model: Mark Mascheroni
Structural Engineer: Robert Silman & Associates
Contractors: Arthur B. Deacon & Sons, Robert Finney

House X
Architect: Peter Eisenman
Associate Architect: Leland Taliaferro
Assistants: Mark Cigolle, Livio Dimitriu, John Nambu, Anthony Perrgola, Noel Quesada
Structural: Robert Silman Associates (Ding Carbonell)
Mechanical: Arthur Spaet & Associates (Arthur Fox)
Landscape: Nicholas Quennell
Cost: Stephen H. Falk
Model: Anthony Pergola
Axonometric Model: Sam Anderson
Photography: Dick Frank

Cannaregio Town Square
Architect: Peter Eisenman
Project Team: David Buege, John Nambu, Joan Ockman
Models: Sam Anderson, Andrew Bartle
Model Photos: Dick Frank

House El even Odd
Architect: Peter Eisenman
Assistants: Mark Mascheroni, Caroline Hancock, Tom Haworth
Collages: Walter Chatham, David Buege, Cary Liu
Models: Tom Hut, John Leeper, Jim Uyeki, John Regan
Coordinator: Eleanor Earle, Judy Geib
Structural Engineer: Robert Silman
Robert Silman Associates
Mechanical Engineer: Marvin Mass, Cosentini Associates
Photography: Dick Frank

Madison Components Plant
Architect: Peter Eisenman

IBA Social Housing
Competition Phase
Architect: Eisenman/Robertson Architects
Partner-in-Charge: Peter Eisenman
Project Architects: Thomas Hut, Thomas Leeser
Drawings: Michelle Andrew
Renderings: Brian Burr
Models: Sam Anderson, John Leeper, Vera Marjanovic
Project Realization Phase
Architects: Eisenman/Robertson Architects; Groetzebach, Plessow & Ehlers
Partners-in-Charge: Peter Eisenman, Dietmar Groetzebach, Gunther Plessow
Associates-in-Charge: Thomas Leeser, Wilfried Hartman
Project Team: Audrey Matlock, Doug Oliver, Frank Chirico
Photos: Reinhard Goerner

Travelers Financial Center
Architect: Eisenman/Robertson Architects; Trott & Bean Architects
Partners-in-Charge: Peter Eisenman, Arthur Baker, Richard Trott
Associates-in-Charge: Richard Morris, Faruk Yorgancioughu, Michael Burkey
Project Architects: Thomas Leeser, Peter Thaler, Ross Woolley
Project Team: Andrea Brown, Wes Jones, Mark Mascheroni, Joanne Rivkin, Scott Sickeler
Structural Engineer: Office of Irwin Cantor
Mechanical Engineer: Cosentini Associates
General Contractor: Turner Construction Company
Model Photographs: Dick Frank
Building Photographs: Wolfgang Hoyt/ESTO

Firehouse for Engine Company 233 and Ladder Company 176
Architect: Eisenman/Robertson Architects
Partner-in-Charge: Peter Eisenman
SeniorArchitect: Arthur Baker
Project Architect: Ross Woolley
Project Team: David Winslow, Mark Wamble
Structural Engineer: Robert Silman
Mechanical Engineer: John Altieri
General Contractor: Bedell Associates

Fuller/Toms Loft
Architects: Peter Eisenman & Faruk Yorgancioglu
Assistants: Richard & Candy Harder, Glen Hamilton
Collaborators: Ragip Erdem, David Winslow, James Brown
Engineering Consultants: John Altieri Associates

Romeo and Juliet Castles
Architect: Eisenman/Robertson Architects
Partner-in-Charge: Peter Eisenman
Project Architects: Thomas Leeser, Renato Rlzzi, Peter Thaler
Drawings: Raleigh Perkins, Susan Knauer, Edward Carroll, Alexis Moser, Carlene Ramus, Joseph Rosa
Graphics: Charles Crawford, James Brown, Leslie Ryan
Models: Hiroshi Maruyama, Raleigh Perkins, Christine Chang, Donna Cohen, Guillaume Ehrman, Rajip Erdem, Mara Graham, Kimberley Hoyt, Marina Kieser, Jonathan Marvel, Michel Mossessian, David Murphee, Fabio Nonis, Peter Robson, Adam Silver, Wolfgang Tschapeller, Charles Barclay, Michael Casey
Model Photos: Dick Frank

Tokyo Opera House
Architects: Eisenman/Robertson Architects; Richard Trott & Partners
Partners-in-Charge: Peter Eisenman, Richard Trott
Project Architects: Thomas Leeser, Hiroshi Maruyama, Benjamin Gianni
Project Team: Manou Ernster, David Goth, Christian Kohl, Mark Schendal, Joseph Tanney, Harvey Burns, David Fratianne, Thomas Lanzelotti, Kevin Miller, Sheri O'Reilly, David Mancino, David Efaw, David Shultis, Kathleen Sullivan, James Samuelson
Model Photographs: Dick Frank

Biocentrum
Architect: Eisenman Architects
Partner-in-Charge: Peter Eisenman
Associate-in-Charge: Thomas Leeser
Project Team: Hiroshi Maruyama, David Biagi, Sylvain Boulanger, Ken Doyno, Judy Geib, Holger Kleine, Christian Kohl, Frederic Levrat, Greg Lynn, Carlene Ramus, Wolfgang Rettenmaier, Madison Spencer, Paul Sorum, Sarah Whiting, David Youse
Mechanical Engineer: Jaros, Baum & Bolles;

Augustine DiGiacomo
Structural Engineer:
Silman Associates, Robert Silman
Landscape Architect:
Hanna/Olin Ltd, Laurie Olin
Artist: Michael Heizer
Color Consultant: Robert Slutzky
Photography: Dick Frank Studios

La Villette
Architects: Eisenman/Robertson Architects;
Jacques Derrida, with Renato Rizzi
Architects-in-Charge: Peter Eisenman,
Jacques Derrida
Project Architects: Thomas Leeser,
Renato Rizzi
Project Team: Franco Alloca,
Paola Marzatico, Hiroshi Maruyama,
Manou Ernster
Model Photos: Dick Frank

University Art Museum
Architect: Eisenman/Robertson Architects
Partner-in-Charge: Peter Eisenman
Associate-in-Charge: Thomas Leeser
Project Architects: Hiroshi Maruyama,
Graeme Morland
Project Team: Michael Duncan,
Manou Ernster, Judy Geib, Fabio Ghersi,
Frances Hsu, Christian Kohl,
Paola Marzatico, Fabio Nonis, Joe Tanney,
Mark Wamble, Sarah Whiting, Gilly Youner
Gold Drawings: Mark Wamble
Landscape Architects:
Hanna/Olin, Philadelphia
Model Photos: Michael Moran

Progressive Corporation Office Building
Architects: Eisenman/Robertson Architects
Partner-in-Charge: Peter Eisenman
Associate-in-Charge: Thomas Leeser
Project Architects: Hiroshi Maruyama,
Fabio Nonis
Graphics and Exhibition Consultants:
Robert Slutzky
Model Photographs: Dick Frank

**Wexner Center for the Visual Arts
and Fine Arts Library**
Architect: Eisenman Architects;
Richard Trott & Partners Architects
Partners-in-Charge: Peter Eisenman,
Richard Trott
Directing Architects: George Kewin,
Michael Burdey
Project Architects: Arthur Baker,
Andrew Buchsbaum, Thomas Leeser,
Richard Morris, James Rudy,
Faruk Yorgancioglu
Project Team: Andrea Brown,
Edward Carroll, Robert Choeff, David Clark,
Chuck Crawford, Tim Decker,
Ellen Dunham, John Durschinger,
Frances Hsu, Wes Jones, Jim Linke,
Michael McInturf, Hiroshi Maruyama,
Mark Mascheroni, Alexis Moser, Harry Ours,
Joe Rosa, Scott Sickeler, Madison Spencer,
Mark Wamble
Landscape Architect:
Hanna Olin Ltd, Laurie Olin
Structural Engineer:
Lantz, Jones & Nebraska Inc., Tom Jones
Mechanical Engineer:
H.A. Williams & Associates
Lighting Design:
Jules Fisher & Paul Marantz Inc.
Civil Engineer: C.F. Bird & P.J. Bull Ltd
Security and Fire: Chapman & Ducibella Inc.
Graphics and Color: Robert Slutzky
Soils Engineer: Dunbar Geotechnical
Audio/Visual: Boyce Nemec
Acoustics: Jaffe Acoustics
Specifications: George Van Neil
Models: Scale Images, Albert Maloof,
Gene Servini
Renderings: Brian Burr
Model Photography: Dick Frank,
Wolfgang Hoyt
Construction Photographs: James Friedman,
Will Shively, D.G. Olshavsky/ARTOG
Final Photographs: Jeff Goldberg/ESTO,
D.G. Olshavsky/ARTOG
General Contractor: Dugan and Meyers,
Jim Smith, Project Manager
Mechanical Contractor: A.T.F. Mechanical
Inc., Bob Weiland, Project Manager
Electrical Contractor: Romanoff Electric,
Sib Goeiz, Project Superintendent
Plumbing Contractor: Radico Inc.,
Frank Czako, Project Manager
Steel Subcontractor: J.T. Edwards,
Jack Edwards, President

Carnegie Mellon Research Institute
Architect: Eisenman Architects
Principal-in-Charge: Peter Eisenman
Associate-in-Charge: Richard N. Rosson
Project Team: Lawrence Blough,
Kelly Hopkin, Richard Labonte, Greg Lynn,
Marisabel Marratt, Mark Wamble, Joe Walter
Project Assistants: Wendy Cox,
Simon Hubacher, Kim Tanzer,
Nicolas Vaucher, Sarah Whiting,
Katinka Zlonicky
Model Photographs: Dick Frank
Landscape Architect: Hanna/Olin Ltd,
Laurie Olin
Mechanical Engineer: Jaros, Baum & Bolles,
Augustine DiGiacomo
Structural Engineer: Ove Arup & Partners,
Guy Nordenson

Guardiola House
Architect: Eisenman Architects
Principal-in-Charge: Peter Eisenman
Associates-in-Charge: George Kewin,
Thomas Leeser
Project Architect: Antonio Sanmartin
Project Team: Nuno Mateus, Jan Kleihues,
Hiroshi Maruyama
Project Assistants:
Begona Fernandez Shaw, Felipe Guardiola,
Lise Anne Couture, Luis Rojo,
Michael McInturf, Madison Spencer,
Simon Hubacher, Maximo Victoria,
Frederic Levrat, Anne Marx, Robert Choeff,
Julie Shurtz, Dagmar Schimkus
Structural Engineer: Gerardo Rodriguez
Photography: Dick Frank

Aronoff Center for Design and Art
Architect: Eisenman Architects;
Lorenz & Williams Inc.
Principal-in-Charge: Peter Eisenman,
Richard Roediger
Associates-in-Charge: George Kewin,
Richard Rosson, Jerome Flynn
Project Architects: Donna Barry, Greg Lynn,
Michael McInturf, Joseph Walter
Project Team: Lawrence Blough,
Kelly Hopkin, Edward Mitchell,
Astrid Perlbinder, Brad Winkeljohn (EA),
Joseph Mitlo, Shari Rotella, Jerome Scott,
James Schriefer, Michael Schuyler (LWI)
Project Assistants: Vincent Costa,
Reid Freeman, Nazli Gonensay,
Martin Houston, Richard Labonte,
Corrine Nacinovic, Jean-Gabriel Neukomn,
Karen Pollock, Joe Schott, Jim Wilson,
Jason Winstanley, Leslie Young (EA)
Construction Manager: Dugan & Meyers Inc,
Francis Dugan, Daniel Dugan,
Andy Englehart, Steve Klinder
Civil Engineer: United Consultants
Landscape Architect: Hargreaves Associates
Engineering: Lorenz & Williams Inc.
Acoustical Design: Jaffe Acoustics
Lighting Design: Fisher Marantz
Audiovisual Design: Boyce Nemec Designs
Color Consultant: Donald Kaufman Color
Photography: Dick Frank

Koizumi Sangyo Office Building
Architects: Eisenman Architects;
K Architects and Associates, Tokyo
Partners-in-Charge: Peter Eisenman,
Kojiro Kitayama
Associate-in-Charge: George Kewin
Project Architects: Hiroshi Maruyama (EA),
Minoru Fujii (KA)
Project Team: Lawrence Blough,
Robert Choeff, Lise Anne Couture,
Begona Fernandez Shaw, Frederic Levrat,
Dagmar Schimkus, Julie Shurtz,
Mark Wamble (EA), Itaru Miyakawa,
Tamihiro Motozawa, Hiroyuki Kubodera,
Kazuhiro Isimaru, Susumu Arasaki,
Yujiro Yamasaki (KA)

Siena Bank Master Plan
Architects: Eisenman Architects;
with Renat Rizzi
Project Architect: Thomas Leeser
Photographer: Dick Frank Studi

Greater Columbus Convention Center
Architects: Eisenman Architects;
Richard Trott and Partners Architects Inc.
Principals-in-Charge: Peter Eisenman,
Richard Trott, Jean Gordon
Associates-in-Charge: Richard Rosson,
Michael Burkey
Project Managers: Tracy Aronoff,
Philip Babb, Thomas Ingledue, Jerome Scott
Project Architects: Mark Wamble,
Thomas Leeser
Project Team: Madison Spenser,
Richard Labonte, Kathleen Meyer,
Dean Maltz, David Trautman,
Lewis Jacobsen, Joe Walter, Nuno Mateus
(EA); Jerry Kehlmeier, David Goth,
Lu Schubert, Kristina Ennis, Tim Decker,
John Meegan, Dave Reltenwald,
Blaide Lewis, James Dean, George Van Neil,
Carol Hummel, Chun Shin, Karen McCoy,
Al Brook (RTPA)
Project Assistants: Yvhang Kong,
John Durschinger, John Curran,
Chiara Scortecci, Ilkka Tarkkanen, Jon Malis,
Andres Viditz-Ward, Giovanni Rivolta,
Francesca Acerboni, Jason Winstanley,
John Juryj, Daniel Perez,
Andres Blanco (EA)
Engineers: Lorenz & Williams Inc.
Principal-in Charge: Richard Roediger
Project Managers: Timothy McCrate
(Structural), John Putnam (Mechanical),
Jack Kolb (Mechanical), Timothy Raberding
(Electrical), Thomas Fischer (Construction
Administration)
Civil Engineer: Moody/Nolan Ltd,
Howard Nolan
Code Consultant: Oregon Group Architects,
Jane Voisard
Roofing: Simpson, Gumperts & Heger Inc,
Kevin Cash
Graphic Design: Mayer/Reed, Michael Reed
Lighting: Jules Fisher & Paul Marantz Inc,
Richard Renfro
Acoustics: Jaffe Acoustics Inc., Mark Holden
Construction Manager:
Turner/Smoot/Zunt, Joel Sloan,
Project Manager
Photography: Dick Frank Studio,
ARTOG/D.G. Olshavsky,
Jeff Goldberg/ESTO

Banyoles Olympic Hotel
Architect: Eisenman Architects
Principal-in-Charge: Peter Eisenman
Associate-in-Charge: George Kewin
Project Designers: Begona Fernandez-Shaw,
Nuno Mateus
Project Team: Ed Mitchell, Anne Peters,
Weiland Vajen
Project Assistants: Lawrence Blough,
John Durschinger, Kelly Hopkin,
Martin Houston, Yuhang Kong,
Richard Labonte, Mari Marratt,
Tom Popoff, Henry Urbach, Joe Walter,
Mark Wamble, Leslie Young
Structural Engineer:
"Static" Ingenieria De Construccion,
Gerardo Rodriguez
Model Photographs: Dick Frank

Cooper Union Housing
Architects: Eisenman Architects;
Thomas Leeser
Principal-in-Charge: Peter Eisenman
Associate-in-Charge: George Kewin
Project Designer: Nuno Mateus
Project Team: Ed Mitchell, Joe Walter,
John Durschinger, Yuhang Kong,
Tom Popoff, Wieland Vajen
Project Assistants: Andreas Blanco,
Lawrence Blough, Reid Freeman,
Begona Fernandez-Shaw, Kelly Hopkin,
Jake Malis, Mari Marratt, Tony Pergola,
Astrid Perlbinder, Anne Peters,
Inigo Rodriguez-San Pedro, Leslie Smith,
Madison Spencer, Ilkka Tarkkanen,
Mark Wamble, Jim Wilson
Structural Consultants:
Severud Associates Consulting Engineers
PC Mechanical, Plumbing, Electrical
Consultants: Jaros, Baum & Bolles
Consulting Engineers
Zoning Consultant: Michael Parley
Code Consultant: Super Structures
Model Photography: Dick Frank

Groningen Music-Video Pavilion
Architect: Eisenman Architects
Principal-in-Charge: Peter Eisenman
Associate-in-Charge: George Kewin
Project Architect: Jorg Gleiter
Project Team: Andrea Stipa,
Anton Viditz-Ward, Reid Freeman

Nunotani Office Building
Architect: Eisenman Architects
Principal-in-Charge: Peter Eisenman
Associate-in-Charge: George Kewin
Project Architects: Mark Wamble,
Tracy Aronoff
Project Team: David Trautman, John Curran
Project Assistants: Thor Thors,
Hans-Georg Berndsen, Karen Pollock,
David Johnson, Evan Yassy,
Gregory Merryweather, Andrea Stipa,
Jason Winstanley, Andre Kikoski
Construction Manager and Contractor:
The Zenitaka Corporation;
Yoshimichi Hama, Director Manager,
Yoshiteru Kagikawa, Director,
Keiichi Kuwana, Deputy Manager
Model Photography: Dick Frank Studio
Building Photography:
Shigeo Ogawa/Shinkenchiku

Atocha 123 Hotel
Architects: Eisenman Architects;
The Austin Company, SA
Principals-in-Charge: Peter Eisenman,
F.E. "Brownie" Higgs
Associate-in-Charge: George Kewin
Project Managers: David Koons,
Jesus Salgado Marques, Luis Guerrero
Project Architects: Gregory Luhan,
Jorg Gleiter, John Curran, Nuno Mateus,
Mark Searls (EA), Antonio de la Morena,
M. Magdalena Velez,
Ramon Jose Farinas (AC)
Project Team: Tracy Aronoff, Mary Marratt,
Andrea Stipa, Joe Walter, Jason Winstanley,
Donald Skinner, John Maze, Tom Gilman,
Andrew Burmeister
Project Assistants: Donna Barry,
Rosa-Maria Colina, Brooks Critchfield,
Angelo Directo, Winka Dubbledam,
John Durschinger, Martin Felsen,
Brad Gildea, Christophe Guinard,
Jan Hinrichs, Brad Khouri, Andre Kikoski,
Robert Kim, Justin Korhammer,
Alexander Levi, Luc Leveque,
Frederic Levrat, James McCrery,
Gregory Merryweather, David Moore,
Maureen Murphy-Ochsner, Karim Musfy,
Alex Nussbaumer, Karen Pollock,
Stefania Rinaldi, Raquel Sendra,
Jody Sheldon, Marc Stotzer,
Masahiro Suzuki, David Swanson,
Thor Thors
Structural Engineer: The Austin Company,
SA, Fernando De La Frost,
Fernando Yandela Terrosa
Contractor: The Austin Company SA
Photography: Dick Frank

Rebstockpark Master Plan
Architect: Eisenman Architects
Consulting Architect:
Albert Speer & Partner GmbH
Landscape Architect: Hanna/Olin Ltd
Consulting Landscape Architect:
Boedeker, Wagenfeld, Niemeyer & Partners
Traffic Planning:
Durth Roos Consulting GmbH
Principals-in-Charge: Peter Eisenman,
Albert Speer, Laurie Olin
Associates-in-Charge:
George Kewin, Gerhard Brand
Project Managers: Norbert Holthausen,
Michael Denkel, Shirley Kressel,
Karina Aicher
Project Architects: Joachim Bothe,
Jorg Gleiter, Nuno Mateus, Mark Wamble,
Matthew White
Project Team: Pornchai Boonsom,
Brad Gildea, Judith Haase,

Justin Korhammer, Luc Levesque,
Gregory Merryweather, Steven Meyer,
Karim Musfy, Andrea Stipa, Marc Stotzer,
Jason Winstanley, Corinna Wydler
Project Assistants: Donna Barry,
Rosa-Maria Colina, John Curran,
John Durschinger, Michael Eastwood,
Carolina Garcia, Nazli Gononsay,
John Juryj, Andre Kikoski, Stephano Libardi,
Greg Lynn, James McCrery, Edward Mitchell,
Jean Nukomn, Karen Pollock, Jon Stephens
Models: Eisenman Architects
Photography: Dick Frank Studios

Alteka Office Building
Architect: Eisenman Architects
Principal-in-Charge: Peter Eisenman
Associate-in-Charge: Richard Rosson
Project Architect: Mark Wamble
Project Team: Gregory Merryweather,
Nazli Gononsay
Project Assistants: Mina Mei-Szu Chow,
Rosa-Maria Colina, Cornelius Deckert,
Robert Kim, Maria Laurent, Frederic Levrat,
Pierre-Olivier Milanini, Hadrian Predock,
Jason Winstanley
Photography: Dick Frank

Emory Center for the Arts
Architect: Eisenman Architects
Principal-in-Charge: Peter Eisenman
Associate-in-Charge: Richard Rosson
Project Manager: Tracy Aronoff
Project Architects: Selim Koder,
Frederic Levrat, Mark Searls
Project Team: Philip Babb,
James Gettinger, Brad Gildea,
Timothy Hyde, Richard Labonte, Ingel Liou,
Gregory Luhan, James Luhur,
James McCrery, Maureen Murphy-Ochsner,
Lindy Roy, David Schatzle, Joseph Walter
Project Assistants: Ted Arleo, Donna Barry,
Federico Beulcke, Sergio Bregante,
Marc Breitler, Winka Dubbeldam,
Daniel Dubowitz, John Durschinger,
David Eisenmann, Abigail Feinerman,
Ralf Feldmeier, Martin Felsen,
Sigrid Geerlings, Robert Holton,
Keelan Kaiser, Patrick Keane, James Keen,
Brad Khouri, Rolando Kraeher, Joseph Lau,
Maria Laurent, Vincent LeFeuvre,
Claudine Lutolf, John Maze, Mark McCarthy,
Steven Meyer, Julien Monfort, David Moore,
Yayoi Ogo, Debbie Park, Axel Rauenbusch,
Ali Reza Razavi, Mirko Reinecke, Tilo Ries,
Stefania Rinaldi, David Ruzicka, Setu Shah,
Tod Slaboden, Giovanni Soleti,
Lucas Steiner, Helene Van gen Hassend,
Marcus Wallner, Benjamin Wayne,
Lois Weinthal, Erin Vali, Irina Verona
Landscape Architect: Hanna/Olin Ltd,
Laurie Olin, Chris Allen, Cora Olgyay
Structural Engineer:
Stanley D. Lindsey & Associates Inc.,
Stanley Lindsey, Tommy Hagood
Mechanical & Electrical Engineer:
Nottingham, Brook & Pennington Inc.,
Charles Pennington, Neil Wych
Acoustical Consultant:
Kirkegaard & Associates Inc.,
Larry Kirkegaard, David Eplee, Brian Cline
Theater and Lighting Design: Theatre
Projects Consultants Inc., Richard Pilbrow,
Robert Long, Peter Lucking, Ben Boltin
Cost Analysis: Donnell Consultants Inc.,
Stewart Donnell, Athol Joffe
Photography: Dick Frank

Max Reinhardt Haus
Architect: Eisenman Architects
Principal-in-Charge: Peter Eisenman
Associate-in-Charge: George Kewin
Project Architects: Edward Mitchell,
Lindy Roy, Richard Labonte
Project Team: Armand Biglari, Brad Gildea,
Norbert Holthausen, Gregory Luhan,
Stefania Rinaldi, David Schatzle,
Jon Stephens
Project Assistants: Federico Beulske,
Mark Bretler, Andrew Burmeister,
Robert Holten, Patrick Keane, Brad Khouri,
Joseph Lau, Vincent LeFeuvre,
Fabian Lemmel, John Maze, Steven Meyer,
Debbie Park, Silke Potting, Benjamin Wade
Landscape Architect: Hanna/Olin Ltd,
Laurie Olin, Shirley Kressel,
Matthew W. White
Color Consultant: Donald Kaufman Color
Structural Engineer: Severud Associates,
Edward M. Messina, Edward DiPaolo
Mechanical Engineer: Jaros, Baum & Bolles,
Augustine A. DiGiacomo,
Kenneth J. Zuar
Wind & Shadow Studies: Spacetec
Datengewinnung, Freiburg, Germany
Cost Estimating: Donnell Consultants Inc.,
Stewart Donnell
Computer Images: Edward Keller
Photography: Dick Frank

Nordliches Derendorf Master Plan
Urban Designers: Eisenman Architects;
Hanna/Olin Landscape Architects
Principals-in-Charge: Peter Eisenman,
Laurie Olin
Associates-in-Charge: George Kewin,
Shirley Kressel
Project Architects: Winka Dubbledam,
Norbert Holthausen, Donna Barry,
Matthew White
Project Team: Edgar Cozzio,
James Gettinger, Brad Gildea, Jorg Lesser,
Jon Stephens, James McCrery
Project Assistants: Barbera Aderbeauer,
Armand Biglari, Frederico Buelcke,
Andy Burmeister, John Durschinger,
Martin Felsen, Patrick Keane, Brad Khouri,
Selim Koder, Fabian Lemmel,
Frederic Levrat, Gregory Luhan,
Maureen Murphy-Ochsner, Stephania
Rinaldi, Lindy Roy, David Schatzle (EA),
Bobbie Huffman, David Rubin,
Howard Supnik, Karen Skafte (HO)
Traffic Planning: Durth Roos Consulting,
Hans-Joachim Fischer
Color Consultants: Donald Kaufman Color,
Donald Kaufman, Taffy Dahl
Computer Modeling: Mathematica Program,
Seamus Moran, Physicist
Photography: Dick Frank, Brian Connelly

Haus Immendorff
Architect: Eisenman Architects
Principal-in-Charge: Peter Eisenman
Associate-in-Charge: George Kewin
Project Architect: Lindy Roy
Project Team: David Schatzle,
Patrick Keane, James Luhur
Project Assistants: Barbara Adabauer,
Ted Arleo, Marc Bretler,
Andrew Burmeister, Chi Yi Chang,
Winka Dubbeldam, David Eisenmann,
Abigail Feinerman, Annette Kahler,
Fabian Lemmel, Jung Kue Liou,
Gregory Luhan, Max Muller,
Mirko Reinecke, Tilo Ries, Lucas Steiner
Construction Manager:
Phillip Holzmann HOG
Structural Engineer:
Severud Associates Consulting Engineers,
PC Mechanical, Plumbing, Electrical
Engineer: Jaros, Baum & Bolles
Photography: Dick Frank

Associates & Collaborators　合伙人/合作者

Francesca Acerboni
Barbara Aderbeauer
Franco Alloca
Tobias Amme
Sam Anderson
Michelle Andrew
Lise Anne Couture
Ted Arleo
Tracy Aronoff, AIA
Phillip Babb, AIA
Arthur Baker
Charles Barclay
Donna Barry
Andrew Bartle
David Beers
Hans-Georg Berndsen
Federico Beulcke
David Biagi
Herve Biele
Armand Biglari
Andres Blanco
Lawrence Blough
Pornchai Boonsom
Joachim Bothe
Sylvain Boulanger
Sergio Bregante
Marc Bretler
Sam Britton
Andrea Brown
James Brown
Andrew Buchsbaum
David Buege
Frederico Buelcke
Andrew Burmeister
Harvey Burns
Francine Cadogan
Edward Carroll
Michael Casey
Ronald Castellano
Christine Chang
Chi-Yi Chang
Frank Chirico
Robert Choeff
David Clark
Donna Cohen
Rosa-Maria Colina
Catherine Colla
Sarah Connoly
Vincent Costa
Wendy Cox
Edgar Cozzio
Charles Crawford
Nestor Crubellati
John Curran
Christophe Dahm
Cynthia Davidson
Ken Doyno
Winka Dubbeldam
Daniel Dubowitz
Michael Duncan
Ellen Dunham
John Durschinger
Arthur Dyess
Michael Eastwood
David Efaw
Guillaume Ehrman
Peter Eisenman, FAIA
David Eisenmann
Rajip Erdem
Manou Ernster
Abigail Feinerman
Ralf Feldmeier
Simon Fellmeth
Martin Felsen
Begona Fernandez-Shaw
Scott Ferris
David Fratianne
Reid Freeman
Jason Frontera
Carolina Garcia
Sigrid Geerlings
Judy Geib
James Gettinger
Fabio Ghersi
Benjamin Gianni
Brad Gildea
Jorg Gleiter
Nazli Gononsay
Mercedes Gonzalez
David Goth
Mara Graham
Susanne Grau
Stephen Griek
Felipe Guardiola
Judith Haase
Daniel Hale
Jan Hartman
Robert Holten
Norbert Holthausen
Robert Holton
Kelly Hopkin
Martin Houston
Kimberley Hoyt
Frances Hsu
Simon Hubacher
Joanne Humphries
Thomas Hut
Timothy Hyde
Diana Ibrahim
Antoinette Jackson
Lewis Jacobsen
Mary Jane McRory
David Johnson
Wes Jones
Jan Jurgens
John Juryj
Annette Kahler
Keelan Kaiser
Rudolph Kammeri
Patrick Keane
James Keen
May Kellner
Therese Kelly
George Kewin
Brad Khouri
Marina Kieser
Andre Kikoski
Rebecca Klapper
Jan Kleihues
Holger Kleine
Susan Knauer
Selim Koder
Christian Kohl
Yvhang Kong
David Koons
Erin Korff
Justin Korhammer
Rolando Kraeher
Kenneth Kraus
Miroslaw Krawczynski
Jung Kue Liou
Richard Labonte
Thomas Lanzelotti
Bernadette Latour
Joseph Lau
Maria Laurent
Vincent Le-Feuvre
Joseph Lechowicz
John Leeper
Jorg Leeser
Thomas Leeser
Fabian Lemmel
Luc Levesque
Frederic Levrat
Andrew Liang
Stephano Libardi
Alexandra Ligotti
Jim Linke
Ingel Liou
Gregory Luhan
James Luhur
Claudine Lutolf
Greg Lynn

Jake Malis
Dean Maltz
David Mancino
Vera Marjanovic
Marisabel Marratt
Hiroshi Maruyama
Jonathan Marvel
Anne Marx
Paola Marzatico
Mark Mascheroni
Nuno Mateus
Audrey Matlock
Dominik Mayer
John Maze
Mark McCarthy
James McCrery
Meegan McFarland
Michael Mcinturf
Gregory Merryweather
Will Meyer
Kathleen Meyer
Steven Meyer
Kevin Miller
Benedicte Mioli
Edward Mitchell
Julian Monfort
David Moore
Graeme Morland
Richard Morris
Alexis Moser
Michel Mossessian
Matthias Muffert
Max Muller
David Murphee
Maureen Murphy-Ochsner
Karim Musfy
Corrine Nacinovic
John Nambu
Jean-Gabriel Neukomn
Fabio Nonis
Jean Nukomn
Alex Nussbaumer
Patrick O'Brien
Sheri O'Oreilly
Joan Ockman
Yayoi Ogo
Doug Oliver
Elizabeth Pacot
Katherina Panagiotou
Debbie Park
Daniel Perez
Tony Pergola
Raleigh Perkins
Astrid Perlbinder
John Peter-Hartman
Anne Peters
Angela Phillips
Karen Pollock
Tom Popoff
Silke Potting
Carlene Ramus
Axel Rauenbusch
Rachel Ravitz
Jurgen Reimann
Mirko Reinecke
Mirko Reinecke
Wolfgang Rettenmaier
Ali Reza Razavi
Tilo Ries
Stefania Rinaldi
Giovanni Rivolta
Miranda Robbins
Peter Robson
Inigo Rodriguez San Pedro
Luis Rofriguez
Christian Rogner
Luis Rojo
Louise Rosa
Joseph Rosa
Richard Rosson, AIA
Lindy Roy
David Ruzicka
Leslie Ryan
James Samuelson
Antonio Sanmartin
Mario Santin
David Schatzle
Mark Schendal
Florian Scheytt
Dagmar Schimkus
Michael Schmidt
Joe SchoK
Chiara Scortecci
Jerome Scott
Mark Searls, AIA
Raquel Sendra
John Seppanen
Setu Shah
Jody Sheldon
Tadao Shimitzu
David Shultis
Julie Shurtz
Scott Sickeler
Stefen Siebrecht
Adam Silver
Tod Slaboden
Leslie Smith
Jeff Smith
Barry Smyth
Giovanni Soleti
Angelika Solleder
Paul Sorum
Lucy Sosa
Michael Speaks
Madison Spencer
Lucas Steiner
Jon Stephens
Andrea Stipa
Marc Stotzer
Kathleen Sullivan
Masahiro Suzuki
David Swanson
Joseph Tanney
Kim Tanzer
Iikka Tarkkanen
Peter Thaler
Sabine Thiel
Thor Thors
Lisa Toms
David Trautman
Wolfgang Tschapeller
Henry Urbach
Weiland Vajen
Erin Vali
Helena Van gen Hassend
Nicholas Vaucher
Irina Verona
Maximo Victoria
Andres Viditz-Ward
Benjamin Wade
Marcus Wallner
Joseph Walter
Mark Wamble
Janine Washington
Benjamin Wayne
Lois Weinthal
Matthew White
Sarah Whiting
Alexander Wiedemann
Jim Wilson.
Brad Winkeljohn
Jason Winstanley
Marcus Witta
Colby Wong
Corinna Wydler
Evan Yassy
Chi Yi Chang
Faruk Yorgancioglu
Gilly Youner
Leslie Young
David Youse
Harry Zernike
Katinka Zlonicky

Chronological List of Buildings & Projects 建筑及项目年表

*Indicates work featured in this book
(see Selected and Current Works).

Liverpool Cathedral Competition
1960
Liverpool, England

Boston City Hall Competition
1961
Boston, Massachusetts
(with Anthony Eardley)

Boston Architectural Center Competition
1963
Boston, Massachusetts
(with Michael Graves)

The American Institute of Architects Headquarters Competition
1964
Washington, DC
(with Michael Graves)

Jersey Corridor Project: A Case Study of a Linear City in the Jersey Corridor between New York and Philadelphia
1964–1966
(with Anthony Eardley and Michael Graves)

Arts Center Competition
1965
University of California
(with Michael Graves)

Manhattan Waterfront Project
1966
New York, New York
Museum of Modern Art
(with Michael Graves)

***House I**
1967–1968
Princeton, New Jersey
Mr and Mrs Bernard M. Barenholz

Townhouse Project
1968
Princeton, New Jersey

***House II**
1969–1970
Hardwick, Vermont
Mr and Mrs Richard Falk

***House III**
1969–1971
Lakeville, Connecticut
Mr and Mrs Robert Miller

***House IV**
1971
Falls Village, Connecticut

House V
1972

***House VI**
1972–1975
Cornwall, Connecticut
Mr and Mrs Richard Frank

Low-Rise High-Density Housing
1973
Staten Island, New York
New York State Urban Development Corporation

House VIII
1973

***House X**
1975
Bloomfield Hills, Michigan
Mr and Mrs Arnold Aronoff

House 11a
1978
Palo Alto, California
Mr and Mrs Forster

***Cannaregio Town Square**
1978
Venice, Italy
Municipal Government of Venice

***House El even Odd**
1980

Eisenman Robertson Architects

Pioneer Courthouse Square Competition
1980
Portland, Oregon

***Madison Components Plant**
1981–1982
Madison, Indiana
Cummins Engine Company

New Brunswick Theological Seminary
1981–1982
New Brunswick, New Jersey
New Brunswick Theological Seminary

***IBA Social Housing**
1981–1985
Berlin, West Germany
Hauert Noack, GmbH & Company

Beverly Hills Civic Center
1982
Beverly Hills, California

***Travelers Financial Center**
1983–1986
Hempstead, New York
Fair Oaks Development and Schottenstein Properties (with Trott & Bean Architects, Columbus, Ohio)

***Firehouse for Engine Company 233 and Ladder Company 176**
1983–1985
Brooklyn, New York
City of New York

***Wexner Center for the Visual Arts and Fine Arts Library**
1983–1989
Columbus, Ohio
Ohio State University

Fin d'Ou T Hou S
1983

***Romeo and Juliet Castles**
1985
Verona, Italy

***Tokyo Opera House**
1985
Tokyo, Japan
City of Tokyo

Cite Unseen 11
1985
Milan, Italy
XV Triennale di Milano

Hardware
1986
Franz Schneider Brakel GmbH & Co.

House
1986
Palm Beach, Florida
Mr Leslie Wexner

Jewlery
1986
Cleto Munari

***University Art Museum**
1986
Long Beach, California
California State University at Long Beach

***Progressive Corporation Office Building**
1986
Cleveland, Ohio
Progressive Corporation

Museum of Futurism
1986
Rovereto, Italy

Eisenman Architects

Tableware
1986
Swid Powell

***Biocentrum**
1987
Frankfurt am Main, West Germany
J.W. Goethe University

EuroDisney Hotel
1987
Paris, France
Disney Development Company

***Fuller/Toms Loft**
1987
New York, New York
Ms Fuller/Mr Toms

***La Villette**
1987
Paris, France
Establissement Public du Parc de la Vilette

***Carnegie Mellon Research Institute**
1987–1989
Pittsburgh, Pennsylvania
Carnegie Mellon University

***Guardiola House**
1988
Cadiz, Spain
D. Javier Guardiola

***Aronoff Center for Design and Art**
1988–present
Cincinnati, Ohio
University of Cincinnati

***Koizumi Sangyo Office Building**
1988–1990
Tokyo, Japan
Koizumi Sangyo Corporation

***Siena Bank Master Plan**
1988
Siena, Italy
Siena Chamber of Commerce

***Greater Columbus Convention Center**
1989–1993
Columbus, Ohio
Greater Columbus Convention Center Authority

***Banyoles Olympic Hotel**
1989
Banyoles, Spain
Consorci Pel Desenvolupament De La Vila Olimpica

***Cooper Union Student Housing**
1989
New York, New York
The Cooper Union

Zoetermeer Houses
1989
The Netherlands
Geerlings Vastgoed B.V., J.G.A. Geerlings

***Groningen Music-Video Pavilion**
1990
Groningen, The Netherlands
Groningen City Festival

***Nunotani Office Building**
1990–1992
Tokyo, Japan
Nunotani Corporation

***Atocha 123 Hotel**
1990–1993
Madrid, Spain
Sociedad Belga de Los Pinares De el Paular

Knoll Textiles
1990
The Knoll Group

***Rebstockpark Master Plan**
1990–present
Frankfurt, Germany
City of Frankfurt, Dieter Bock and Buropark an der Frankfurter Messe GdR

***Alteka Office Building**
1991
Tokyo, Japan
Alteka Corporation

***Emory Center for the Arts**
1991
Atlanta, Georgia
Emory University

Advanta Haus
1992
Berlin, Germany
Advanta Management AG, Dieter Bock

***Max Reinhardt Haus**
1992
Berlin, Germany
Advanta Management AG, Dieter Bock
Ostinvest, Klaus-Peter Junge and Klaus Dieter

***Nordliches Derendorf Master Plan**
1992
Dusseldorf, Germany
City of Dusseldorf Planning Department

Zurich Insurance Headquarters Study
1992
Frankfurt, Germany
Buropark an der Frankfurter Messe

Friedrichstrasse Competition
1992–1993
Berlin, Germany
Advanta Management AG and Ostinvest

***Haus Immendorff**
1993
Dusseldorf, Germany
Professor Jorg Immendorff

Madgeburg, Damaschkeplatz II
1993
Madgeburg, Germany
Hauert and Noack GmbH & Co.

Cities of Artificial Excavation
1993–1994
Montreal, Canada
Canadian Centre for Architecture

Tours Regional Music Conservatory and Contemporary Arts Center Competition
1993–1994
Tours, France
City of Tours
(with Jean Yves Barrier Architect)

Architecture of Display
1995
New York, New York
Comme des Garçons, The Architectural League

Celebration Fire Station
1995
Orlando, Florida
Disney Development Company

United Nations Headquarters
1995
Geneva, Switzerland

Awards & Exhibitions 获奖及展览情况

*Indicates catalog

奖项

建筑设计荣誉奖
进步建筑杂志（Progressive Architecture）
艺术中心
埃默里大学，亚特兰大，佐治亚州
1993

建筑项目奖章（Architectural Projects Honor Award）
美国建筑师学会，纽约分会
艺术中心
埃默里大学，亚特兰大，佐治亚州
1993

荣誉奖（Citation）
美国建筑师学会，纽约分会
布谷办公楼
布谷公司，东京，日本
1993

国家奖章
美国建筑师学会
韦克斯纳视觉艺术中心
俄亥俄大学，哥伦布，俄亥俄州
1993

建筑设计奖
进步建筑杂志
Alteka 办公楼
东京，日本
1992

建筑设计奖
进步建筑杂志
建筑、艺术系设计与规划
辛辛那提大学，辛辛那提，俄亥俄州
1991

第五界威尼斯国际建筑双年展
建筑、艺术系设计与规划
辛辛那提大学，辛辛那提，俄亥俄州
1991

一等奖
国际邀请竞赛
Rebstockpark 总体规划
法兰克福，德国
1991

国家奖章
美国建筑师学会
小泉三洋大楼
小泉三洋公司
东京，日本
1991

建筑设计奖
进步建筑杂志
巴塞罗那奥林匹克旅馆
巴塞罗那，西班牙
1990

建筑设计荣誉奖
进步建筑杂志
卡内基·梅隆研究协会
卡内基·梅隆大学
匹兹堡，宾西法尼亚州
1990

建筑项目奖章
进步建筑杂志
瓜迪奥拉别墅，加的斯，西班牙
1989

建筑项目奖章
美国建筑师学会，纽约分会
卡内基·梅隆研究协会
卡内基·梅隆大学
匹兹堡，宾西法尼亚州
1989

建筑项目奖章
美国建筑师学会，纽约分会
瓜迪奥拉别墅，加的斯，西班牙
1989

一等奖
国际邀请竞赛
巴塞罗那奥林匹克旅馆
巴塞罗那，西班牙
1989

一等奖
国际邀请竞赛
大哥伦布会议中心
哥伦布，俄亥俄州
1989

金奖
第五届世界建筑双年展
Interarch
卡纸住宅
1989

国家奖章
美国建筑师学会，纽约分会
富勒/汤姆斯阁楼
纽约市，纽约州

室内设计荣誉奖
美国建筑师学会，纽约分会
IBA 社会住宅
柏林，德国
1988

建筑设计荣誉奖
进步建筑杂志
大学艺术博物馆
长滩加利福尼亚大学
长滩，加利福尼亚
1987

特别奖
国际邀请竞赛
生物研究中心
法兰克福，前西德
1987

建筑设计奖
进步建筑杂志
韦克斯纳视觉艺术中心
俄亥俄州立大学，哥伦布，俄亥俄州
1985

设计杰出奖
纽约城市艺术委员会
233 机械公司和 176 梯子公司的消防站
纽约
1985

石狮奖
第三届国际建筑双年展
罗密欧与朱丽叶城堡
1985

一等奖
国际邀请竞赛
韦克斯纳视觉艺术中心
俄亥俄州立大学，哥伦布，俄亥俄州
1985

特别奖
西柏林国际竞赛
南 Friedrichstadt
西柏林，前西德
1981

建筑设计荣誉奖
进步建筑杂志
11a 住宅
1979

建筑设计荣誉奖
进步建筑杂志
X 住宅
1976

展览

人工挖掘之城（Cities of Artificial Excavation）
彼得·埃森曼建筑作品，1978—1988
马德里，西班牙
1995 年 3 月

***人工挖掘之城**
彼得·埃森曼建筑作品，1978—1988
加拿大建筑中心（Center Canadien d ‘Architecture）
蒙特利尔
1999 年 3—6 月

艺术家的色彩：伊门多夫住宅
Galerie fur Architektur Renate Kammer and Angelika Hinrichs
汉堡，德国
1993 年 12 月—1994 年 1 月

M Emory 项目
哈佛大学设计研究院
剑桥，马萨诸塞州
1993 年 12 月—1994 年 1 月

GA International 93：
埃默里大学艺术中心
GA 画廊
东京，日本
1993 年 4 月—5 月
K2 画廊
大阪，日本
1993 年 5 月—6 月

***彼得·埃森曼作品中的格栅、剥落、描图、折叠**
圣保罗艺术博物馆
圣保罗，巴西
1993 年 5 月—6 月

彼得·埃森曼：Why Pink?
Sadock & Uzzan Galerie
巴黎，法国
1993 年 3 月—6 月

***建筑艺术博物馆**
Cantonale 艺术博物馆
卢加诺，瑞士
1992 年 9 月—11 月

法兰克福 Rebstockpark 总体规划
Finanzbehorde Eingang Gansemarkt
汉堡，德国
1992 年 7 月

***1991 年威尼斯第五届国际建筑双年展**
现代艺术中心
辛辛那提，俄亥俄州
1992 年 4 月—5 月
里夫画廊
哥伦布，俄亥俄州
1992 年 6 月—8 月
克里夫兰现代艺术中心
克里夫兰，俄亥俄州
1992 年 9 月—10 月
代顿艺术学院
代顿，俄亥俄州
1992 年 12 月—1993 年 1 月

***世界建筑三年展**
奈良，日本
1992 年 5 月—6 月

***展现法兰克福**
Aedes 画廊
柏林，德国
1991 年 12 月

1991 年威尼斯第五届国际建筑双年展
威尼斯，意大利
1991 年 7 月—8 月

II Tesoro Dell'Architettura
Palazzo Medici-Riccardi
佛罗伦斯，意大利
1990 年 3 月

Snakes and Ladders
Max Protetch 画廊
纽约市，纽约州
1989 年

纽约建筑 1970—1990 年
Deutsche 建筑博物馆
MOPU 画廊
马德里，斯维勒，德国，莫斯科
1988 年

一个建筑的缺席
当代艺术中心
辛辛那提，俄亥俄州
1986年10月—1987年1月

***移动的箭，爱神以及其他爱神：**
一个建筑的缺席
约翰·尼古拉斯展览馆
纽约
1986年4—6月

***移动的箭，爱神以及其他爱神：**
建筑协会
伦敦
1986年2—3月

La Casa Domestica
第17届三年展
米兰，意大利
1986年1月

Le Affinita' Elettive
第16届三年展
米兰，意大利
1985年12月

Nerot Mitzvah：犹太宗教之光的当今概念
"开始，结束，再次开始：烛台
伊斯兰博物馆
1985年9—10月

第三届国际建筑双年展
威尼斯，意大利
1985年7—8月

批判的边缘：对美国当今建筑的反思
齐默曼博物馆，Rutgers大学
新不伦瑞克，新泽西
1985年3—6月

Find'Ou T Hou S
建筑展览协会
伦敦，英国
1985

Cite Un Seen 1
建筑博物馆（Architeckturmuseum）
法兰克福，德国
1984

国际双年展（Bauausstellung）
柏林，德国
1984

跨越乌托邦：美国建筑的转换态度
电影
麦克布莱克伍德工厂
现代博物馆
纽约
1983年12月

讽刺（Follies）：20世纪后期建筑景观
Leo Castelli展廊
纽约
1983年10—11月

家之建筑
曼德维尔艺术展廊，加利福尼亚大学
圣迭戈，加利福尼亚州
1983

回应
Palacio de las Alhajas
马德里，西班牙
1982

米兰三年展（Triennale）
米兰，意大利
1982

建筑模型作品
当代艺术中心
日内瓦，瑞士
Nouveau Musee，Lyons，法国
1982

当今建筑潮流
国家美术馆，亚历山大Soutos博物馆
雅典，希腊
1982

美国自治（Autonomia Americana）埃森曼/海杜克
Stichting建筑博物馆

绘画的时代
威尼斯双年展
威尼斯，意大利
1980

自足建筑
哈佛大学
Frogg博物馆
剑桥，马萨诸塞州
1980

商业住宅
Leo Castelli展廊
纽约
1980

10mmagini per Venezia
威尼斯和鹿特丹
1980

Assonometria
Antonia Jannone
米兰，意大利
1979

Assenza - Presenza
博洛尼亚，意大利
1978

Numberals
纽黑文，康涅狄格州
1978

Abraham，埃森曼，海杜克，罗西
纽约
1977

X住宅
普林斯顿，新泽西州
1977

美国建筑绘画200年
纽约市，纽约州
1977

Biennale Di Venezia
威尼斯，意大利
1976

五位建筑师
那不勒斯，意大利
1973

纽约五（The New York Five）
艺术网
伦敦，英国
1975

五位建筑师
普林斯顿大学建筑学院
普林斯顿，新泽西州
1974

住区的另一个机遇：低层与高层的交替布局
现代艺术博物馆
纽约市，纽约州
1973

Birch Burdette Long Memorial 绘画竞赛
建筑联盟
纽约市，纽约州
1973

当代
罗马，意大利
1973

第15届三年展
米兰，意大利
1973

年轻的纽约建筑师
哥伦比亚大学
纽约市，纽约州
波士顿建筑中心
波士顿，马萨诸塞州
1972

* **博物馆的建筑艺术**
现代艺术博物馆
纽约市，纽约州
1968

* **新城市：建筑与城市改造**
现代艺术博物馆

40之下40（Forty Under Forty）
建筑联盟
纽约市，纽约州
1966

白，金，灰
洛杉矶，加利福尼亚大学
加利福尼亚州
1974

Bibliography 参考文献

*Indicates major critical writings

Books and Articles by Peter Eisenman

"Aspiring to Disagree: Report of the Jury, Twenty-Second Awards Program." *Progressive Architecture* (Stamford, Connecticut, January 1975), pp. 42–65. (Transcription of the jury's comments, including Peter Eisenman as juror.)

Cities of Artificial Excavation: The Work of Peter Eisenman, 1978–1988. New York: Rizzoli International Publications/Montreal: Centre Canadien d'Architecture, 1994. (A catalog for the exhibition of the same name at the Centre Canadien d'Architecture, May 2 to June 19, 1994.)

Eisenman, Peter (guest co-editor). "The City as Artifact." *Casabella* (special issue, nos. 359–60, Milan, November/December 1971), cover (in Italian and English).

Eisenman, Peter (guest editor with Robert A.M. Stern). "White and Gray: Eleven Modern American Architects." *Architecture + Urbanism* (no. 52, Tokyo, April 1975), pp. 25–180 (in Japanese; English summary pp. 2–4).

*Eisenman, Peter (with Jeff Kipnis and Nina Hofer). *Fin D'ou T Hou S,* Folio IV. London: The Architectural Association, 1985.

*Eisenman, Peter et al. *Recente Projecten — Peter Eisenman Recent Projects.* Nijmegen, The Netherlands: SUN Publishing Co., 1989 (in Dutch and English, edited by Arie Graafland).

Eisenman, Peter with Anthony Vidler and Raphael Moneo. *The Wexner Center for the Visual Arts: Ohio State University.* New York: Rizzoli International Publications, 1989.

Eisenman, Peter. "A Review of Alison and Peter Smithson's *Ordinariness and Light.*" *Architectural Forum* (vol. 133, New York, May 1971), pp. 76–80.

Eisenman, Peter. "Aratacism: On the Theoretical Ruins of Arata Isozaki." *Arata Isozaki: Works 30: Architectural Models, Prints, Drawings.* Rikuyo-sha Publishing, Inc., Tokyo, 1992, pp. 186–97 (in Japanese and English).

Eisenman, Peter. "Architecture and the Problem of the Rhetorical Figure." *Architecture + Urbanism* (special feature on Eisenman/Robertson Architects, no. 202, Tokyo, July 1987), cover, pp. 17–22 (in Japanese and English). (Travelers Financial Center, pp. 23–32; Firehouse for Engine Co. 233, pp. 33–48; IBA Social Housing, Berlin, pp. 49–70; Tableware, Calendar, Hardware, Jewelry, pp. 71–80.)

*Eisenman, Peter. "Architecture and the Return of Ornamentation." *Architecture + Urbanism* (no. 211, Tokyo, April 1988), pp. 11–14 (in Japanese).

Eisenman, Peter. "Architecture as a Second Language: The Limits of Representation." Unpublished manuscript, 1985.

Eisenman, Peter. "Architecture as a Second Language: The Texts of Between." In Marco Diani and Catherine Ingraham (eds), *Threshold: Restructuring Architectural Theory.* Evanston: Northwestern University Press, 1989, pp. 69–73. (Reprinted from *Threshold, Journal of The School of Architecture, University of Illinois at Chicago.*)

*Eisenman, Peter. "Architecture as a Second Language: The Texts of Between." *Threshold: Journal of the School of Architecture, University of Illinois at Chicago* (vol. IV, Illinois, Spring 1988), pp. 71–5.

Eisenman, Peter. "Architettura Concettuale: dal Livello Percettivo della Forma ai Suoi Significati Nascoti." In *Contemporanea, Incontri Internazionali d'Arte.* Florence: STIAV Press, 1973, pp. 317–19. (Catalog for the exhibition, in Italian and English; Italian translation of "Notes on Conceptual Architecture.")

*Eisenman, Peter. "Architettura e figura retorica." *Eidos* (no. 1, Asolo, 1987), pp. 12–19. (Italian translation by Fabio Nonis of "Architecture and the Problem of the Rhetorical Figure.")

Eisenman, Peter. "Arquitectura Concettuale: Dal Livello Percettivo della Forma ai Suoi Significati Nascoti." *Casabella* (no. 386, Milan, February 1974), pp. 25–7. (Italian translation of "Notes on Conceptual Architecture II"; in Italian and English.)

Eisenman, Peter. "Aspects du Modernism: La Maison Domino ou le Signe Autoreferentiel." *Les Cahiers de la Recherche Architecturale 12* (Paris, November 1982), pp. 58–65. (French translation of "Aspects of Modernism: The Maison Domino and the Self-referential Sign.")

*Eisenman, Peter. "Aspects of Modernism: Maison Dom-ino and the Self-Referential Sign." *Oppositions 15/16* (Cambridge, Winter/ Spring 1980), pp. 119–28, and *Arsenale Editrice* (Venezia: 1989).

Eisenman, Peter. "Banff Transcripts." *Section A* (vol. 2, nos. 3/4, Montreal, September, 1984), pp. 20–6.

*Eisenman, Peter. "Behind the Mirror: On the Writings of Philip Johnson." *Oppositions 10* (Cambridge, Fall 1977), pp. 1–13.

Eisenman, Peter. "Biology Center for the J.W. Goethe University of Frankfurt, Frankfurt am Main, 1987." *Assemblage* (no. 5, Cambridge, February 1988), pp. 29–50.

Eisenman, Peter. "Blue Line Text." *Deconstruction: Omnibus Volume.* New York: Rizzoli International Publications, 1989, pp. 150–1. (Reprinted from *Architectural Design: Contemporary Architecture.*)

Eisenman, Peter. "Cannaregio." *IAUS Catalog 3: Idea as Model*. New York: Rizzoli International Publications, 1981, p. 120.

Eisenman, Peter. "Cardboard Architecture." *Architecture and Urbanism* (no. 11, Tokyo, November 1973), pp. 185–9.

Eisenman, Peter. "Cardboard Architecture: House I." (Non-published text, August 15, 1972.)

Eisenman, Peter. "Castelli di Carte: Due Opere di Peter Eisenman." *Casabella* (no. 374, Milan, February 1973), pp. 17–31 (Italian translation of "Houses of Cards", in Italian and English).

Eisenman, Peter. "Castillos de Romeo y Julieta." *Arquitectura 270* (Madrid, January/ February 1988, pp. 66–79 (in Spanish).

Eisenman, Peter. "CHORA L WORKS." *Arquitectura 270* (Madrid, January/February 1988), pp. 53–65 (in Spanish).

Eisenman, Peter. "Como casas de naipes." *El Paseante* (no. 8, Madrid, 1988), pp. 96–109 (in Spanish).

Eisenman, Peter. "Conceptual Architecture II: Double Deep Structure I." *Architecture + Urbanism* (no. 3, Tokyo, March 1974), pp. 83–8.

Eisenman, Peter. "Dall Ogetto alla Relazionita: la Casa del Fascio di Terragni." In L. Ferrario & D. Pastore, *Giuseppe Terragani: La Casa del Fascio.* Rome: Instituto Mides, 1982, pp. 54–8. (Italian translation of "From the Object to Relationship: the Casa del Fascio.")

Eisenman, Peter. "Dall'oggetto alla Relazionalita: la Casa del Fascio di Terragni." *Casabella* (no. 344, Milan, January 1970), pp. 38–41. (Italian translation of "From Object to Relationship: The Casa del Fascio.")

Eisenman, Peter. "Das Carnegie Mellon Research Institute: Uberlegungen des Architekten." *Archithese* (Zurich, February 1989), pp. 31–6.

Eisenman, Peter. "Das Guardiola–Haus: Uberlegungen des Architekten." *Archithese* (1:89, Zurich, January/February 1989), pp. 21–5 (in German, translated by Simon Hubacher).

*Eisenman, Peter. "Das Symbol." In *Das Neue Berlin.* Berlin: Gebr. Mann Verlag, 1987, pp. 86–90 (in German).

Eisenman, Peter. "Divisions of Excess: The Emptiness of Hiromi Fujii."

Eisenman, Peter. "El Fin de lo Clasico: El Fin del Comienzo, El Fin del Fin." *Arquitecturas Bis* (Barcelona, March 1984), pp. 29–37. (Spanish translation of "The End of the Classical.")

Eisenman, Peter. "En Terror Firma: Auf den Faehrten der Grotexte." *Archithese* (1:89, Zurich, January/February 1989), pp. 29–30 (in German). (Translation of "En Terror Firma: In Trails of Grotextes.")

Eisenman, Peter. "En Terror Firma: In Trails of Grotextes." *Deconstruction: Omnibus Volume.* New York: Rizzoli International Publications, 1989, pp. 152–3. (Reprinted from *Form, Being, Absence.*)

*Eisenman, Peter. "En Terror Firma: In Trails of Grotextes." *Form, Being, Absence/ Architecture and Philosophy: Pratt Journal of Architecture* (New York, 1988), pp. 111–21.

Eisenman, Peter. "En Terror Firma: In Trails of Grotextes." In Arie Graafland (ed.), *Recent Projects — Peter Eisenman.* Nijmegen, The Netherlands: SUN Publishing Co., 1989, pp. 19–24 (in Dutch and English).

Eisenman, Peter. "En Terror Firma: In trails of Grotextes." *Architectural Design: Decontructivist Architecture II* (vol. 59, London, 1989), pp. 40–43. (Reprinted from *Form, Being, Absence.*)

Eisenman, Peter. "Fin d'Ou T Hou S." In *Follies: Architecture for the Late Twentieth Century Landscape.* New York: Rizzoli International Publications, 1983, pp. 54–6. (Catalog from exhibition at Leo Castelli Gallery.)

Eisenman, Peter. "Fin d'Ou T Hou S." In *Follies: Arquitectura Para el Paisaje de Finales del Siglo XX.* Madrid: MOPU Arquitectura, 1984, pp. 54–7. (Spanish translation.)

Eisenman, Peter. "Firehouse for Engine Company 233, Ladder Company 176, Brooklyn, New York." *Architectural Design 54* (London, November/December 1984), pp. 14–15.

Eisenman, Peter. "Folding in Time: The Singularity of Rebstock." In Greg Lynn (ed.), *Folding in Architecture.* Architectural Design Profile No. 102. London: Academy Group Ltd./Architectural Design, 1993, pp. 22–5. (Rebstockpark Master Plan, Alteka Office Building, Center for the Arts, pp. 26–35.)

*Eisenman, Peter. "From Golden Lane to Robin Hood Gardens; Or If You Follow the Yellow Brick Road, It May Not Lead to Golder's Green." *Architectural Design* (vol. 42, no. 9, London, September 1972), pp. 557–73, 588–92, and *Oppositions 1* (Cambridge, September 1973), pp. 27–56.

*Eisenman, Peter. "From Object to Relationship II: Giuseppe Terragni." *Perspecta 13–14 : The Yale Architectural Journal* (Cambridge: MIT Press, 1971), pp. 36–75.

Eisenman, Peter. "Genuinamente Ingles: la Destruction de la Caja." *Arquitectura* (no. 211, Madrid, March/April 1978), pp. 54–71. (Spanish translation of "Real and English: Destruction of the Box.")

Eisenman, Peter. "Giuseppe Terragni: Casa Giuliani-Frigerio, Como." *Lotus 42* (Milan, 1984), pp. 69–71 (in Italian and English).

Eisenman, Peter. "Hollein's Cave(at): The Haas Haus." *Architecture + Urbanism* (Tokyo, January 1992), pp. 122–3 (in English and Japanese).

Eisenman, Peter. "House El Even Odd." *Architecture + Urbanism* (no. 123, Tokyo, December 1980), pp. 96–8 (in Japanese and English).

Eisenman, Peter. "House II." *IAUS Catalog 3: Idea as Model.* New York: Rizzoli International Publications, 1981, pp. 34–5.

*Eisenman, Peter. "House III: To Adolph Loos and Bertolt Brecht." *Progressive Architecture* (Stamford, Connecticut, May 1974), p. 92.

Eisenman, Peter. "House VI." *Progressive Architecture* (Stamford, Connecticut, June 1977), pp. 57–9.

Eisenman, Peter. "House VI: Frank Residence." Text for catalog and exhibition *Houses for Sale* at the Leo Castelli Gallery. New York: Rizzoli International Publications, 1980, pp. 172–6.

*Eisenman, Peter. "House VI: The Frank Residence." *GA Document Special Issue: 1970–1980* (Tokyo, 1980), pp. 172–3 (in Japanese and English).

Eisenman, Peter. "House X." *At Home with Architecture: Contemporary Views of the House.* San Diego: Mandeville Art Gallery, 1983, pp. 18–21 (catalog).

Eisenman, Peter. "House X." *IAUS Catalog 3: Idea as Model.* New York: Rizzoli International Publications, 1981, pp. 82–3.

Eisenman, Peter. "Il Futuro della Tradizione: Una Ricerca di *Perspecta 12,* 'Rivista Minore'." *Casabella* (no. 345, Milan, February 1970), pp. 28–33. (Italian translation of "The Big Little Magazine.")

Eisenman, Peter. "Il Wexner Center for Visual Arts a Columbus, Ohio." In Pippo Ciorra (ed.), *Botta, Eisenman, Gregotti, Hollein: musei* (Milano, 1991), pp. 65–81, cover (in Italian).

Eisenman, Peter. "In My Father's House are Many Mansions." *IAUS Catalog 12: John Hejduk: Seven Houses.* New York: Rizzoli International Publications, 1980, pp. 8–20.

Eisenman, Peter. "In Zeit einfalten: Die Singularitat des Rebstock-Gelandes." *Frankfurt Rebstockpark: Folding in Time.* Munich: Prestel-Verlag/Frankfurt: Deutsches Architekturmuseum, 1992 (in German). (Catalog for the exhibition of the Rebstockpark Master Plan project at the Deutsches Architekturmuseum in Frankfurt.)

Eisenman, Peter. "Indicencies: In the Drawing Lines of Tadao Ando." *Tadao Ando: Details.* Yukio Futagawa (ed.), Tokyo: A. D. A. Edita, 1991, pp. 6–9. (in Japanese and English).

*Eisenman, Peter. "It Signifies Nothing ... It Is All Dissimulation." Unpublished manuscript, 1985.

*Eisenman, Peter. "L'inizio la fine e poi di nuovo l'inizio-alcune osservazioni sull'idea di scaling." In Gabriella Belli e Franco Rella (eds), *La Citta' e le Forme.* Milan: Mazzotta, pp. 93–8 (in Italian).

Eisenman, Peter. "La Ciudad de la Excavacion Artificial." *Arquitectura* (Madrid, January/February 1984), pp. 42–3. (Spanish translation of "The City of Artificial Excavation.")

Eisenman, Peter. "La Ciudad de la Excavacion Artificial." *Arquitecturas Bis* (Barcelona, December 1983), pp. 21–3. (Spanish translation of "The City of Artificial Excavation.")

Eisenman, Peter. "La Futilidad de los Objectos." *Arquitectura* (Madrid, January/ February 1984), pp. 33–56. (Spanish translation of "The Futility of Objects.")

Eisenman, Peter. "La Maison Dom-ino e il Segno Autoreferenziale." *Sulle Trace di Le Corbusier.* San Marco: Arsenale Press, 1989, pp. 21–35.

Eisenman, Peter. "Laboratorios in Frankfurt." *Arquitectura 270* (Madrid, January/February 1988) (in Spanish).

*Eisenman, Peter. "Le Rappresentazioni del Dubbio: Nel Segno del Segno." *Rassegna 9* (Milan, March 1982), pp. 69–74 (in Italian); also "The Representations of Doubt: At the Sign of the Sign." *Rassegna 9,* (English original, unpaginated).

*Eisenman, Peter. "Lecturas de MiMESis: Malinterpretadas no Significan NADA." *Mies Van der Rohe: Su Architectura v Sus Discipulos.* Madrid: La Direccion General para la Vivienda y Arquitectura, 1987, pp. 92–104. (Exhibition catalog. Spanish translation of "miMISes READING: does not mean A THING", German version.)

Eisenman, Peter. "Meier's Smith House: Letter to the Editor." *Architectural Design.* (London, August 1971), p. 520.

*Eisenman, Peter. "Metaphysics, Mystique, and Power." *AA Files* (no. 12, London, Summer 1986), p. 107. (A review of John Hejduk's *Mask of Medusa.*)

*Eisenman, Peter. "miMISes READING: does not mean A THING." In *Mies Reconsidered: His Career, Legacy. and Disciples.* Chicago: Art Institute of Chicago/New York: Rizzoli Publications, 1986, pp. 86–98 (catalog).

*Eisenman, Peter. "miMISes READING: does not mean A THING." *Mies van der Rohe: Vorbild und Vermaechtnis.* Frankfurt: Deutsches Architekturmuseum, pp. 85–95. (Exhibition catalog. German translation of "miMISes READING: does not mean A THING.")

*Eisenman, Peter. "Misreading Peter Eisenman." *Houses of Cards.* New York: Oxford University Press, 1987, pp. 167–86.

Eisenman, Peter. "Modern Architecture 1919–1939: Polemics." Princeton University Library, 1968 (catalog with introduction and annotations for an exhibition of architectural books; unpublished).

*Eisenman, Peter. "Moving Arrows, Eros and Other Errors." *Arquitectura* (no. 270, Madrid, January/February 1988), pp. 67–81 (in Spanish and English). (Reprinted from *The Culture of Fragments.*)

*Eisenman, Peter. "Moving Arrows, Eros and Other Errors." *The Culture of Fragments,* Precis 6. New York: Columbia Graduate School of Architecture, Planning and Preservation, Spring 1987, pp. 139–43.

*Eisenman, Peter. "Moving Arrows, Eros and other Errors." *Chiasmos II: Strategier* (Copenhagen, December 1988), pp. 29–32. (Danish translation by Carsten Juel-Christiansen.)

Eisenman, Peter. "Moving Arrows, Eros, and Other Errors: An Architecture of Absence." III Architectural Biennale, Venice, 1985. (Original text to accompany exhibition panels.)

Eisenman, Peter. "Moving Arrows, Eros, and Other Errors: An Architecture of Absence." *SD* (special feature on Peter Eisenman, Tokyo, March 1986), pp. 53–7. (Japanese translation of text for Venice Biennale.)

Eisenman, Peter. "Notes on a Conceptual Architecture." *Formalism, Realism, Contextualism* (Tokyo, 1979). (Japanese translation of "Notes on a Conceptual Architecture.")

Eisenman, Peter. "Notes on Conceptual Architecture IIA." *Environmental Design Research Association* (vol. II, Stroudsburg, Pennsylvania, 1973), pp. 319–22.

Eisenman, Peter. "Notes on Conceptual Architecture IIA." *On Site* (no. 4, New York, 1973), pp. 41–4.

Eisenman, Peter. "Notes on Conceptual Architecture: Towards a Definition." *Casabella* (nos. 359–360, Milan, November/December 1971), pp. 48–58 (in Italian and English).

*Eisenman, Peter. "Notes on Conceptual Architecture: Towards a Definition." *Design Quarterly* (special double issue, nos. 78–9, 1970), pp. 1–5, cover.

Eisenman, Peter. "Oltre Lo Sguardo: L'Architettura nell'Epoca dei Media Elettronici." ("Visions' Unfolding: Architecture in the Age of Electronic Media.") *Domus* (Milan, January 1992), pp. 17–24 (in Italian and English).

Eisenman, Peter. "Peter Eisenman." *Der Postmoderne Salon: Architekten uber Architecten* (Berlin, 1991), pp. 88–95.

Eisenman, Peter. "Peter Eisenman." *Yale Seminars in Architecture* vol. 2. New Haven: Yale University Press, 1982, pp. 49–87.

Eisenman, Peter. "Peter Eisenman: The Affects of Singularity." *A.D.* (vol. 62, November/December 1992).

Eisenman, Peter. "Post-Functionalism." *Arquitecturas/Bis 22* (Barcelona, 1978), pp. 6–12. (Spanish translation of "Postfunctionalism.")

*Eisenman, Peter. "Post-Functionalism." *Oppositions 6* (Cambridge, Fall 1976), pp. i–iii (editorial).

*Eisenman, Peter. "Post/EI Cards: A Reply to Jacques Derrida." *Assemblage* (no. 12, Cambridge, August 1990), pp. 14–17.

Eisenman, Peter. "Postfunktionalismus." In Gerald R. Blomeyer & Barbara Tietze, *Opposition zur Moderne: Aktuelle Positionen in der Architekur.* Weisbaden: Friedr. Vieweg & Sohn Verlagsgesellschaft, 1980. (German translation of "Post-Functionalism.")

*Eisenman, Peter. "Real and English: Destruction of the Box I." *Oppositions 4* (Cambridge, May 1974), pp. 5–34.

Eisenman, Peter. "Recent Works." *Architectural Design: Deconstruction II* (no. 1/2, London, 1989), pp. 40–61.

*Eisenman, Peter. "Representations of the Limit: Writing a 'Not Architecture'." In Daniel Libeskind, *Chamberworks: Architectural Meditations on Themes from Heraclitus.* London: Architectural Association Folio, 1983, pp. 6–8.

Eisenman, Peter. "Residence, Critique of Weekend House by Philosopher, Sociologist, and Architect Himself." *Progressive Architecture* (Stamford, Connecticut, June 1977), pp. 57–67.

Eisenman, Peter. "Roosevelt Island Competition." *Controspazio* (no. 4, Rome, December 1975), pp. 30–1.

*Eisenman, Peter. "Sandboxes: House 11a." *Architecture + Urbanism* (special issue, no. 112, Tokyo, January 1980), pp. 221–4, cover (in Japanese and English).

Eisenman, Peter. "Semiotica e Architettura (la Casa del Fascio), *Casabella* (Milan, October, 1977), p. 25 (in Italian and English).

Eisenman, Peter. "Shore Birds, or The Rocket's Red Glare." Introduction to Brad Collins & Diane Kasprowicz (eds), *Gwathmev Siegel: Buildings and Projects 1982–1992.* New York: Rizzoli International Publications, 1993, pp. 4–13.

Eisenman, Peter. "Sports of Our Climate." *New York Times* (July 16, 1983), opposite editorial page.

Eisenman, Peter. "Stadium Ghosts." *New York Times* (April 2, 1980), opposite editorial page.

*Eisenman, Peter. "Strong Form, Weak Form." *Architecture in Transition: Between Deconstruction and New Modernism.* Munich: Prestel, 1991, pp. 33–45 (in German and English).

*Eisenman, Peter. "TEXt AS Zero, or The Destruction of Narrative." In Lars Lerup, *Planned Assaults.* Centre Canadienne d'Architecture/MIT Press, 1987, pp. 93–8.

*Eisenman, Peter. "The Authenticity of Difference: Architecture and the Crisis of Reality." *Center: A Journal for Architecture in America, University of Texas at Austin* (vol. 4, New York, 1988), pp. 50–7.

Eisenman, Peter. "The Author's Affect: Passion and the Moment of Architecture." *Anyone* (New York, 1991) pp. 200–11.

Eisenman, Peter. "The Author's Affect: Passion and the Moment of Architecture." *Critical Space* (June 1992), pp. 213–19 (in Japanese). (Translation of English text published in *Anyone*.)

Eisenman, Peter. "The Beginning, the End, and the Beginning Again." *Nerot Mitzvah: Contemporary Ideas for Light in Jewish Ritual.* Catalog for the Israel Museum, Jerusalem, 1986, p. 62.

*Eisenman, Peter. "The Beginning, the End, and the Beginning Again: Some Notes on an Idea of Scaling." *SD* (special issue on Peter Eisenman, Tokyo, March 1986), pp. 6–7, cover (in Japanese).

Eisenman, Peter. "The Big Little Magazine: *Perspecta 12*" and "The Future of the Architectural Past." *Architectural Forum* (vol. 131, no. 3, New York, October 1969), pp. 74–5, 104.

Eisenman, Peter. "The Blue Line Text." *Architectural Design: Contemporary Architecture.* (London, January/February 1989), pp. 6–9.

Eisenman, Peter. "The Carnegie Mellon Research Institute." *GA Documents 23* (Tokyo, April, 1989), pp. 82–4.

Eisenman, Peter. "The City as Memory and Immanence." *Zone* (Toronto 1986), pp. 440–1.

Eisenman, Peter. "The City as Memory and Immanence." *Zone* (vol. 1, New York, Fall 1985), pp. 440–1.

Eisenman, Peter. "The City of Artificial Excavation." (Extract) *Architectural Design Profile* (London, 1983), pp. 24–7.

*Eisenman, Peter. "The City of Artificial Excavation." *Architectural Design* (London, nos. 1–2, January 1983), pp. 91–3.

Eisenman, Peter. "The End of the Classical." *Montana State University Architectural Review* (vol. 3, Bozeman, Montana, Spring 1984), pp. 2–11. (Reprinted from *Perspecta 21.*)

*Eisenman, Peter. "The End of the Classical." *Perspecta 21: The Yale Architectural Journal* (Cambridge, Summer 1984), pp. 154–72.

Eisenman, Peter. "The End of the Classical: The End of the Beginning, the End of the End." *Chiasmos II: Strategier* (Copenhagen, December 1988), pp. 9–21. (Danish translation by Carsten Juel-Christiansen.)

Eisenman, Peter. "The Futility of Objects." *Harvard Architecture Review* (vol. 3, Cambridge, Winter 1984), pp. 65–82.

*Eisenman, Peter. "The Futility of Objects." *Lotus 42* (Milan, February 1984), pp. 63–75. (A condensed version of "The Futility of Objects" in Italian.)

*Eisenman, Peter. "The Graves of Modernism." *Oppositions 12* (Cambridge, Spring 1978), pp. 36–41.

*Eisenman, Peter. "The House of the Dead as the City of Survival." *IAUS Catalog 2: Aldo Rossi in America: 1976–1979.* New York: Institute for Architecture and Urban Studies, 1979, pp. 4–15 (introduction).

*Eisenman, Peter. "The Houses of Memory: The Texts of Analogy." Introduction to Aldo Rossi, *The Architecture of the City.* Cambridge: MIT Press, 1982, pp. 3–12.

Eisenman, Peter. "The OSU Center for the Visual Arts" and "Fin D'ou T Hou S." *Architectural Design* (vol. 55, nos. 1/2, London, January 1985), pp. 44–55.

Eisenman, Peter. "The Story of AND O." *Tadao Ando: The Yale Studio & Current Works.* New York: Rizzoli International Publications, 1989, pp. 137–9.

Eisenman, Peter. "Three Texts for Venice." *Domus* (no. 611, Milan, November 1980), pp. 9–11, cover (in Italian and English).

Eisenman, Peter. "Towards an Understanding of Form in Architecture." *Architectural Design* (London, October 1963), pp. 457–8.

*Eisenman, Peter. "Transformations, Decompositions and Critiques: House X." *Architecture + Urbanism* (special issue, no. 112, Tokyo, January 1980), pp. 14–151, cover (in Japanese and English).

Eisenman, Peter. "Unfolding Events: Frankfurt Rebstockpark and the Possibility of a New Urbanism." *Unfolding Frankfurt.* Berlin: Ernst & Sohn, 1991, pp. 8–17 (in German and English).

Eisenman, Peter. "University Art Museum, California State University, Long Beach, California." *GA Document 18: GA International '87* (Tokyo, April 1987), pp. 13–15.

Eisenman, Peter. "University Campus, Long Beach, California: The Museum Rediscovered." *Lotus 50* (Milan, 1986), pp. 128–35 (in Italian and English).

Eisenman, Peter. "Vioes que se desdobram: A arquitetura na Epoca da Midia Eletronica." *Oculum 3* (Sao Paulo, March, 1993), pp. 14–22 (in Portuguese and English).

Eisenman, Peter. "Wettbewerbsprojeckt Rebstock in Frankfurt, 1991." *Werk, Bauen + Wohnen* (March 1992).

Eisenman, Peter. "Yale Seminars in Architecture, Peter Eisenman." *Architecture + Urbanism* (no. 166, Tokyo, July 1984), pp. 19–26. (Japanese translation of Yale Seminars in Architecture.)

Eisenman, Peter. "Zum Forshungsinstut der Carnegie Mellon Universitat und der Verwaltungsgebaude der Oxford Development Company." *Dekonstruktion? Dekonstructivismus?...* (Lengerich, 1990), pp. 110–16.

*Eisenman, Peter. *House X.* New York: Rizzoli International Publications, 1982.

*Eisenman, Peter. *Houses of Cards.* New York: Oxford University Press, 1987.

Eisenman, Peter. Introduction to *Writings* by Philip Johnson. New York: Oxford University Press, 1979, pp. 10–25. (Edited version of "Behind the Mirror.")

*Eisenman, Peter. *La Fine del Classico.* Venice: Cluva, 1987. (Italian translations of essays and articles. Introduction by Franco Rella, postscript by Renato Rizzi, translated by Renato Rizzi and Daniela Toldo.)

*Eisenman, Peter. *Moving Arrows. Eros, and Other Errors: An Architecture of Absence,* Box 3. London: The Architectural Association, 1986.

Eisenman, Peter. Preface to *IAUS Catalog 3: Idea as Model.* New York: Rizzoli International Publications, 1981, p. 1.

*Eisenman, Peter. *The Formal Basis of Modern Architecture.* Dissertation for the Degree of Doctor of Philosophy. Cambridge: University of Cambridge, 1963 (unpublished).

Eisenman, Peter. *The Ohio State Center for the Visual Arts Competition.* New York: Rizzoli International Publications, 1984.

Frankfurt Rebstockpark: Folding in Time. Munich: Prestel-Verlag/Frankfurt: Deutsches Architekturmuseum, 1992 (in German). (A catalog for the exhibition of the Rebstockpark Master Plan at the Deutsches Architekturmuseum in Frankfurt.)

La Diversitat Cultural en el Dialeg Nord-Sud. Barcelona: Departament de Cultura de la Generalitat de Catalunya/Institut Catala de Bibliografia, 1991, pp. 111–14 (in Catalan).

Peter Eisenman & Frank Gehry. New York: Rizzoli International Publications, 1991. (A catalog for the Fifth International Exhibition of Architecture at the 1991 Venice Biennale.)

Unfolding Frankfurt. Berlin: Ernst & Sohn, 1991 (in German and English). (A catalog for the exhibition of Rebstockpark project at Aedes Gallery, Berlin.)

Forthcoming Books

Eisenman, Peter (with Jacques Derrida and Jeff Kipnis). *CHORA L WORKS.* London: Architectural Association.

*Eisenman, Peter. *Giuseppe Terragni: Transformations, Decompositions, Critiques.* New York: Rizzoli International Publications.

Rakatansky, Mark (ed.). *Collected Essays of Peter Eisenman.* Princeton, NJ: Princeton Architectural Press, 1994.

Articles on Eisenman Architects

"A Skyscraper For A Post-Phallic Age." *Harper's Magazine* (July 1993), p. 12.

Abrams, Janet. "Misreading Between the Lines." *Blueprint* (no. 14, London, February 1985), pp. 16–17.

"Am Rebstock soll eine neue Trabantenstadt entstehen." *Wiesbadener Kurier* (December 13, 1992) (in German).

American Institute of Architecture, New York Chapter. "Nunotani Headquarters Building", "Center for the Arts." *New York Architecture Volume 6.* New York: AIA New York Chapter, 1993, pp. 9, 14–15, 35, 38–39.

Anderson, Kurt. "A Crazy Building in Columbus: Peter Eisenman, Architecture's

Bad Boy, Finally Hits His Stride." *Time* (November 20, 1989).

"Anything Goes." *New Criterion* (New York, January 1992), pp. 2–3.

"Architektur Ist Ein Schwaches Medium." *Taz Hamburg* (Hamburg, July 7, 1992), p. 19 (in German).

"Arquitecto americano expoe maquetes exoticas no Masp." *Folha de S. Paulo* (May 18, 1993), Section D, p. 3.

Armaly, Fareed. "Plan." *Intelligente Ambiente II* (1994), pp. 50–67.

Baird, George. "The Return of Peter Eisenman." *GSD News* (1994), pp. 21–3.

"BAC Craft Awards Highlight Craftsmanship." *Bricklayers and Allied Craftsman* (February 1993).

"Bahnhof Friedrichstrasse." *Lotus 80* (1994), pp. 118–22.

Barris, Roann. "Eisenman and the Erosion of Truth." *20/1 Art & Culture* (School of Architecture & Design, University of Illinois at Chicago, 1:1, Spring 1990), pp. 20–37.

Beckelmann, Jurgen. "Der Optimismus der Architekten." *Frankfurter Rundschau* (December 12, 1992).

Benjamin, Andrew (ed.). *Re:Working Eisenman.* London: Academy Editions/Ernst & Sohn, 1993.

Benson, Robert. "Convention." *Inland Architect* (July/August 1993), pp. 53–8.

Benson, Robert. "Eisenman's Architectural Challenge." *New Art Examiner* (Summer 1990), pp. 27–30.

Benson, Robert. "Wexing Eloquent in Columbus." *Inland Architect* (May/June 1990), pp. 34–43, cover.

"Berlin's Reinhardt Homage: Why Not Eisenman?" *International Herald Tribune* (December 1992) pp. 19–20.

"Berlin: A New Twist." *ARTnews* (April 1993).

Bischoff, Michael. "Warum Le Corbusier die Welt nicht verandert hat." *Frankfurter Allgemeine Zeitung* (January 4, 1992).

Blackford, Darris. "Architect, Critics Give Convention Center High Marks." *Columbus Dispatch* (February 27, 1993).

Blackford, Darris. "Great Expectations Convention Center's Curtain Going Up." *Columbus Dispatch* (February 28, 1993).

Blackford, Darris. "Open House Draws Crowd." *Columbus Dispatch*

Blackford, Darris. "Those Who Built It Like the Center." *Columbus Dispatch* (February 28, 1993), p. 1.

Bletter, Rosemary Haag. "Five Architects — Eisenman, Graves, Gwathmey, Hejduk, Meier." *Journal of the Society of Architectural Historians* (New York, May 1979), pp. 205–7.

Blundo, Joe. "If the New Convention Center Were a House ..." *Columbus Dispatch* (March 7, 1993).

Bottiger, Helmut. "Wider das Kunstfeindliche Subentionstheater." *Frankfurter Rundschau* (Frankfurt, July 29, 1992).

Bouw, Matthjis "Peter Eisenman — Towards a Supple Geometry." *Wiederhall 16* (1993), pp. 40–55.

Campbell, Robert. "The Fascinating and Abstract World of Peter Eisenman." *Boston Globe* (May 27, 1986), p. 67.

Castro, Ricardo. "Cryptic Philosophy, New Direction in Architecture." *Gazette* (Montreal, March 1994), p. 2

"Center's Architecture Draws International Recognition." *Columbus Dispatch* (March 7, 1993), pp. 70, 86.

Ciorra, Pippo (ed.). *Peter Eisenman: opere e progetti.* Serie Documenti di Architettura, no. 71. Milan: Electa, 1993.

"College of Design, Architecture, Art and Planning, University of Cincinnati, Ohio." *Zodiac 7* (Milan, March/August 1992), pp. 96–109 (in English and Italian).

Color of an Artist: Haus Immendorff. Hamburg: Artfound Print Co./Galerie fur Architektur Renate Kammer und Angelika Hinrichs, owners, 1993 (in German and English). (A catalog for the exhibition of the Haus Immendorff project at Galerie fur Architektur Renate Kammer und Angelika Hinrichs, Hamburg.)

Colquhoun, Alan. "The Competition for the Center for the Visual Arts at Ohio State University." *The Ohio State Center for the Visual Arts Competition.* New York: Rizzoli International Publications, 1984, pp. 132–5.

"Convention Center Earns Foundation Award." *Columbus Dispatch* (October 8, 1993).

Cooleybeck, Patrick and John Abela. "Do It For Van Gogh." *Dimensions: Journal of University of Michigan College of Architecture* (vol. 6, Ann Arbor, 1992), pp. 92–101.

"Costruire L' Impossibile" *Cultura* (November 1992) p. 18.

Cuomo, Alberto. "Architettura e Negativita." *Controspazio* (anno X, Rome, July/August 1978), pp. 42–7 (in Italian).

Dal Co, Francesco. "Ten Architects in Venice." *Architecture + Urbanism* (no. 121, Tokyo, October 1980), pp. 26–33 (in English and Japanese).

Dal Co, Francesco. "Ten Architects in Venice." In *10 Immagini Per Venezia.* Venice: Officina Edizioni, pp. 55–65 (in Italian). (Catalog of the exhibition in Venice, April 1–30, 1980.)

Dannatt, Adrian. "Any Old How." *Building Design* (London, March 6, 1992), p. 14.

Darley, Gillian. "Ideas Sit Heavily on the Wall." (March 15, 1993).

Davidson, Cynthia. "A Game of Eisenman Seeks." *Architecture + Urbanism* (Tokyo, September 1991), pp. 12–13, cover (in Japanese and English).

Davis, Douglas. "Modernism Revisited." *The Ohio State Center for the Visual Arts Competition.* New York: Rizzoli International Publications, 1984.

Davis, Douglas. "The Death of Semiotics (in Late Modern Architecture), the Corruption of Metaphor (in Post-Modernism), the Birth of Punctum (in Neomania)." *Art Forum* (May, 1984), pp. 56–63.

Dawson, Layla. "Peter Eisenman: The Man Who Discovered Folding Architecture." *Pace Magazine* (vol. 47, Hong Kong, October 1992).

de Mendonca, Denise Xavier. "Peter Eisenman — Blue Line Text; Edificio-Sede da Nunotani Toquio, Japoa; Centro de Convencoes, Greater Columbus, Ohio; Concurso Nordliches Derendorf, Dusseldorf, Alemanha"; "Instabilidade e Turbulencia." *Arquitectura e Urbanismo* (April/May 1993), pp. 46–66, cover.

"Del mundo mecanico al electronico." *Arquitectura* (Buenos Aires, December 12, 1992) (in Spanish).

"Der Turmbau am Zirkus." *Frankfurter Allgemeine* (Frankfurt, November 17, 1992), p. 1.

Derrida, Jacques. "A Letter to Peter Eisenman." *Assemblage* (no. 12, August 1990), pp. 7–13.

Derrida, Jacques. "A Letter to Peter Eisenman." *Critical Space* (no. 6, Tokyo, 1992), pp. 98–106 (in Japanese).

Derrida, Jacques. "Why Peter Eisenman Writes Such Good Books." *Architecture + Urbanism: EISENMANAMNESIE* (extra edition, Tokyo, August 1988), pp. 113–24, cover (in Japanese and English). (Translation of the French original in Jacques Derrida, *Psyche: L'invention de l'autre,* Paris: Galilee, 1987, pp. 495–508.)

Derrida, Jacques. "Why Peter Eisenman Writes Such Good Books." In Arie Graafland (ed.), *Peter Eisenman: Recent Projects,* Nijmegen: SUN, 1989, pp. 169–82 (in Dutch and English). (Reprinted from *Architecture + Urbanism: EISENMANAMNESIE,* August 1988, pp. 113–24.)

Derrida, Jacques. "Why Peter Eisenman Writes Such Good Books." In Marco Diani and Catherine Ingraham (eds). *Threshold: Restructuring Architectural Theory*. Evanston: Northwestern University Press, 1989, pp. 99–105. (Reprinted from *Architecture + Urbanism: EISENMANAMNESIE*, August 1988, pp. 113–24.)

di Forti, Massimo. "Costruire l'impossibile." *Il Messaggero* (Rome, November 3, 1992), p. 18 (in Italian).

Dibar, Carlos L. "La deconstruccion de Nunotani: Un edificio de oficinas en Tokio." *El Cronista Arquitectura & Disegno* (Buenos Aires, December 2, 1992), pp. 1–2, 8, cover (in Spanish).

"Die Ordnung Des Scheinbar Regellosen." *Frankfurter Allgemeine Zeitung* (December 1993).

Doubilet, Susan. "The Divided Self." *Progressive Architecture* (March 1987), pp. 81–92, cover. (Berlin, Travelers, Artifacts.)

"Dusseldorf: der Kunstier Bau." *Ambiente* (Munich, October 1993), p. 18 (in German). (Haus Immendoff)

Dyer Szabo, Brenda. "Contrasti e Armonia: An Alliance of Oppositions." *Habitat Ufficio* (no. 58, Milan, June/July 1992), pp. 50–9 (in Italian and English).

Egelkraut, Ortrun. "Flurstuck Am Zirkus 1." *Berliner Zeitung* (Berlin, November 16, 1992) (in German).

Egelkraut, Ortrun. "Planung ohne Grund." *Berliner Zeitung* (Berlin, December 18, 1992) (in German).

"Ein neuer Stadtteil am Rebstock." *Gallus Drehscheibe* (February 1993) (in German).

"Ein Projekt im Namen Max Reinhardt." *Der Tages Spiegel* (Berlin, July 29, 1992) (in German).

"Eisenman diz que Sao Paulo tem perfil do seculo 20." *Folha de S. Paulo* (May 19, 1993), p. 2.

"Eisenman und Rebstock." *Arkitektur* (April 1992) (in German).

Eisenman, Peter. "People Who Live in Glass Houses Should Not Throw Stones." In Color of An Architect: Peter Eisenman, Haus Immendorff, Hamburg: Artfound Print Co./Galerie fur Architektur Renate Kammer und Angelika Hinrichs, owners, 1993, pp. 2–11, (in German and English).

"Eisenman, Provocador." *Folha de S. Paulo* (May 19, 1993), p. 17.

"Er Labt Wande Explodieren." *Hamburger Abendblatt* (Hamburg, June 30, 1992) (in German).

Evans, Robin. "Not to Be Used for Wrapping Purposes — Peter Eisenman: Fin d'Ou T Hou S." *AA Files* (London, Autumn 1985), pp. 68–78.

"Excavating Eisenman." *Architecture* (June 1994), pp. 57–63.

"Experimentos arquitectonicos." *Arquitectura* (Buenos Aires, January 23, 1993), p. 6.

"Experimentos arquitectonicos." *Arquitectura* (Buenos Aires, January 23, 1993) (in Spanish).

"Falten Statt Hamburger." *Hamburger Morgen Post* (Hamburg, July 2, 1992) (in German).

"'Faltungen' auf Rebstockgelande." *Frankfurter Allgemeine Zeitung* (Frankfurt, December 16, 1992) (in German).

Filho, Antonio Goncalves. "Masp abriga a polemica arquitectura de Eisenman." *Folha de S. Paulo* (May 17, 1993), Section 4, p. 8.

Filler, Martin. "Peter Eisenman: Polemical Houses." *Art in America* (vol. 68, no. 9, New York, November 1980), pp. 126–33.

Findsen, Owen. "Exhibit Spotlights Architects' Designs for DAAP, Disney." *Cincinnati Enquirer.*

Fischer, Adelheid. "Towers of Power." *Express* (Memphis, Summer 1992), pp 21–5.

Forster, Kurt. "Eisenman/Robertson's City of Artificial Excavation." *Archetype* (vol. 2, no. 2, Melbourne, Spring 1981), pp. 84–5.

Forster, Kurt. "Monuments to the City." *Harvard Architectural Review* (Spring 1984), pp. 107–21.

Forster, Kurt. "Traces and Treason of a Tradition." *The Ohio State Center for the Visual Arts Competition.* New York: Rizzoli International Publications, 1984, pp. 135–40.

Fox, Catherine. "Emory's Sights Set Clearly on World-Class Arts Facility." *Atlanta Journal* (Living Section, Atlanta, October 9, 1992).

Frampton, Kenneth. "Apropos Eisenman." *Domus* (no. 688, Milan, September 1987) (in Italian and English).

Frampton, Kenneth. "Eisenman Revisited: Running Interference." *Architecture + Urbanism: EISENMANAMNESIE* (extra edition, Tokyo, August 1988) pp. 57–69, cover (in Japanese and English). (Reprinted from *Domus*, no. 688, September 1987.)

Frampton, Kenneth. "Zabriskie Point: The Somnambulist Trajectory." *Casabella* (nos. 586/587, Milan, January/February 1992), pp. 8–13 (in Italian).

Frederic Levrat. "Peter Eisenman." *L'Architecture d'Aujourd'hui* (no. 279, Paris, February 1, 1992), pp. 98–115, cover (in French and English).

Freedman, Adele. "Turning Architecture from the Sun toward the Moon." *Toronto Globe & Mail* (February 23, 1993), p. C13.

Fuji, Hirome. "From Conception to Decomposition." *Architecture + Urbanism* (no. 112, Tokyo, January 1980), pp. 249–52 (in Japanese and English).

Galloway, David. "Berlin's Reinhardt Homage: Why Not Eisenman?" *International Herald Tribune* (December 19–20, 1993), p. 18.

Galloway, David. "Berlin's Reinhardt Homage: Why Not Eisenman?" *International Herald Tribune* (nos. 19/20, December 1992), p. 16.

Gandelsonas, Mario. "From Structure to Subject." *Oppositions 17* (Cambridge, Summer 1979), pp. 6–29.

Gandelsonas, Mario. "From Structure to Subject: The Formation of an Architectural Language." In Eisenman, Peter, *House X.* New York: Rizzoli International Publications, pp. 7–31.

Gandelsonas, Mario. "Linguistics in Architecture." *Casabella* (no. 374, Milan, 1973), pp. 17–30 (in Italian and English).

Gandelsonas, Mario. "On Reading Architecture." *Progressive Architecture* (vol. 53, March 1972), pp. 68–88.

Gass, William. "House VI." *Progressive Architecture* (June 1977), pp. 60, 62, 64.

Ghirardo, Diane. "Peter Eisenman: il Camouflage dell'avanguardia." *Casabella* (1994), pp. 22–7.

Gilson, Nancy. "Buildings that Challenge." *Columbus Dispatch* (Arts Section, June 3, 1992).

Giovannini, Joseph, "Greater Convention Center." *Domus 750* (June 1993), pp. 25–33.

Giovannini, Joseph. "Beyond Convention." *Architecture* (May 1993), pp. 52–63, cover.

Gironnay, Sophie. "L'archeologie fictive revele la pensee et les aeuvres souvent troublantes de Peter Eisenman; Le Deconstructivisme." *Le Devoir* (1994), p. C1.

Gironnay, Sophie. "Provocateur et Architecte." *Le Devoir* (1994).

Giurgola, Romaldo. "Five on Five: the Discreet Charm of the Bourgeoisie." *Architectural Forum* (vol. 138, no. 4, May 1973), pp. 46–57.

Glusberg, Jorge. "Centro Wexner." *Projeto* (no. 163, Buenos Aires, 1993), pp. 50–3.

"Go to Venice and See the Whole World." *Independent* (London, September 11, 1991).

Goldberger, Paul. "A Remembrance of Visions Pure And Elegant." *New York Times* (January 3, 1993), Section 2, p. 29.

Goldberger, Paul. "Architecture's 'Big Five' Elevate Form." *New York Times* (November 26, 1973), p. 33.

Goldberger, Paul. "The Museum that Theory Built." *New York Times* (Arts and Leisure Section, November 5, 1989), pp. C1, C38.

Graafland, Arie. "Decentering van Structuraliteit in het Ontwerp — Peter Eisenman's Decompositie." *Esthetische Theorie en Ontwerp: Architectuur. Macht en Lichaam* (Delft, Uitgeverij SUN, November 1984), pp. 213–32, cover (in Dutch).

Graafland, Arie. "Peter Eisenman: Architecture in absentia." In Arie Graafland (ed.), *Peter Eisenman: Recent Projects,* Nijmegen: SUN, 1989, pp. 95–126 (in Dutch and English).

Green, Jonathan. "Algorithms for Discovery." *The Wexner Center for the Visual Arts, The Ohio State University.* New York: Rizzoli International Publications, 1989, pp. 28–31.

Gregory, Anne. "New Convention Center Makes Sense." *Upper Arlington News* (Arlington, November 11, 1992).

Griddings, Scalings, Tracings and Foldings in the Work of Peter Eisenman. Sao Paulo: Editora Pini, Ltda, 1993 (in Portuguese and English). (A catalog for the exhibition of the same title at the Museu de Arte de Sao Paulo, Brazil.)

Gutman, Robert. "House VI." *Progressive Architecture* (June 1977), pp. 57–67.

Hacker, Marc. "With a Certain Laughter and Dance." *Investigation in Architecture/ Eisenman Studios at the GSD: 1983–85.* Cambridge: Harvard Graduate School of Design, 1986, pp. 26–42.

Hales, Linda. "American Architects at Venice Biennale." *Washington Post* (Washington, DC, September 19, 1991).

Hallett, Joe. "Columbus Beholds Unconventional Center." *The Blade* (Toledo, February 28, 1993), p. 3.

Heller, Fran. "New York Architect Finds Responsive 'Home' in Ohio." *Cleveland Jewish News* (October 30, 1992), pp. 42–3.

Henderson, Justin. "Unconventional Wisdom." *Interiors* (June 1993), pp. 82–5.

"Hochkonjunktur Der Hochhausprojekte." *Gallus Drehscheibe* (1993).

Hofer, Nina. "Fin d'Ou T Hou S." *Architectural Follies.* New York: Leo Castelli Gallery, October 22 – November 19, 1983.

Huxtable, Ada Louise. "The Troubled State of Modern Architecture." *New York Review of Books* (May 1, 1980), pp. 22–9.

Irace, Fulvio. "Eisenman House VI: Architettura di Geometrie Visibili/Analysis of Eisenman's House VI." *MODO* (no. 12, Milan, September 1978), pp. 27–31 (in Italian).

Janik, Detlev. "Die Ordnung des Scheinbar Regellosen." *Frankfurter Rundschau* (September 11, 1993).

Jansen, Karin. "Manege frei: Peter Eisenman." *Die Wochenschau,* Wettbewerbe.

Jencks, Charles. "The Perennial Architectural Debate, The Eisenman Paradox: Elitism, Populism and Centrality." *Architectural Design Profile.* London: 1983, pp. 4–23.

Jencks, Charles. *Post Modern Architecture,* revised edition. London: Academy Editions, 1978, pp. 8, 64, 66, 100, 101, 118, 121, 122, 126, 127.

Jodidio, Philip, (ed.). *Contemporary American Architects.* Cologne: Benedikt Taschen GmbH, 1993, pp. 56–61 (in English, German, French). (Wexner Center for the Arts, Columbus Convention Center.)

Jodidio, Philip. "Jour de Cristal." *Connaissance des Arts* (April 1993), pp. 59–63.

Johnson, Philip. "Philip Johnson on Eisenman and Gehry." *Architectural Design* (London, January/February 1992), pp. 26–31.

Johnson, Philip. "Philip Johnson on Peter Eisenman." *Architecture + Urbanism: EISENMANAMNESIE* (extra edition, Tokyo, August 1988), pp. 9–11, cover (in Japanese and English).

Johnson, Philip. Introduction to *Peter Eisenman and Frank Gehry.* New York: Rizzoli International Publications, 1991, pp. 2–3. (Catalog for the Fifth International Exhibition of Architecture at the 1991 Venice Biennale.)

Kahn, Eve M. "New From the Deconstructivist Guru." *Wall Street Journal* (May 12, 1993), p. A12.

Kamin, Blair. "America's Oddest Architect." *Chicago Tribune* (April 19, 1993), Section 13 (Sunday Arts), cover, pp. 4–5, 28.

Kipnis, Jeffrey. "Architecture Unbound: Consequences of the Recent Work of Peter Eisenman." *SD* (special issue on Peter Eisenman, Tokyo, March 1986), pp. 26–33 (in Japanese and English).

Kipnis, Jeffrey. "Star Wars III: The Battle at the Center of the Universe." *Investigation in Architecture/Eisenman Studios at the GSD: 1983–85.* Cambridge: Harvard Graduate School of Design, 1986, pp. 42–7.

Kipnis, Jeffrey. "The Law of ana-. On Choral Works." In Arie Graafland (ed.), *Peter Eisenman: Recent Projects,* Nijmegen: SUN, 1989, pp. 145–60 (in Dutch and English).

Kipnis, Jeffrey. "The Ohio State University Center for the Visual Arts and the Architecture of Modification." *Casabella* (488/9, Milan, January/February 1984), pp. 96–9 (in English).

Kitayama, Kirjo. "Peter Eisenman Tokyo, Koizumi Sangyo and Nunotani Headquarters." *Lotus 76,* pp. 87–95.

Klemmer, Clemens. "Gefaltete Architektur." *Neue Burcher Zeitung* (September 19, 1992), p. 9 (in German).

Krauss, Rosalind. "Death of a Hermaneutic Phantom: Materialization of the Sign in the Work of Peter Eisenman." *Architecture + Urbanism* (no. 112, Tokyo, January 1980), pp. 189–219.

Krauss, Rosalind. "Death of a Hermeneutic Phantom: Materialization of the Sign in the Work of Peter Eisenman." In Peter Eisenman, *Houses of Cards,* New York: Oxford University Press, 1987, pp. 166–84.

Kwinter, Sanford. "Der Genius Der Materie: Projekt Fur Cincinnati." *Arch +* (December 1993).

Kwinter, Sanford. "The Genius of Matter: Eisenman's Cincinnati Project." In *Peter Eisenman and Frank Gehry.* New York: Rizzoli International Publications, 1991, pp. 8–9. (Catalog for the Fifth International Exhibition of Architecture at the 1991 Venice Biennale.)

Lacy, Bill. "Peter Eisenman Building — Before and After Earthquake." *Architectural Record* (New York, January 1992), p. 48 (cartoon).

Lauer, Heike. "Living and Working at Rebstockpark." *Archigrad 1/1992: Planning and Building on the 50th Parallel* (Frankfurt, 1992), pp. 37–41.

LeCuyer, Annette. "Designs on the Computer." *Architectural Review* (January 1995), pp. 76–9.

Lemos, Peter. "The Triumph of the Quill." *Village Voice* (May 3, 1983), pp. 96, 99.

Lerup, Lars. "House X by Peter Eisenman." *Design Book Review* (San Francisco, Summer 1983), pp. 44–8.

Levesque, Luc. "Le theatre desoperation pliable: Concours de circonstances ..." *Inter art actuel* (vol. 53, Quebec, 1992), pp 34–7 (in French).

Libeskind, Daniel. "Peter Eisenman and the Myth of Futility." *Harvard Architectural Review* (vol. 3, Winter 1984), pp. 61–4.

Libeskind, Daniel. "The Lamentable Center: A Response to Peter Eisenman." Unpublished article, 1984.

Litt, Steven. "Daring Design Shouts Gotcha!" *Plain Dealer* (Cleveland, April 25, 1993).

Litt, Steven. "Eisenman on Design." *Plain Dealer* (Cleveland, October 28, 1992).

Litt, Steven. "Exhibit Explores Master Architects' Visionary Works." *Plain Dealer* (September 20, 1992).

"Living Room." *Architecture* (August 1992), p. 101.

Lubowski, Bernd. "Heimkehr und spate Versohnung ." *Berliner Morgenpost* (Berlin, July 29, 1992) (in German).

Lubowski, Bernd. "Senat sperrt sich gegen Max-Reinhardt-Museum." *Berliner Morgenpost* (Berlin, November 12, 1992) (in German).

Lubowski, Bernd. "Making shop drawings with the spirit of Japanese craftmanship." *Nikkei Architecture* (Tokyo, February/March 1992), pp. 254–8 (in Japanese).

Ludlow, Randy. "Columbus Proud of New Lady." *Cincinnati Post* (March 3, 1993).

Lynn, Greg (guest editor). "Rebstock Park Masterplan", "Alteka Office Building", "Center for the Arts, Emory University." In *Folding in Architecture,* Architectural Design Profile no. 102. London: Academy Group Ltd, 1993, pp. 26–35.

MacLeod, Douglas. "ACS Kompendium '93." (November 1993).

Macrae-Gibson, Gavin. "The Anxiety of the Second Fall: House El Even Odd, Peter Eisenman." In *The Secret Life of Buildings: An American Mythology for Modern Architecture.* Cambridge: MIT Press, 1985, pp. 30–51.

Maier, Susanne. "Hausverbot im Theater sienes Vaters!" *B.Z.* (July 29, 1992), p. 32 (in German).

"Manhattan Waterfront Design Project: Peter Eisenman and Michael Graves." In *The New City: Architecture and Urban Renewal.* New York: Museum of Modern Art, 1967, pp. 36–41. (Catalog including Manhattan Waterfront design project with Michael Graves.)

Maruyama, Hiroshi. "Of a mis-Leading, for a mis-Leading." *SD* (special issue on Peter Eisenman, Tokyo, March 1986), pp. 7–8 (in Japanese and English).

"Max Reinhardt Monument Challenges 'Business as Usual'." *Architectural Record* (March 1993).

McGuigan, Cathleen (with Maggie Malone). "Stone, Steel and Cyberspace." *Newsweek* (February 27, 1995), p. 73.

Mendini, Alessandro. "Dear Peter Eisenman." *Domus* (no. 611, Milan, November 1980), p. 1 (editorial, in Italian and English).

Merkel, Jayne. "Looking for the Future: The Center for the Visual Arts Competition at Ohio State University." *Inland Architect* (December 1983), pp. 110–16.

Mitarheiter, N.Z. & Petersohn, Harlmut. "Ein City-Kristall mit Signatur des Jahrhunderts." (Berlin, December 12, 1982) p. 20 (in German).

Mitarheiter, N.Z. & Petersohn, Harlmut. "Nach oben ist der Alex offen." (November 18, 1992) (in German).

Mitarheiter, N.Z. & Petersohn, Harlmut. "Nunotani Building." *Nikkei Architecture* (no. 442, Tokyo, September 1992), pp. 206–11 (in Japanese).

Mitarheiter, N.Z. & Petersohn, Harlmut. "Nunotani Building." *Shinkenchiku* (67:10, Tokyo, October 1992), pp. 187–206 (in Japanese).

Moenninger, Michael. "Die geplants Unordnung." *Frankfurter Allgemeine Zeitung* (February 22, 1993).

Moenninger, Michael. "Marke Eigenbau: Der Architekt Peter Eisenman." *Frankfurter Allgemeine Magazin* (Heft 681, March 19, 1993), pp. 14–23, cover (in German).

Mollard, Beth. "Nightmare on High Street: Center's Design Causes Delay." *Columbus Dispatch* (Columbus).

Moneo, Rafael. "Unexpected Coincidences." In *The Wexner Center for the Visual Arts, The Ohio State University.* New York: Rizzoli International Publications, 1989, pp. 40–5.

Moore, Rowan. "Eisenman Plays the Centre Forward." *Blueprint* (April 14, 1993), p.14.

Muschamp, Herbert. "This Time, Eisenman Goes Conventional." *New York Times* (March 2, 1993).

Nasar, Jack. "Convention Center's Interior an Exciting Space." *Columbus Dispatch* (March 14, 1993), p. 5C.

Newkirk, Margaret. "Our Strange New Convention Center: Haute or Hoot?" *The Other Paper* (Columbus, November 12–18, 1992), cover.

Newkirk, Margaret. "News Notes." *Oculus* (55:4, December 1992).

"No Construyo Solo Edificios; Construyo Ideas." *El Pais* (November 1993).

"Nunotani Building." *Japan Architect 9* (January 1993), pp. 150–3 (in Japanese and English).

"Ohio State University Center for the Visual Arts." *Section A* (vol. 2, no. 2, Montreal, April/May 1984), pp. 22–5.

Paschke, Regina. "Viel Ehre mit Profit." *Die Tageszeitung* (Berlin, July 29, 1992) (in German).

"Pastel City Tinged with Menace." *Globe and Mail* (March 1994), pp. 14, 15.

Pelissier, Alain. "Peter Eisenman: L'Espace Autre." *Techniques et Architecture* (vol. 360, Paris, June/July 1985), pp. 26–55 (in French).

Perrella, Stephen. "Columbus Convention Center Opens." *Newsline,* Columbia University Architecture, Planning, Preservation (New York, March/April 1993), p. 8.

"Peter Eisenman Il Progetto Del Vuoto." *Risk* (1994), p. 19.

"Peter Eisenman in Buenos Aires." *Clarin Arquitectura* (special supplement, Buenos Aires, December 5, 1992), pp. 1–5, cover.

"Peter Eisenman." *Archimade* (no. 44, June 1994), pp. 5–31.

"Peter Eisenman: Emory Center For the Arts." *GA International 36* (Tokyo, 1993), pp. 32–3.

"Peter Eisenman: Max Reinhardt Haus, Berlin." *AA Files 25* (London, March 1993), pp. 8–13.

"Peter Eisenman; Oltre lo sguardo." *Domus* (no. 734, Milan, January 1992), pp. 17–24 (in Italian and English).

"Peter, los filosofos y tambien los criticos." *Arquitectura* (Buenos Aires, December 12, 1992).

Pinon, Helio. "La Forma de la Forma." *Arquitectura de los Neovanguardias* (Barcelona, 1984), pp. 117–65 (in Spanish).

Pione, Carolyn. "Columbus Convention Center 'Not Your Typical Box'." *Cincinnati Enquirer* (April 11, 1993), p. D-5.

Pione, Carolyn. "Columbus Convention Center: New Frontier in Architecture?" *Alliance Review* (March 11, 1993), p. 12, cover.

Pommer, Richard. "The New Architectural Supremacists." *Artforum* (vol. 15, no. 2, October 1976), pp. 38–43.

Pontbriand, Chantal. *Parachute.* Revue D'Art Contemporain Inc., 1993.

Post, Nadine. "Cockeyed Optimism in Columbus." *Engineering News-Record* (February 8, 1993), pp. 22–4, 58, cover, editorial.

Post, Nadine. "Throwing Construction A Curve: No More Mister Highbrow." *Engineering News-Record* (June 21, 1993), pp. 24–7, cover.

Rajchman, John. "Perplications" *Critical Space* (no. 6, Tokyo, 1992), pp. 113–29 (in Japanese).

Rajchman, John. "Perplications: On the Space and Time of Rebstockpark." *Unfolding Frankfurt*. Berlin: Ernst & Sohn, 1991, pp. 20–77, (in German and English). (Catalog for the exhibition of the Rebstockpark project at Aedes Gallery.)

"Regenswasser von den Dachern in den Weiher." *Frankfurter Allgemeine Zeitung* (December 9, 1992) (in German).

"Reinhardt-Haus vom Senat abgelehat." *Berliner Zeitung* (November 13, 1992), p.1.

Rella, Franco. "Figure nel labirinto. La matamorfosi di una metafora." In *La Fine del Classico*. Venice, 1987, pp. 9–25.

Riskind, Jonathan. "Convention Center: Beauty or Beast?" *Columbus Dispatch* (July 15, 1992), p. 58.

Robertson, Jaquelin. "Five on Five: Machines in the Garden." *Architectural Forum* (vol. 138, no. 4, May 1973), pp. 46–57.

Rochon, Lisa. "Nunotani's Headquarter Building." *International Contract* (Toronto, May 1993), pp. 46–9.

Roffo, Analia. "La arquitectura no resulve problemas sociales." *Buenos Aires* (February 4, 1993), p. 8.

Rosen, Cheryl. "Columbus Readies Daring New Facility." *Business News Travel* (December 1, 1992), cover.

Runge, Irene. "Konsulat Wachtiger als Schauspielhaus." *Neues Deutschland* (November 13, 1992) (in German).

Salvato, Al. "New 'Gem' to Brighten UC Campus (Architects Praise DAAP Building Plan)." *Cincinnati Post* (June 1994).

Schmidt, Martinus. "Erben fuhlen sich bruskiert." *Neuezeit* (Berlin, November 13, 1992) (in German).

Schubert, Peter. "Krimi um 'Filetgrundstuck'." *Berliner Morgenpost* (Berlin, November 14, 1992) (in German).

Schwarz, Ullrich. "Peter Eisenman: Dazwischen." *Der Architezkt* (January, 1993).

Schweitzer, Eva. "150 Meter hoch an der Weidendammer Brucke?" *Berliner Morgenpost* (December 16, 1992) (in German).

Seabrook, John. "The David Lynch of Architecture." *Vanity Fair* (January 1991), pp. 74–9, 125–9.

"Seismische Architektur Das Nunotani Building in Tokio." *Architektur, Innenarchitektue. Technischer Ausbau* (Leinfelden-Echterdingen, Germany: Verlagsanstalt Alexander Koch GmbH, April, 1993), pp. 36–9.

Sola-Morales , Ignasi de. "Del Objeto a la Dispersion: Arquitectura Artificialis. Proyecto de Peter Eisenman para le Friedrichstrasse de Berlin." *Arquitecturas/Bis 42* (Barcelona, December 1983), pp. 16–21 (in Spanish).

Sola-Morales, Ignasi de. "Cuatro Notas Sobre la Arquitectura Reciente de Peter Eisenman." *El Croquis* (41, Madrid, December 1989), pp.16–23, cover (in Spanish and English).

Sola-Morales, Ignasi de. "Peter Eisenman: Arquitectura del Presente." *La Vanguardia* (Barcelona, September 16, 1986), p. 33 (in Spanish).

Solomon, Nancy B. "Flexible Theaters." *Architecture* (August 1992), pp. 95–102.

Somol, R.E. "O-O." *Progressive Architecture* (special issue on Eisenman/Wexner Center, October 1989), p. 88, cover.

Somol, R.E. "Peter Eisenman: Wexner Center for the Visual Arts, Columbus/ Ohio." *Domus* (Milan, January 1990), pp. 38–47 (in Italian and English).

Somol, Robert. "Accidents Will Happen." *Architecture + Urbanism* (Tokyo, September 1991), pp. 4–7, cover (in Japanese and English).

Sorkin, Michael. "Architecture: Solid Geometry: Coming off a Theoretical Tangent, Architect Peter Eisenman Puts a New Spin on Design." *House and Garden* (October 1989), pp. 62–6.

Speicher, Stephan. "Baukunst und Investorengluck." *Frankfurter Allgemeine Zeitung* (December 12, 1992) (in German).

Stache, Rainer. "Aus fur dieses haus? Schade!" *Berliner Morgenpost* (December 16, 1992) (in German).

Stein, Jerry. "CAC Salutes 'Superstar' Architects." *Cincinnati Post* (April 4, 1992), p. 7B.

Steinhausen, Ansgar. "Zwischen Stadtautobahn und Gleisanlagen." *Badische Zeitung* (January 18, 1993), p. 11.

Tafuri, Manfredo. "Les Bijoux Indiscrets." Introduction to *Five Architects NY*. Rome: Officina Edizioni, 1976, pp. 7–34. (Introduction to Italian translation.)

Tafuri, Manfredo. "Meditations of Icarus." Unpublished manuscript. (Later published in Eisenman, Peter, *Houses of Cards*, New York: Oxford University Press, 1987, pp. 167–87.)

Tafuri, Manfredo. "Peter Eisenman: The Meditations of Icarus." In Peter Eisenman, *Houses of Cards,* New York: Oxford University Press, 1987, pp. 167–87.

Taki, Koji. "Dialogue on Peter Eisenman." *Architecture + Urbanism* (no. 112, Tokyo, January 1980), pp. 245–48 (in Japanese and English).

Taylor, John. "Mr. In-Between, Deconstructing with Peter Eisenman." *New York Magazine* (October 17, 1988), pp. 46–52.

Taylor, Mark. "Eisenman's Coup." *Progressive Architecture* (special issue on Peter Eisenman, October 1989), p. 89, cover.

Terrahe, Antje. "In Raum und Zeit entFalten: Wohnraum schaffen, ohne ins Umland zu Wuchern." *Frankfurter Rundschau* (December 12, 1992), p. 8 (in German).

"The 39th Annual P/A Awards: Architectural Design Award, Alteka Office Building." *Progressive Architecture* (January 1992), pp. 63–5.

"The 40th Annual Progressive Architecture Design Citation, Center for the Arts." *Progressive Architecture* (January 1993), pp. 78–81.

"The Site Report—Nunotani Tokyo NC Building." *Nikkei Architecture* (3:2, Tokyo: 1992), pp. 254–8 (in Japanese).

"Tokio Scragerklotz." *Ambiente* (January/ February 1993), p. 20.

Trelstad, Julie M. "Convention Center Breaks Convention." *I.D.* (March/April 1993), p. 19.

"Turning Architecture from the Sun to the Moon." *Globe and Mail* (February 1993), p. C13.

"Un progetto per Siena: Il concorso internazionale per piazza matteotei — la Lizza." Bernardo Secchi & Chiara Merlini (eds), Milan: Electa, 1992, pp. 80–5 (in Italian and English).

"Una Scomposta Nervrosi." *Architettura* (March 1993), p. 32.

Unger, Oscar. "Ein neuer Stadtteil zeischen Autobahn und Rebstockweiher." *Frankfurter Neue Presse* (December 9, 1992) (in German).

"Unspeakable or Unreadable." *Blueprint* (no. 84, London, February 1992), p. 12. (Review of *Anyone*.)

Vallongo, Sally. "Columbus Exhibition Contrasts Top Architects." *The Blade* (July 19, 1992).

van Dijk, Hans. "Autonomia Americana. Eisenman/Hejduk." *Wonen TA/BK* (21/22, Hilversun, The Netherlands, 1981) (in Dutch). (Catalog of the exhibition at the Stichting Architectuur Museum.)

van Dijk, Hans. "Eisenman/Hejduk: Architectuur Halverwege Amerika en Europa." *Wonen TA/BK* (21/22, Hilversun, The Netherlands, November 1980), pp. 7–10 (in Dutch).

van Dijk, Hans. "Peter Eisenman en het Berlangan Naar de Klassieke Rede." ("Peter Eisenman and the Desire for Classical Reason.") *Wonen TA/BK* (Hilversun, The Netherlands, February 1977), pp. 4–10 (in Dutch).

"Verjungende Falten." *Taz Hamburg.* (July 2, 1992), p. 19 (in German).

Vidler, Anthony. "After the End of the Line." *Architecture + Urbanism: EISENMANAMNESIE* (extra edition, Tokyo, August 1988), pp. 147–61, cover (in Japanese and English).

Vidler, Anthony. "Counter-Monuments in Practice: The Wexner Center for the Visual Arts." In *The Wexner Center for the Visual Arts, The Ohio State University.* New York: Rizzoli International Publications, 1989, pp. 32–8.

Vidler, Anthony. *The Architectural Uncanny: Essays in the Modern Unhomely.* Cambridge, Massachusetts: MIT Press, 1992.

"Warum Le Corbusier Die Welt Nicht Verandert Hat." *Ereignisse Und Gestalten* (January 1992).

Weibel, Peter. "Intelligente Wesen in Einem Intelligenten Universum." *Intelligente Ambiente* (1994), pp. 6–26.

"Wettbewerbsprojekt Rebstock in Frankfurt." *Werk, Bauen + Wohnen* (Frankfurt, March 1992), pp. 36–9 (in German).

"Wexner Center for the Visual Arts." *Museo d'arte e architectura.* Lugano: Edizioni Charta s.r.l., 1992, pp. 94–101. (Catalog for the exhibition at the Museo Cantonale d'arte.)

"What's On In February & March Halls of Fame." *Elle Decor* (February/March 1993).

Whiteman, John. "Site Unseen — Notes on Architecture and the Concept of Fiction: Peter Eisenman: Moving Arrows, Eros and Other Errors." *AA Files* (no. 12, London, Summer 1986), pp. 76–84.

Wise, Michael Z. "Reinhardt's Restitution." *Forward* (New York, February 26, 1993), pp. 1, 11.

Wise, Michael. "United Berlin Divided Over Its 'Eiffel Tower'." *Los Angeles Times* (March 29, 1993), pp. F1, F9.

Witte, Joachim. "Droht Zweite Enteignung Max Reinhardt?" (November 13, 1992) (in German).

Wittmann, Jochen. "Konkurrenz fur das Brandenburger Tor." *Der Standard Immobilien* (January 7, 1933), p. 1.

Wurm, Fabian. "Elogen auf Eisenman." *Baumeister* (February 1993).

Yatsuka, Hajime. "The Adventure in the Labyrinth of the Knight for Purity." *SD* (special issue on Peter Eisenman, Tokyo, March 1986), pp. 70–1 (in Japanese and English).

Zellner, Lisa K. "Convention Center Group Takes Show On Road." *Upper Arlington News* (Arlington, November 4, 1992).

Zohlen, Gerwin. "Im Dienst der Erben." *Suddeutsche Zeitung* (December 12, 1992) (in German).

Zuck, Barbara. "Vision Serves Geldin Well as Director of Wexner Center." *New York Times* (June 1994), p. 2B.

Acknowledgments 致 谢

The work in this book was made possible:

through the efforts of my staff, including particularly Richard Rosson, Sergio Bregante, Setu Shah, Cynthia Davidson, Diano Ibrahim, Megan McFarland, Bernadette Latour, Megan McFarland, Benedicte Mioli, Rudolph Kammeri, and Alexander Wiedemann;

through all the Clients who commissioned this work; and

through the co-operative efforts of Engineers and Consultant.

We would like to thank The Images Publishing Group, Paul Latham, and Alessina Brooks for their invitation to participate in this monograph series.

Photography Credits

Dick Frank Studio
House, I, House II, House III, House IV, House VI, Cannaregio Town Square, Madison Components Plant, Romeo and Juliet Castles, Tokyo Opera House, Biocentrum, La Villette, Carnegie Mellon Research Institute, Guardiola House, Aronoff Center for Design and Art, Koizumi Sangyo Office Building, Siena Bank Competition, Banyoles Olympic Hotel, Cooper Union Housing, Atocha 123 Hotel, Rebstockpark Master Plan, Alteka Office Building, Emory Center for the Arts, Max Reinhardt Haus, Nordliches Derendorf Master Plan, Haus Immendorff

Ron Forth
Aronoff Center for Design and Art

David Morton
House III

D.G. Olshavasky/ARTOG
Wexner Center for the Visual Arts and Fine Arts Library, Greater Columbus Convention Center

Jeff Goldberg/ESTO
Fuller/Toms Loft, Wexner Center for the Visual Arts and Fine Arts Library, Greater Columbus Convention Center

Jochen Littkemann
Max Reinhardt Haus

Mark C. Schwarts
Progressive Corporation Office Building

Masao Ueda
Koizumi Sangyo Office Building

Michael Moral
University Art Museum

Reingard Gorner
IBA Social Housing

Shigwo Ogawa/Shinkenchiku
Nunotani Office Building

Van der Vlugt & Claus
Groningen Music-Video Pavilion

Wolfgang Hoyt/ESTO
Firehouse for Engine Company No 233 and Ladder Company No 176, Travelers Financial Center

Rendering Credits

Ed Keller/*Straylight Imaging*
Max Reinhardt Haus

Index 索 引

Bold page numbers refer to projects included in Selected and Current Works.